PROGRESS IN

Nucleic Acid Research and Molecular Biology

Volume 26

PROGRESS IN
Nucleic Acid Research and Molecular Biology

Volume 26
DNA: Multiprotein Interactions

edited by

WALDO E. COHN

Biology Division
Oak Ridge National Laboratory
Oak Ridge, Tennessee

1981

ACADEMIC PRESS
A Subsidiary of Harcourt Brace Jovanovich, Publishers

New York   London   Toronto   Sydney   San Francisco

COPYRIGHT © 1981, BY ACADEMIC PRESS, INC.
ALL RIGHTS RESERVED.
NO PART OF THIS PUBLICATION MAY BE REPRODUCED OR
TRANSMITTED IN ANY FORM OR BY ANY MEANS, ELECTRONIC
OR MECHANICAL, INCLUDING PHOTOCOPY, RECORDING, OR ANY
INFORMATION STORAGE AND RETRIEVAL SYSTEM, WITHOUT
PERMISSION IN WRITING FROM THE PUBLISHER.

ACADEMIC PRESS, INC.
111 Fifth Avenue, New York, New York 10003

United Kingdom Edition published by
ACADEMIC PRESS, INC. (LONDON) LTD.
24/28 Oval Road, London NW1 7DX

LIBRARY OF CONGRESS CATALOG CARD NUMBER: 63–15847

ISBN 0-12-540026-8

PRINTED IN THE UNITED STATES OF AMERICA

81 82 83 84   9 8 7 6 5 4 3 2 1

Contents

LIST OF CONTRIBUTORS xi
PREFACE xv
ABBREVIATIONS AND SYMBOLS xvii
SOME ARTICLES PLANNED FOR FUTURE VOLUMES xxi

Introduction: DNA–Multiprotein Interactions in Transcription, Replication, and Repair   1
R. K. Fujimura

I. Replicative DNA Polymerase and Its Complex
Chairman and Summarizer: David Korn

Summary   5
David Korn

Enzyme Studies of ϕX174 DNA Replication   9
Ken-ichi Arai, Naoko Arai, Joseph Shlomai, Joan Kobori, Laurien Polder, Robert Low, Ulrich Hübscher, LeRoy Bertsch, and Arthur Kornberg

The DNA Replication Origin (*ori*) of *Escherichia coli*: Structure and Function of the *ori*-Containing DNA Fragment   33
Yukinori Hirota, Masao Yamada, Akiko Nishimura, Atsuhiro Oka, Kazunori Sugimoto, Kiyozo Asada, and Mitsuru Takanami

Replication of Linear Duplex DNA *in Vitro* with Bacteriophage T5 DNA Polymerase   49
R. K. Fujimura, S. K. Das, D. P. Allison, and B. C. Roop

Mechanisms of Catalysis of Human DNA Polymerases α and β 63
David Korn, Paul A. Fisher, and Teresa S.-F. Wang

Structural and Functional Properties of Calf Thymus DNA Polymerase δ 83
Marietta Y. W. Tsang Lee, Cheng-Keat Tan, Kathleen M. Downey, and Antero G. So

II. Mechanisms of Transcription
Chairman: Elliot Volkin
Summarizer: Robert K. Fujimura

Summary   99
R. K. Fujimura

Regulatory Circuits of Bacteriophage Lambda   103
S. L. Adhya, S. Garges, and D. F. Ward

III. Chromatin Transcription and Replication
Chairman and Summarizer: Ronald L. Seale

Summary   121
Ronald L. Seale

Site of Histone Assembly   123
Ronald L. Seale

Chromatin Replication in *Tetrahymena pyriformis*   135
A. T. Annunziato and C. L. F. Woodcock

Role of Chromatin Structure, Histone Acetylation, and the Primary Sequence of DNA in the Expression of SV40 and Polyoma in Normal or Teratocarcinoma Cells   151
G. Moyne, M. Katinka, S. Saragosti, A. Chestier, and M. Yaniv

IV. Control of Transcription in Eukaryotes
Chairman and Summarizer: William J. Rutter

Summary   171
William J. Rutter

V. Mechanisms of DNA Repair
Chairman: Richard F. Kimball

Repair Replication Schemes in Bacteria and Human Cells   181
Philip C. Hanawalt, Priscilla K. Cooper, and Charles Allen Smith

Recent Developments in the Enzymology of Excision Repair of DNA   197
Errol C. Friedberg, Corrie T. M. Anderson, Thomas Bonura, Richard Cone, Eric H. Radany, and Richard J. Reynolds

Multiprotein Interactions in Strand Cleavage of DNA Damaged by UV and Chemicals   217
Erling Seeberg

In Vitro Packaging of Damaged Bacteriophage T7 DNA — 227
Warren E. Masker, Nancy B. Kuemmerle, and Lori A. Dodson

The Inducible Repair of Alkylated DNA — 237
John Cairns, Peter Robins, Barbara Sedgwick, and Phillipa Talmud

VI. Functions Induced by Damaged DNA
Chairman and Summarizer: Evelyn M. Witkin

Summary — 247
Evelyn M. Witkin

Inducible Error-Prone Repair and Induction of Prophage Lambda in Escherichia coli — 251
Raymond Devoret

DNA and Nucleoside Triphosphate Binding Properties of recA Protein from Escherichia coli — 265
K. McEntee, G. M. Weinstock, and I. R. Lehman

Molecular Mechanism for the Induction of "SOS" Functions — 281
Michio Oishi, Robert M. Irbe, and Lee M. E. Morin

Induction and Enhanced Reactivation of Mammalian Viruses by Light — 303
Larry E. Bockstahler

Comparative Induction Studies 315
Ernest C. Pollard, D. J. Fluke, and Deno Kazanis

VII. Conclusion

Concluding Remarks 325
Ernest C. Pollard

INDEX 327

CONTENTS OF PREVIOUS VOLUMES 331

List of Contributors

Numbers in parentheses indicate the pages on which the authors' contributions begin.

S. L. ADHYA (103), *Laboratory of Molecular Biology, National Cancer Institute, Bethesda, Maryland 20205*

D. P. ALLISON (49), *Biology Division, Oak Ridge National Laboratory, Oak Ridge, Tennessee 37830*

CORRIE T. M. ANDERSON (197), *Laboratory of Experimental Oncology, Department of Pathology, Stanford University, Stanford, California 94305*

A. T. ANNUNZIATO[1] (135), *Department of Zoology, University of Massachusetts, Amherst, Massachusetts 01003*

KEN-ICHI ARAI (9), *Department of Biochemistry, Stanford University School of Medicine, Stanford, California 94305*

NAOKO ARAI (9), *Department of Biochemistry, Stanford University School of Medicine, Stanford, California 94305*

KIYOZO ASADA (33), *Institute for Chemical Research of Kyoto University, Uji-shi, Japan*

LEROY BERTSCH (9), *Department of Biochemistry, Stanford University School of Medicine, Stanford, California 94305*

LARRY E. BOCKSTAHLER (303), *Bureau of Radiological Health, Food and Drug Administration, Rockville, Maryland 20857*

THOMAS BONURA (197), *Laboratory of Experimental Oncology, Department of Pathology, Stanford University, Stanford, California 94305*

JOHN CAIRNS (237), *The Imperial Cancer Research Fund, London N.W. 7, England*

A. CHESTIER (151), *Department of Molecular Biology, Pasteur Institute, 75015 Paris, France*

RICHARD CONE (197), *Laboratory of Experimental Oncology, Department of Pathology, Stanford University, Stanford, California 94305*

PRISCILLA K. COOPER (181), *Department of Biological Sciences, Stanford University, Stanford, California 94305*

S. K. DAS (49), *Biology Division, Oak Ridge National Laboratory, and The University of Tennessee–Oak Ridge Graduate School of Biomedical Sciences, Oak Ridge, Tennessee 37830*

RAYMOND DEVORET (251), *Radiobiologie Cellulaire, Laboratoire d'Enzymologie, C.N.R.S., 91190 Gif-sur-Yvette, France*

[1] Present address: Department of Cellular Biology, Scripps Clinic and Research Foundation, 10666 N. Torrey Pines Road, La Jolla, California 92037.

LIST OF CONTRIBUTORS

LORI A. DODSON (227), *Biology Division, Oak Ridge National Laboratory, Oak Ridge, Tennessee 37830*

KATHLEEN M. DOWNEY (83), *Howard Hughes Medical Institute Laboratory, Departments of Medicine and Biochemistry and Center for Blood Diseases, University of Miami School of Medicine, Miami, Florida 33101*

PAUL A. FISHER (63), *Laboratory of Experimental Oncology, Department of Pathology, Stanford University School of Medicine, Stanford, California 94305*

D. J. FLUKE (315), *Zoology Department, Duke University, Durham, North Carolina 27706*

ERROL C. FRIEDBERG (197), *Laboratory of Experimental Oncology, Department of Pathology, Stanford University, Stanford, California 94305*

R. K. FUJIMURA (1, 49, 99), *Biology Division, Oak Ridge National Laboratory, Oak Ridge, Tennessee 37830*

S. GARGES (103), *Laboratory of Molecular Biology, National Cancer Institute, Bethesda, Maryland 20205*

PHILIP C. HANAWALT (181), *Department of Biological Sciences, Stanford University, Stanford, California 94305*

YUKINORI HIROTA (33), *National Institute of Genetics, Yata 1,111 Mishima-shi, Japan*

ULRICH HÜBSCHER (9), *Department of Biochemistry, Stanford University School of Medicine, Stanford, California 94305*

ROBERT M. IRBE (281), *The Public Health Research Institute of New York, New York, New York 10016*

M. KATINKA (151), *Department of Molecular Biology, Pasteur Institute, 75015 Paris, France*

DENO KAZANIS (315), *Zoology Department, Duke University, Durham, North Carolina 27706*

JOAN KOBORI (9), *Department of Biochemistry, Stanford University School of Medicine, Stanford, California 94305*

DAVID KORN (5, 63), *Laboratory of Experimental Oncology, Department of Pathology, Stanford University School of Medicine, Stanford, California 94305*

ARTHUR KORNBERG (9), *Department of Biochemistry, Stanford University School of Medicine, Stanford, California 94305*

NANCY B. KUEMMERLE (227), *Biology Division, Oak Ridge National Laboratory, Oak Ridge, Tennessee 37830*

MARIETTA Y. W. TSANG LEE (83), *Howard Hughes Medical Institute Laboratory, Departments of Medicine and Biochemistry and Center for Blood Diseases, University of Miami School of Medicine, Miami, Florida 33101*

LIST OF CONTRIBUTORS   xiii

I. R. LEHMAN (265), *Department of Biochemistry, Stanford University School of Medicine, Stanford, California 94305*
ROBERT LOW (9), *Department of Biochemistry, Stanford University School of Medicine, Stanford, California 94305*
WARREN E. MASKER (227), *Biology Division, Oak Ridge National Laboratory, Oak Ridge, Tennessee 37830*
K. MCENTEE (265), *Department of Biochemistry, Stanford University School of Medicine, Stanford, California 94305*
LEE M. E. MORIN (281), *The Public Health Research Institute of New York, New York, New York 10016*
G. MOYNE[2] (151), *Department of Molecular Biology, Pasteur Institute, 75015 Paris, France*
AKIKO NISHIMURA (33), *National Institute of Genetics, Yata 1,111 Mishima-shi, Japan*
MICHIO OISHI (281), *The Public Health Research Institute of New York, New York, New York 10016*
ATSUHIRO OKA (33), *Institute for Chemical Research of Kyoto University, Uji-shi, Japan*
LAURIEN POLDER (9), *Department of Biochemistry, Stanford University School of Medicine, Stanford, California 94305*
ERNEST C. POLLARD (315, 325), *Zoology Department, Duke University, Durham, North Carolina 27706*
ERIC H. RADANY (197), *Laboratory of Experimental Oncology, Department of Pathology, Stanford University, Stanford, California 94305*
RICHARD J. REYNOLDS[3] (197), *Laboratory of Experimental Oncology, Department of Pathology, Stanford University, Stanford, California 94305*
PETER ROBINS (237), *The Imperial Cancer Research Fund, London N.W. 7, England*
B. C. ROOP (49), *Biology Division, Oak Ridge National Laboratory, Oak Ridge, Tennessee 37830*
WILLIAM J. RUTTER (171), *Department of Biochemistry and Biophysics, University of California, San Francisco, California 94143*
S. SARAGOSTI (151), *Department of Molecular Biology, Pasteur Institute, 75015 Paris, France*
RONALD L. SEALE (121, 123), *Scripps Clinic and Research Foundation, La Jolla, California 92037*

[2] Present address: Institut de Recherches, Scientifiques sur le Cancer, B.P. 8, 94800 Villejuif, France.
[3] Present address: Department of Physiology, Harvard School of Public Health, Boston, Massachusetts 02115.

BARBARA SEDGWICK (237), *The Imperial Cancer Research Fund, London N.W. 7, England*

ERLING SEEBERG (217), *Norwegian Defence Research Establishment, Division for Toxicology, N-2007, Kjeller, Norway*

JOSEPH SHLOMAI (9), *Department of Biochemistry, Stanford University School of Medicine, Stanford, California 94305*

CHARLES ALLEN SMITH (181), *Department of Biological Sciences, Stanford University, Stanford, California 94305*

ANTERO G. SO (83), *Howard Hughes Medical Institute Laboratory, Departments of Medicine and Biochemistry, University of Miami School of Medicine, Miami, Florida 33101*

KAZUNORI SUGIMOTO (33), *Institute for Chemical Research of Kyoto University, Uji-shi, Japan*

MITSURU TAKANAMI (33), *Institute for Chemical Research of Kyoto University, Uji-shi, Japan*

PHILLIPA TALMUD (237), *The Imperial Cancer Research Fund, London N.W. 7, England*

CHENG-KEAT TAN (83), *Howard Hughes Medical Institute Laboratory, Departments of Medicine and Biochemistry and Center for Blood Diseases, University of Miami School of Medicine, Miami, Florida 33101*

TERESA S.-F. WANG (63), *Laboratory of Experimental Oncology, Department of Pathology, Stanford University School of Medicine, Stanford, California 94305*

D. F. WARD (103), *Laboratory of Molecular Biology, National Cancer Institute, Bethesda, Maryland 20205*

G. M. WEINSTOCK (265), *Department of Biochemistry, Stanford University School of Medicine, Stanford, California 94305*

EVELYN M. WITKIN (247), *Department of Biological Sciences, Douglass College, Rutgers University, New Brunswick, New Jersey 08903*

C. L. F. WOODCOCK (135), *Department of Zoology, University of Massachusetts, Amherst, Massachusetts 01003*

MASAO YAMADA (33), *National Institute of Genetics, Yata 1,111 Mishima-shi, Japan*

M. YANIV (151), *Department of Molecular Biology, Pasteur Institute, 75015 Paris, France*

Preface

As the Preface to Volume 1 (1963) indicated, this serial publication was conceived by its two original editors (the late J. H. Davidson and the undersigned) as "a continuing periodical assessment or reassessment of those areas of knowledge in the field of nucleic acids [and, a bit later, in the field of molecular biology] that have arisen or advanced notably since the publication in 1960 of the last of the three volumes of *The Nucleic Acids: Chemistry and Biology*, edited by Chargaff and Davidson." The preponderant interest in chemistry, including enzymology, at that time is shown by the fact that all but one of the eleven articles in Volume 1 dealt with the properties of nucleic acids or with the enzymes that form them. While this basic interest in the chemistry of nucleic acids and of their precursors and analogues has been retained in the 25 volumes that have since appeared, the growth of interest in the functioning and involvement of nucleic acids in biological phenomena—such as genetics, virology, and immunology—has led to an increase in contributions in those areas. The general thrust has been to explain the phenomena in terms of chemical mechanisms, these in turn stemming from the chemical properties of the nucleic acids themselves.

Whatever nucleic acids do in replication or specification of protein (translation) is done in conjunction with proteins, whether the presumably structural proteins of the ribosomes, or the polymerases or ligases or "factors" involved in transcription or translation, and we find many papers in these volumes dealing with protein–nucleic acid interactions. One aspect of the involvement of complexes of nucleic acids and proteins was explored intensively at a recent symposium entitled "DNA–Multiprotein Interactions in Transcription, Replication, and Repair," which was divided into the following sessions: Replicative DNA Polymerase and Its Complex; Mechanisms of Transcription; Chromatin Transcription and Replication; Control of Transcription in Eukaryotes; Mechanisms of DNA Repair; and Functions Induced by Damaged DNA. Those selected by the organizers to present papers were among those research leaders whom we would have been pleased to invite to present extended papers in their respective fields, so it was arranged to present the proceedings in this special, separate volume. Its form, therefore, differs somewhat from that of the typical volume, which would contain a small number of longer contributions on diverse subjects, but if each section is viewed as a multiauthored survey of its particular area, the difference is not as great as a cursory inspection of the Table of Contents might suggest.

Comments and suggestions from readers are desired. As stated in an earlier Preface, "We seek to provide a forum for discussion . . . and we welcome suggestions . . . as to how this end may best be served."

WALDO E. COHN

Abbreviations and Symbols

All contributors to this Series are asked to use the terminology (abbreviations and symbols) recommended by the IUPAC-IUB Commission on Biochemical Nomenclature (CBN) and approved by IUPAC and IUB, and the Editor endeavors to assure conformity. These Recommendations have been published in many journals (*1, 2*) and compendia (*3*) in four languages and are available in reprint form from the Office of Biochemical Nomenclature (OBN), as stated in each publication, and are therefore considered to be generally known. Those used in nucleic acid work, originally set out in section 5 of the first Recommendations (*1*) and subsequently revised and expanded (*2, 3*), are given in condensed form (I-V) below for the convenience of the reader. Authors may use them without definition, when necessary.

I. Bases, Nucleosides, Mononucleotides

1. *Bases* (in tables, figures, equations, or chromatograms) are symbolized by Ade, Gua, Hyp, Xan, Cyt, Thy, Oro, Ura; Pur = any purine, Pyr = any pyrimidine, Base = any base. The prefixes S-, H_2, F-, Br, Me, etc., may be used for modifications of these.

2. *Ribonucleosides* (in tables, figures, equations, or chromatograms) are symbolized, in the same order, by Ado, Guo, Ino, Xao, Cyd, Thd, Ord, Urd (Ψrd), Puo, Pyd, Nuc. Modifications may be expressed as indicated in (1) above. Sugar residues may be specified by the prefixes r (optional), d (=deoxyribo), a, x, l, etc., to these, or by two three-letter symbols, as in Ara-Cyt (for aCyd) or dRib-Ade (for dAdo).

3. *Mono-, di-, and triphosphates of nucleosides* (5') are designated by NMP, NDP, NTP. The N (for "nucleoside") may be replaced by any one of the nucleoside symbols given in II-1 below. 2'-, 3'-, and 5'- are used as prefixes when necessary. The prefix d signifies "deoxy." [Alternatively, nucleotides may be expressed by attaching P to the symbols in (2) above. Thus: P-Ado = AMP; Ado-P = 3'-AMP] cNMP = cyclic 3':5'-NMP; Bt_2cAMP = dibutyryl cAMP, etc.

II. Oligonucleotides and Polynucleotides

1. Ribonucleoside Residues

(a) Common: A, G, I, X, C, T, O, U, Ψ, R, Y, N (in the order of I-2 above).

(b) Base-modified: sI or M for thioinosine = 6-mercaptopurine ribonucleoside; sU or S for thiouridine; brU or B for 5-bromouridine; hU or D for 5,6-dihydrouridine; i for isopentenyl; f for formyl. Other modifications are similarly indicated by appropriate *lower-case* prefixes (in contrast to I-1 above) (*2, 3*).

(c) Sugar-modified: prefixes are d, a, x, or l as in I-2 above; alternatively, by *italics* or boldface type (with definition) unless the entire chain is specified by an appropriate prefix. The 2'-*O*-methyl group is indicated by *suffix* m (e.g., -Am- for 2'-*O*-methyladenosine, but -mA- for 6-methyladenosine).

(d) Locants and multipliers, when necessary, are indicated by superscripts and subscripts, respectively, e.g., -m_2^6A- = 6-dimethyladenosine; -s^4U- or -^{4}S- = 4-thiouridine; -ac^4Cm- = 2'-*O*-methyl-4-acetylcytidine.

(e) When space is limited, as in two-dimensional arrays or in aligning homologous sequences, the prefixes may be placed *over the capital letter*, the suffixes *over the phosphodiester symbol*.

2. Phosphoric Residues [left side = 5', right side = 3' (or 2')]

(a) Terminal: p; e.g., pppN... is a polynucleotide with a 5'-triphosphate at one end; Ap is adenosine 3'-phosphate; C > p is cytidine 2':3'-cyclic phosphate (*1, 2, 3*); p < A is adenosine 3':5'-cyclic phosphate.

(b) Internal: hyphen (for known sequence), comma (for unknown sequence); unknown sequences are enclosed in parentheses. E.g., pA-G-A-C(C_2,A,U)A-U-G-C > p is a sequence with a (5') phosphate at one end, a 2':3'-cyclic phosphate at the other, and a tetranucleotide of unknown sequence in the middle. (**Only codon triplets should be written without some punctuation separating the residues.**)

3. Polarity, or Direction of Chain

The symbol for the phosphodiester group (whether hyphen or comma or parentheses, as in 2b) represents a 3'-5' link (i.e., a 5'...3' chain) unless otherwise indicated by appropriate numbers. "Reverse polarity" (a chain proceeding from a 3' terminus at left to a 5' terminus at right) may be shown by numerals or by right-to-left arrows. Polarity in any direction, as in a two-dimensional array, may be shown by appropriate rotation of the (capital) letters so that 5' is at left, 3' at right when the letter is viewed right-side-up.

4. Synthetic Polymers

The complete name or the appropriate group of symbols (see II-1 above) of the repeating unit, **enclosed in parentheses if complex or a symbol,** is either (a) preceded by "poly," or (b) followed by a subscript "n" or appropriate number. **No space follows "poly"** (2, 5).

The conventions of II-2b are used to specify known or unknown (random) sequence, e.g.,

polyadenylate = poly(A) or A_n, a simple homopolymer;

poly(3 adenylate, 2 cytidylate) = poly(A_3C_2) or $(A_3,C_2)_n$, an *irregular* copolymer of A and C in 3:2 proportions;

poly(deoxyadenylate-deoxythymidylate) = poly[d(A-T)] or poly(dA-dT) or $(dA-dT)_n$ or $d(A-T)_n$, an *alternating* copolymer of dA and dT;

poly(adenylate,guanylate,cytidylate,uridylate) = poly(A,G,C,U) or $(A,G,C,U)_n$, a random assortment of A, G, C, and U residues, proportions unspecified.

The prefix copoly or oligo may replace poly, if desired. The subscript "n" may be replaced by numerals indicating actual size, e.g., $A_n \cdot dT_{12-18}$.

III. Association of Polynucleotide Chains

1. *Associated* (e.g., H-bonded) chains, or bases within chains, are indicated by a *center dot* (not a hyphen or a plus sign) separating the *complete* names or symbols, e.g.:

poly(A) · poly(U)   or $A_n \cdot U_m$
poly(A) · 2 poly(U)   or $A_n \cdot 2U_m$
poly(dA-dC) · poly(dG-dT)   or $(dA-dC)_n \cdot (dG-dT)_m$.

2. *Nonassociated* chains are separated by the plus sign, e.g.:

2[poly(A) · poly(U)] → poly(A) · 2 poly(U) + poly(A)
or $2[A_n \cdot U_m] \to A_n \cdot 2U_m + A_n$.

3. Unspecified or unknown association is expressed by a comma (again meaning "unknown") between the completely specified chains.

Note: In all cases, each chain is completely specified in one or the other of the two systems described in II-4 above.

IV. Natural Nucleic Acids

RNA	ribonucleic acid or ribonucleate
DNA	deoxyribonucleic acid or deoxyribonucleate
mRNA; rRNA; nRNA	messenger RNA; ribosomal RNA; nuclear RNA
hnRNA	heterogeneous nuclear RNA
D-RNA; cRNA	"DNA-like" RNA; complementary RNA

mtDNA	mitochondrial DNA
tRNA	transfer (or acceptor or amino-acid-accepting) RNA; replaces sRNA, which is not to be used for any purpose
aminoacyl-tRNA	"charged" tRNA (i.e., tRNA's carrying aminoacyl residues); may be abbreviated to AA-tRNA
alanine tRNA or tRNA^{Ala}, etc.	tRNA normally capable of accepting alanine, to form alanyl-tRNA, etc.
alanyl-tRNA or alanyl-tRNA^{Ala}	The same, with alanyl residue covalently attached. [*Note:* fMet = formylmethionyl; hence tRNA^{fMet}, identical with tRNA$_f^{Met}$]

Isoacceptors are indicated by appropriate subscripts, i.e., tRNA$_1^{Ala}$, tRNA$_2^{Ala}$, etc.

V. Miscellaneous Abbreviations

P_i, PP_i	inorganic orthophosphate, pyrophosphate
RNase, DNase	ribonuclease, deoxyribonuclease
t_m (not T_m)	melting temperature (°C)

Others listed in Table II of Reference 1 may also be used without definition. No others, with or without definition, are used unless, in the opinion of the editor, they increase the ease of reading.

Enzymes

In naming enzymes, the 1978 recommendations of the IUB Commission on Biochemical Nomenclature (4) are followed as far as possible. At first mention, each enzyme is described *either* by its systematic name *or* by the equation for the reaction catalyzed *or* by the recommended trivial name, followed by its EC number in parentheses. Thereafter, a trivial name may be used. Enzyme names are not to be abbreviated except when the substrate has an approved abbreviation (e.g., ATPase, but not LDH, is acceptable).

REFERENCES*

1. *JBC* **241**, 527 (1966); *Bchem* **5**, 1445 (1966); *BJ* **101**, 1 (1966); *ABB* **115**, 1 (1966), **129**, 1 (1969); and elsewhere.†
2. *EJB* **15**, 203 (1970); *JBC* **245**, 5171 (1970); *JMB* **55**, 299 (1971); and elsewhere.†
3. "Handbook of Biochemistry" (G. Fasman, ed.), 3rd ed. Chemical Rubber Co., Cleveland, Ohio, 1970, 1975, Nucleic Acids, Vols. I and II, pp. 3–59.
4. "Enzyme Nomenclature" [Recommendations (1978) of the Nomenclature Committee of the IUB]. Academic Press, New York, 1979.
5. "Nomenclature of Synthetic Polypeptides," *JBC* **247**, 323 (1972); *Biopolymers* **11**, 321 (1972); and elsewhere.†

Abbreviations of Journal Titles

Journals	Abbreviations used
Annu. Rev. Biochem.	ARB
Arch. Biochem. Biophys.	ABB
Biochem. Biophys. Res. Commun.	BBRC

*Contractions for names of journals follow.

†Reprints of all CBN Recommendations are available from the Office of Biochemical Nomenclature (W. E. Cohn, Director), Biology Division, Oak Ridge National Laboratory, Box Y, Oak Ridge, Tennessee 37830, USA.

Biochemistry	Bchem
Biochem. J.	BJ
Biochim. Biophys. Acta	BBA
Cold Spring Harbor Symp. Quant. Biol.	CSHSQB
Eur. J. Biochem.	EJB
Fed. Proc.	FP
Hoppe-Seyler's Z. physiol. Chem.	ZpChem
J. Amer. Chem. Soc.	JACS
J. Bacteriol.	J. Bact.
J. Biol. Chem.	JBC
J. Chem. Soc.	JCS
J. Mol. Biol.	JMB
Nature, New Biology	Nature NB
Nucleic Acid Research	NARes
Proc. Nat. Acad. Sci. U.S.	PNAS
Proc. Soc. Exp. Biol. Med.	PSEBM
Progr. Nucl. Acid Res. Mol. Biol.	This Series

Some Articles Planned for Future Volumes

tRNA Splicing in Lower Eukaryotes
J. ABELSON AND G. KNAPP

Ribosomal RNA: Structure and Interactions with Proteins
R. BRIMACOMBE

Metabolism and Function of Cyclic Nucleotides
W. Y. CHEUNG

Accuracy of Protein Synthesis: A Reexamination of Specificity in Codon–Anticodon Interaction
H. GROSJEAN AND R. BUCKINGHAM

Mechanism of Interferon Action
G. SEN

The Regulatory Function of the 3'-Region of mRNA and Viral RNA Translation
U. LITTAUER AND H. SOREQ

Participation of Aminoacyl-tRNA Synthetases and tRNAs in Regulatory Processes
G. NASS

Queuine
S. NISHIMURA

Viral Inhibition of Host Protein Synthesis
A. SHATKIN

Ribosomal Proteins: Structure and Function
A. R. SUBRAMANIAN

RNA-Helix Destabilizing Proteins
W. SZER AND J. O. THOMAS

Introduction: DNA-Multiprotein Interactions in Transcription, Replication, and Repair

| R. K. FUJIMURA[1]
| Biology Division
| Oak Ridge National Laboratory
| Oak Ridge, Tennessee

Arthur Kornberg once said[2] that enzymes have at least three faces, a "catalytic" face, a "regulatory" or "allosteric" face, and a "social" face, the last being the one that recognizes neighboring cellular constituents. This symposium tried to deal with these three aspects of enzymes that interact with DNA. As the title suggests, functional aspects of the interaction were stressed over structural aspects. One objective was to inform scientists in related fields of the latest developments in other fields. Some up-to-the-minute findings were reported, so that even the people directly involved in each respective field learned something new. For the first time, we also had poster sessions in which about two dozen posters were presented to encourage exchanges of nascent ideas. They served the function of stimulating discussion so well that they were left up for three days instead of one as originally planned.

Because of the breadth of the subject and the limited time available, only a few aspects of replication, transcription, and repair were taken up by the formal presentations. They were divided into six sessions as shown in the Table of Contents.

The organizing committee[1] asked Waldo Cohn, the editor of this series, to publish these proceedings, and he and the publisher have agreed. The symposium on mRNA in 1976 was published in a similar

---

[1] Chairman of the Organizing Committee. I thank members of the Organizing Committee of the conference for their advice and suggestions during the process of selecting speakers. They are: R. Julian Preston, Audrey L. Stevens, Paul A. Swenson, John R. Totter, and Wen-Kuang Yang of the Biology Division of the Oak Ridge National Laboratory, and Ronald L. Seale of Scripps Clinic and Research Foundation. We express our appreciation for a grant from the National Science Foundation (Grant No. PCM-8000282) to the University of Tennessee, to bring speakers from abroad. The conference was sponsored by the Biology Division of the Oak Ridge National Laboratory and was supported by the Office of Health and Environmental Research, U. S. Department of Energy.

[2] In "Reflections on Biochemistry" (A. Kornberg, B. L. Horecker, L. Cornudella, and J. Oro, eds.), p. 250. Pergamon, New York, 1976.

manner. Learning from that experience, we have avoided some time-consuming aspects of the endeavor by leaving the decision of whether or not to publish his talk up to each speaker and by not incorporating discussions or unscheduled talks.

Many of the presentations were reviews of recent advances in the field. Most of the speakers or their laboratories have contributed significantly to advances in their respective fields. I hope readers will agree that these contributions are of value to people working on various aspects of DNA–protein interactions.

I. Replicative DNA Polymerase and Its Complex

Chairman and Summarizer: DAVID KORN

Summary DAVID KORN	5
Enzyme Studies of ϕX174 DNA Replication KEN-ICHI ARAI, NAOKO ARAI, JOSEPH SHLOMAI, JOAN KOBORI, LAURIEN POLDER, ROBERT LOW, ULRICH HÜBSCHER, LEROY BERTSCH, AND ARTHUR KORNBERG	9
The DNA Replication Origin (ori) of *Escherichia coli*: Structure and Function of the *ori*-Containing DNA Fragment YUKINORI HIROTA, MASAO YAMADA, AKIKO NISHIMURA, ATSUHIRO OKA, KAZUNORI SUGIMOTO, KIYOZO ASADA, AND MITSURU TAKANAMI	33
Replication of Linear Duplex DNA *in Vitro* with Bacteriophage T5 DNA Polymerase R. K. FUJIMURA, S. K. DAS, D. P. ALLISON, AND B. C. ROOP	49
Mechanisms of Catalysis of Human DNA Polymerases α and β DAVID KORN, PAUL A. FISHER, AND TERESA S.-F. WANG	63
Structural and Functional Properties of Calf Thymus DNA Polymerase δ MARIETTA Y. W. TSANG LEE, CHENG-KEAT TAN, KATHLEEN M. DOWNEY, AND ANTERO G. SO	83

Summary

DAVID KORN

Laboratory of Experimental
  Oncology
Department of Pathology
Stanford University School of
  Medicine
Stanford, California

This session included five presentations concerned with prokaryotic DNA replication and with the structure and catalytic properties of the phage T5 DNA polymerase and of DNA polymerases α, β, and δ from mammalian tissues.

The first paper, delivered by K.-I. Arai, summarized a large amount of recent work from the laboratory of Arthur Kornberg that focused on the replication *in vitro* of several model phage DNA genomes by purified protein components. The most complex of these systems is that involved in the replicative sequence of ϕX174, SS $\rightarrow$ RF $\rightarrow$ SS. With respect to the first phase of this sequence, SS $\rightarrow$ RF, the requirements for the formation of a faithful initiation complex proved to be unexpectedly complicated and have been the object of intensive scrutiny for many years. At present, it is believed that the formation of the initiation complex on ϕX174 DNA requires coating of the template with DNA-binding protein (DBP) and the sequential participation of *Escherichia coli* proteins i, n, n', n", dnaB, dnaC, and "primase."[1] The n' protein has been shown recently to be a ϕX174 ssDNA-dependent ATPase, whose activity is not abolished by DBP, and which appears to recognize a specific hairpin structure, composed of about 55 nucleotides, that is presumed to be the origin for complementary-strand synthesis. The dnaB protein has been dubbed a "mobile promoter"; i.e., on a circular ssDNA template and without DBP, the dnaB protein plus primase will form multiple, but nonrandom, primers that are capable of being recognized and elongated by the holoenzyme of *E. coli* DNA polymerase III. In the presence of DBP, in contrast, the "general priming reaction" is prevented, and all the other proteins noted above are required for the "specific priming reaction" at the origin for complementary strand synthesis. The dnaB protein is known to be a DNA-

---

[1] Primase, the polymerase that synthesizes the primer for DNA polymerase, is a product of the *dnaG* gene.

dependent ATPase. It is currently believed that the protein, in the presence of ATP, is modified into a form with very high affinity for ϕX174 DNA, and that the tightly bound protein–ATP complex provides a recognition signal for primase. It is further believed that upon hydrolysis of the bound ATP, the resulting protein–ADP complex adopts a different "conformation" that has low affinity for DNA and thus dissociates from the template. Using restriction mapping techniques, the Kornberg group has obtained evidence that the protein n' recognition site resides between genes F and G, within both HaeIII fragment z1 and the HhaI fragment 6 of the ϕX174 genome, and that the movement of the dnaB protein occurs in a direction ($5' \rightarrow 3'$ on the template strand) opposite to that of the DNA polymerase. Arai then proceeded to summarize additional features of the SS → RF → SS sequence, most of which have been published. The group has succeeded in coupling these reactions to obtain a rapid and efficient multiplication of RF and the new synthesis of ssDNA progeny.

Yukinori Hirota presented the results of recent studies of the physical mapping of the *E. coli* origin of replication (*ori*), which he and his colleagues had earlier succeeded in cloning in a nonreplicating DNA vector. Techniques of exonuclease and restriction nuclease digestion and modification were used to identify the essential sequence and topological information contained in the *ori* segment. Particularly imaginative was the use of three independent biological criteria to test for the preservation or elimination of ori function: (*a*) insertion in an F plasmid and testing its ability to replicate in an Hfr host; (*b*) insertion into phage λ in place of the "*c*" region and testing its ability to replicate in an iλ lysogen; and (*c*) insertion into Col E1 and testing replicability in a polA host. These studies reveal that the minimum length of DNA necessary for ori function is a 245- to 255-base-pair segment that is stringently bounded on its left-hand end by the sequence TATTTTA and on its right-hand end by the sequence CAC; single-base changes within these sequences are sufficient to destroy ori function. The intervening region of about 245 base-pairs is complex in that it is possible to identify multiple alternative potential conformers containing hairpins, loops, etc. The region contains multiple interspersed repeats of the sequence GATC. By use of restriction enzymes to generate deletions and insertions within the intervening region, Hirota and his colleagues obtained results supporting the interpretation that only a relatively few, interspersed specific sequences are essential to ori function; and that what is required is the preservation of complex topological relationships within this region, relationships that appear to be preserved in spite of substantial deletions at some restriction sites, but to be prevented by relatively minimal degrees of sequence perturbation at other such sites. These data support a number of provocative speculations about the function of the *ori* region; among other things, they are compatible with the long-standing hypothesis of *in vivo* interaction of the replication origin with the bacterial cell membrane.

Robert Fujimura presented an up-to-date review of the major structural and catalytic properties of the phage T5 DNA polymerase, an enzyme comprised of a single polypeptide of 96,000 daltons. Work by his group indicates that it is capable of performing a very extensive strand-displacement synthesis on nicked, circular duplex DNAs, but only a limited strand-displacement synthesis on nicked linear duplex DNA (e.g., phage T5 DNA); that the enzyme possesses an intrinsic $3' \rightarrow 5'$-exonuclease activity, the properties of which are consistent with postulated redactory function; and that the polymerase appears to be extremely processive, as assayed with synthetic poly(dA) · oligo(dT) primer-templates. This laboratory has recently obtained a partially purified factor from infected or uninfected endoI polA *E. coli* that appears to enable the T5 polymerase to utilize intact linear duplex T5 DNA, but not intact circular duplex PM2 DNA. The stimulatory fraction appears to be entirely free of detectable endonuclease activity.

David Korn reviewed an extensive series of catalytic and kinetic studies of human DNA polymerases α and β that have illuminated a number of the specific molecular signals that appear to regulate the specific interaction of these polymerases with their DNA substrates. DNA polymerase β, which is capable of performing a limited strand-displacement reaction on nicked, duplex linear or circular DNA molecules, can be shown to interact in a catalytically significant way exclusively with base-paired 3'-primer termini, and the strength of the binding interaction is essentially identical for 3'-terminal OH, P, and H (dideoxy) residues. The polymerase does not interact with intact or flush-ended duplex DNA molecules, and it manifests only a weak interaction with single-stranded, heteropolymeric DNA that appears to be independent of 3' termini and noncompetitive with primer-template. In sharp contrast to these results, DNA polymerase α, which is incapable of utilizing a nicked duplex DNA substrate, demonstrates measurable binding affinity only for single-stranded polydeoxynucleotide (template), and none at all for intact, nicked, or blunt-ended duplex DNA molecules. Kinetic analysis of the single-strand interaction reveals positive cooperativity and indicates that each polymerase molecule possesses a minimum of two strongly interacting template binding sites. Binding to template is independent of the presence of 3'-primer termini; in contrast, the polymerase appears to be capable of interacting with such termini only in concert with, or subsequent to, the binding of single-strand, and the termini must present a minimal degree (two to five base-pairs) of complementarity to template sequence. A further series of studies with synthetic homo- and heterodeoxypolymers and "hook" polymers indicate that polymerase α appears to respond allosterically to base-composition-mediated signals that are generated by its template-binding interactions. These data support the provocative interpretation that polymerase α may be intrinsically capable of recognizing specific start and stop signals that may be crucially involved in the regulation of mammalian DNA replication.

The final paper of this session, by Antero So, described the recent successes obtained by his group in accomplishing the complete purification and structural characterization of a novel species of mammalian DNA polymerase activity, DNA polymerase δ. The enzyme has an estimated molecular weight of about 150,000, polymerase properties that are generally identical to those of DNA polymerase α, and an intrinsic $3' \rightarrow 5'$-exonuclease activity. By gel electrophoresis in dodecyl sulfate, the polymerase appears to be comprised of two major polypeptides of 60,000 and 49,000 daltons, respectively. It is not yet clear how these subunits are associated stoichiometrically to produce the approximately 150,000 dalton catalytically active polymerase. The data presented by So are generally in agreement with those published by Samuel Wilson (NIH) that describe an apparently identical polymerase species from murine myeloma. Wilson and his colleagues distinguished this species from polymerase α per se by designating the two species DNA polymerases $\alpha 1$ and $\alpha 2$, respectively. After a number of years of considerable controversy regarding the legitimacy of the existence of the polymerase δ species, the recent studies by So and his co-workers, together with the publication from Wilson's laboratory, would seem to establish beyond reasonable doubt that the entity is real. Moreover, the Wilson report contains some partial peptide mapping data that show no significant homologies between the myeloma polymerases α and δ. The possible physiological functions of this newest recognized member of the mammalian DNA polymerase family remain to be determined—as indeed they do for polymerases α, β, and γ as well.

Enzyme Studies of ϕX174 DNA Replication

KEN-ICHI ARAI
NAOKO ARAI
JOSEPH SHLOMAI
JOAN KOBORI
LAURIEN POLDER
ROBERT LOW
ULRICH HÜBSCHER
LEROY BERTSCH AND
ARTHUR KORNBERG

Department of Biochemistry
Stanford University School of Medicine
Stanford, California

Replication of the single-stranded DNA of phage ϕX174 can be divided into three stages (*1*). In stage I, the single-stranded circle is converted into a duplex replicative form (RF-I) by the action of the DNA replication enzymes of the *Escherichia coli* host cell. In the second stage, multiple copies of RF-I are generated using the newly synthesized RF-I as a template; this process requires two additional proteins, the phage-encoded gene *A* protein and host rep protein (*2*, *3*). In the final stage, further production of RF-I ceases and only the viral strand is synthesized for envelopment into the capsid of a progeny phage. This last stage requires not only the gene *A* and rep proteins (*4*, *5*), but also the products of the phage genes *B*, *D*, *F*, *G*, and *H*.

The enzymology of *E. coli* replication proteins employed in ϕX DNA replication is a guide to their action in replicating the host chromosome. This contribution deals with two aspects of this process. The first is the mechanism of discontinuous replication required in the first stage, the conversion of the single-strand (SS) viral circle coated with single-strand-binding (SSB) protein into RF—the SS → RF system. We find that production of the complementary strand is initiated at a specific region of the viral strand template by a mobile replication promoter. This mobile replication promoter can move processively on the template in a direction opposite to primer and DNA chain growth. The second aspect is reconstitution of the RF replication system with

Progress in Nucleic Acid Research
and Molecular Biology, Vol. 26

ISBN 0-12-540026-8

In contrast, in φX, primers have been located at many regions of the chromosome (18, 22), whether coupled with DNA replication or uncoupled from it, indicating that priming is initiated at many sites. However, a unique property of protein n' suggests that complementary strand replication of φX also starts at a specific site. Protein n' [probably identical to factor Y (23)] contains an intrinsic ATPase activated specifically by φX DNA. Unlike the low levels of activity seen with M13 and G4 DNA effectors, it is not abolished by coating the φX DNA with SSB protein (10). This suggests that protein n' recognizes a φX DNA region(s) not coated with SSB protein. The specific DNA sequence recognized by protein n' was determined after fragmentation of φX DNA by restriction endonuclease *Hae*III, followed by digestion with exonuclease VII (Fig. 2).

A 55-nucleotide fragment within the untranslated region between genes F and G, a location analogous to that of the complementary strand origin of G4 DNA (24), specifically supported the ATPase activity of protein n' (25). Within this fragment, there is a 44-nucleotide sequence with a potential for forming a relatively stable hairpin structure.

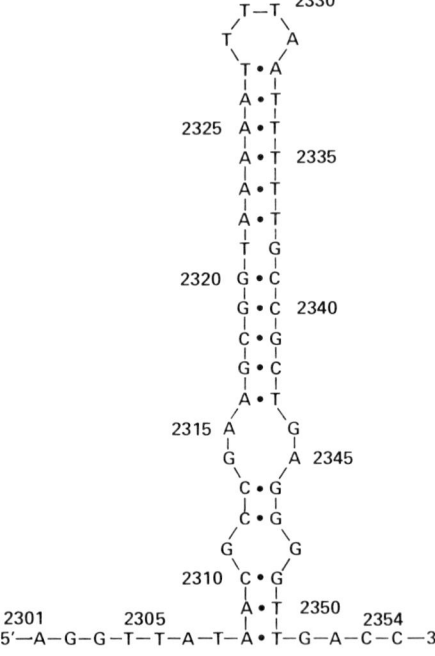

FIG. 2. Potential secondary structure in the recognition locus of protein n' (25).

III. Only dnaB Protein, among the Prepriming Proteins, Is Required with Uncoated φX DNA

An important clue to the mechanism of initiation of complementary strand replication was obtained only after extensive purification of each protein. At an earlier point in reconstitution studies (in 1974), the priming system was dependent on proteins n, i, dnaC, dnaB, and primase (Table II) (26). SSB protein and spermidine were also absolutely needed. With a more highly purified system (in 1978), proteins n' and n" were added to the list, and all the prepriming proteins and primase were strictly required. However, spermidine was no longer needed and, furthermore, omission of spermidine completely abolished the requirement for SSB protein. A further analysis of the requirements showed that only dnaB protein and primase were needed to initiate DNA synthesis on uncoated φX DNA (27). This simple priming system is active on any uncoated single-stranded phage DNA and also on poly(dT) and is inhibited simply by coating DNA with SSB protein. These results indicate that, on uncoated single-strand DNA, dnaB protein and primase act as a nonspecific general priming system synthesizing multiple primers that are then elongated by DNA polymerase III holoenzyme (Fig. 3).

The general priming system is active only on single-stranded regions, since it is completely inhibited by SSB protein. Simply coating the phage DNA with SSB protein restores template specificity. SSB-

TABLE II
REQUIREMENTS FOR φX174 SS → RF-II REACTION

Component omitted	DNA synthesis[a] (pmol)	
	1974	1978
None (complete)	120	210
dnaB protein	1	1
dnaC protein	8	2
Protein i	15	3
Proteins n + n"	5	2
Protein n'	—	7
Primase	8	4
DNA polymerase III holoenzyme	2	<1
Single-strand-binding protein	18	51
Spermidine	2	205
Spermidine and single-strand-binding protein	1	180

[a] Data are from references 26 and 27.

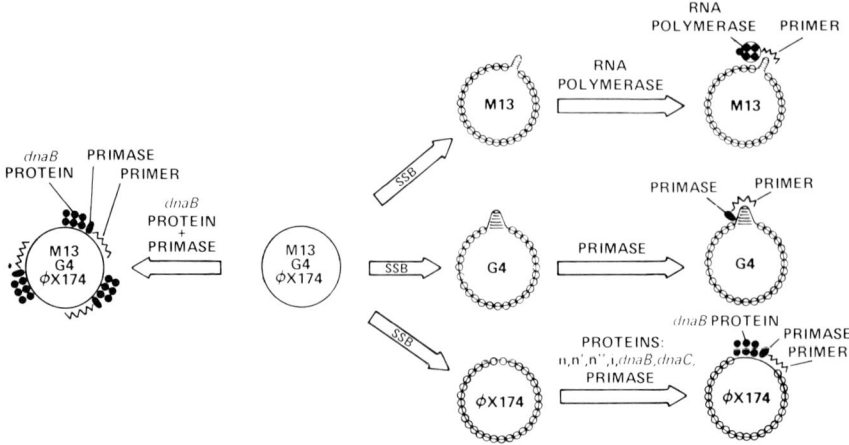

FIG. 3. Scheme for nonspecific priming by dnaB protein and primase on uncoated DNA and the specific priming system for M13, G4, and φX DNAs coated with single-strand-binding protein (SSB) (27).

coated φX DNA is reactivated for the dnaB protein-primase priming system by the action of the five additional prepriming proteins. Of key interest is how dnaB protein and primase initiate priming on single-stranded, uncoated regions of DNA and how the prepriming proteins reactivate SSB-coated φX DNA.

IV. A General Priming System Employing dnaB Protein and Primase

The general priming action of dnaB protein and primase was demonstrated by the synthesis of RNA primers on uncoated ssDNA. RNA synthesis was completely dependent on dnaB protein, primase, and DNA template and was inhibited by SSB protein. The amount of RNA synthesized was surprisingly large; as much as 30% of template was covered with primers (27). dNTPs were also incorporated if ATP or ATP(γS) was present (K. Arai, S. Yasuda and A. Kornberg, unpublished). Analysis of the RNA products revealed several important features (Fig. 4). First, multiple primers are synthesized on each template; their sizes range from 10 to 60 nucleotides. Thus the system has an intrinsic property to synthesize short primers. Second, the pattern of primers is not random, but rather characteristic for each template, indicating that particular regions of DNA are preferentially recognized by dnaB protein and primase. Third, SSB protein inhibits the

ENZYME STUDIES OF φX174 DNA REPLICATION

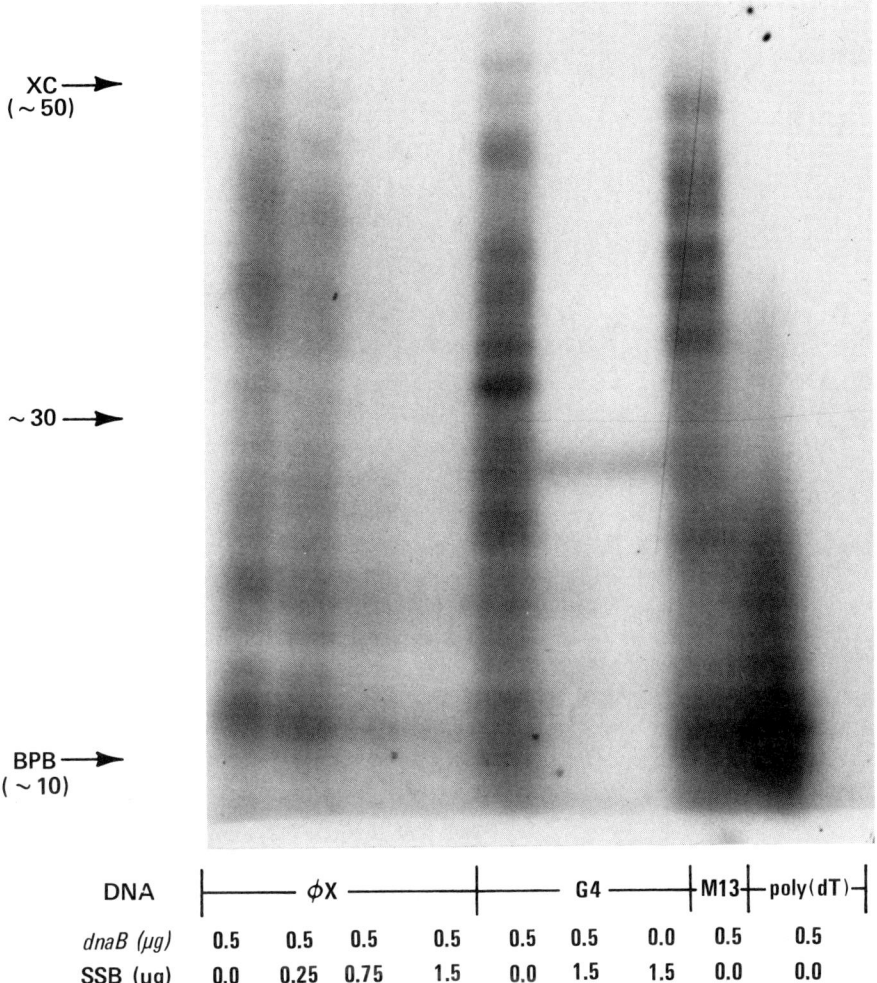

FIG. 4. Electrophoresis of RNA primer transcripts. The reaction mixtures (25 μl) contained [α-^{32}P]rNTPs, phage DNA, 0.15 μg of primase, and additional components as indicated. The poly(dT) reaction mixture contained [α-^{32}P]ATP and poly(dT). Incubation was for 30°C for 30 minutes. RNA transcripts were fractionated by electrophoresis in 15% polyacrylamide gel containing 7 M urea and located by autoradiography (27). Approximate lengths are given on the left in nucleotides. XC, xylene cyanol; BPB, bromophenol blue.

general priming system, yet allowing, on G4 DNA, the synthesis of the specific primer identical to that produced by primase in the absence of dnaB protein. These results indicate that the general priming system initiates priming at single-stranded regions, and its activity on poly(dT) further supports this conclusion.

Hybridization of RNA primers to various restriction fragments of φX DNA have shown that virtually all regions are transcribed (K. Arai, unpublished). Moreover, the pattern of primers synthesized by the specific priming system on coated DNA was not distinguishable in terms of size and distribution from that of the nonspecific (general) priming system on uncoated φX DNA. Clearly, recognition of DNA regions for priming is carried out primarily by dnaB protein and primase; the binding protein and other prepriming proteins do not appear to participate significantly in this process.

V. Formation of (φX DNA) · (dnaB Protein) · Primase Complexes

Formation of a (dnaB protein) · primase · DNA complex as a result of interaction of dnaB protein and primase was demonstrated by the binding of radioactive dnaB protein and primase to DNA (Table III). With or without ATP, very little (dnaB protein) · DNA complex could be isolated. However, the presence of both ATP and primase significantly increased the stability of a (dnaB protein) · DNA complex. ADP was not effective, but, with ATP present in the reaction mixture, ADP could stabilize the complex during gel filtration. The dnaB protein

TABLE III
BINDING OF dnaB PROTEIN TO φX174 DNA[a]

Additions		dnaB protein bound (molecule/circle)
Reaction mixture	Gel filtration	
None	None	0.04
ATP	ATP	0.33
Primase, ATP	ATP	1.30
Primase, ATP	ADP	1.10
Primase, ADP	ADP	0.16
Primase, four rNTPs	ATP	1.86
Primase, four rNTPs, SSB	ATP	0.00
p[NH]ppA	p[NH]ppA	1.94
p[NH]ppA, SSB	p[NH]ppA	0.00

[a] Labeling of dnaB protein was performed according to Rice and Means (28). ^{3}H-labeled dnaB protein (1.5 μg) was mixed with 1.5 nmol (as nucleotide) of φXDNA in the presence or the absence of additional components as indicated. The amount of primase and single-strand-binding protein (SSB) was 0.4 μg and 4 μg, respectively. After incubation at 30°C for 5 minutes, the reaction mixtures were filtered through BioGel A-5m (600 μl column) at room temperature and the ^{3}H complexed with DNA was determined.

was also bound to DNA in the presence of primase and the four rNTPs, the condition for synthesis of multiple primers. SSB protein completely inhibited this process. Similar experiments with labeled primase demonstrated the formation of a DNA · primase complex in the presence of dnaB protein and four rNTPs. These results suggest that dnaB protein and primase interact directly on DNA and form a (dnaB protein) · primase complex. The dnaB protein binds very tightly without primase if p[NH]ppA[2], a nonhydrolyzable analog of ATP, is used, providing SSB protein is absent. This observation suggests that dnaB protein, complexed with ATP and without hydrolysis, binds tightly to DNA and provides a recognition signal for primase action. It is therefore essential to understand the nature of the interaction of dnaB protein with single-stranded DNA.

VI. dnaB Protein as a Replication Promoter

The reaction mechanism for the DNA-dependent ATPase action by dnaB protein revealed by DNA binding and ATP binding studies is outlined in Fig. 5. First, dnaB protein forms a binary complex with

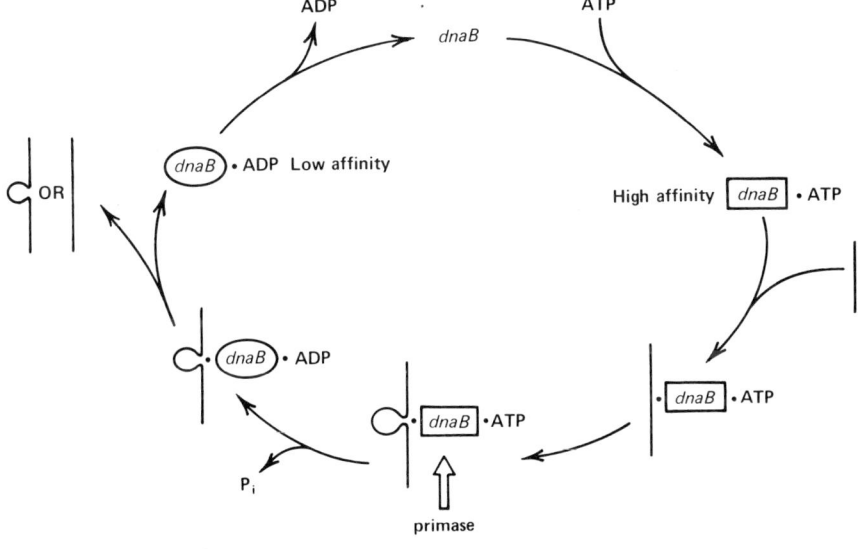

FIG. 5. Scheme for DNA-dependent ATPase reaction of dnaB protein.

[2] Adenosine 5'-[β,γ-imido]triphosphate, sometimes also abbreviated AMP-P[NH]P as in Seeberg, this volume. See *PNAS* **74**, 2222 (1977). [Ed.]

ATP with a strong affinity for ssDNA. The dnaB · ATP complex then binds DNA to form the ternary complex, DNA · dnaB · ATP, creating a unique secondary structure in the DNA. (The dnaB protein probably wraps single-stranded DNA around its oligomeric form.) Primase then recognizes the unique DNA structure complexed with dnaB protein. Thus, in its role in general priming, dnaB protein can be regarded as a replication promoter for primase action. Hydrolysis of ATP is not required up to this point; synthesis of multiple primers proceeds without ATP hydrolysis. By hydrolysis of ATP, the (dnaB protein) · ATP complex is converted to a (dnaB protein) · ADP complex, which has a low affinity for DNA. Dissociation from DNA enables another cycle of binding and ATPase action to ensue, in effect, a distributive mechanism.

Conformational changes in dnaB protein at the ssDNA binding site, induced by an interconversion of the liganded ATP and ADP, must play a key role. ATP utilization by dnaB protein has features in common with GTP utilization in peptide elongation on ribosomes (29) and with supercoiling of duplex DNA catalyzed by "DNA gyrase" (30). A possible model for general priming is shown in Fig. 6a. In general priming, dnaB protein "engineers" the DNA to produce a secondary structure within its domain. The system has the intrinsic property of synthesizing only short primers, inasmuch as the priming by pri-

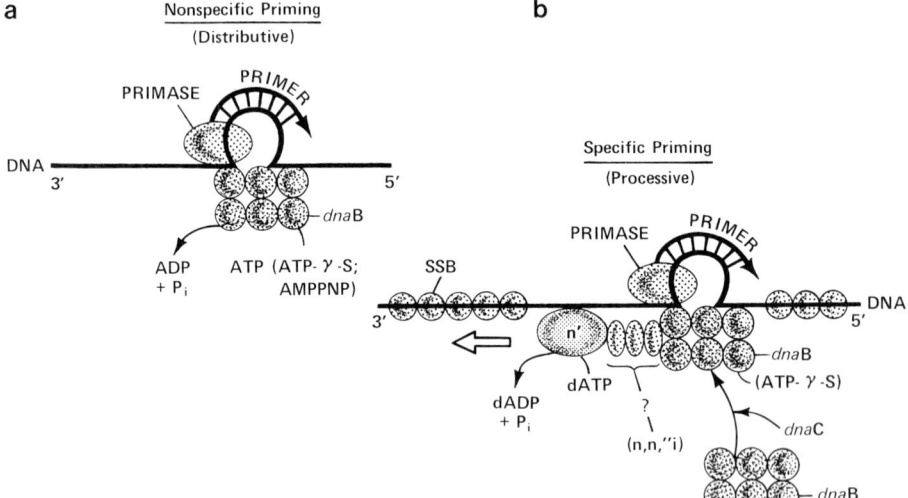

FIG. 6. Hypothetical mechanism for nonspecific priming (a) and specific priming (mobile replication promoter) (b). ATP-γ-S is adenosine 5'-[γ-thio]triphosphate, sometimes abbreviated ATP[γS].

mase takes place only in the DNA domain determined by dnaB protein. Synthesis of multiple primers does not require ATP hydrolysis as the binding of dnaB protein can take place at multiple sites on ssDNA.

VII. Processivity of dnaB Protein in the Replication Intermediate

Approximately one molecule of dnaB protein is bound to SSB protein-coated ϕX DNA by the action of prepriming proteins and ATP (12, 17, 18, K. Arai, unpublished), even without primase present. The stability of the (dnaB protein) · (ϕX DNA) complex formed with the

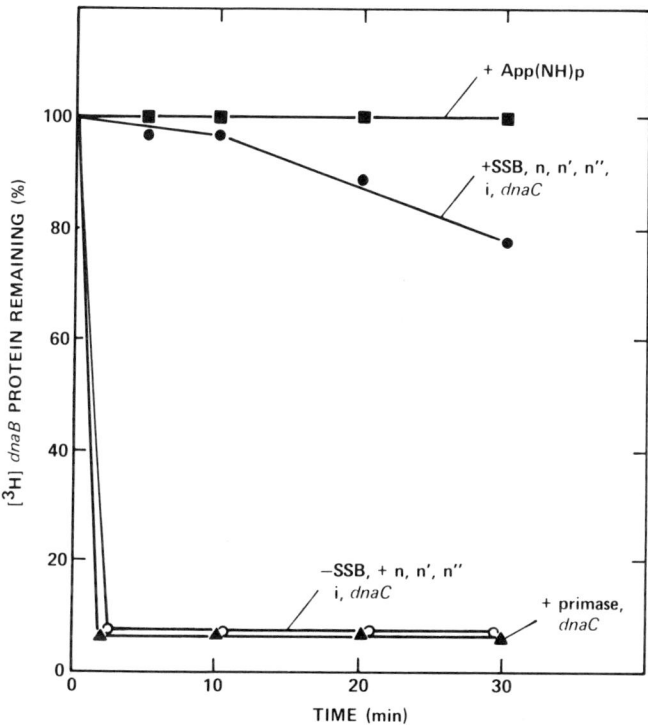

FIG. 7. Exchangeability of dnaB protein bound to ϕX174 DNA. ^{3}H-labeled dnaB protein (0.75 μg) was mixed with ϕX DNA (750 pmol as nucleotide) in the presence of the components as indicated. After incubation at 30°C for 20 minutes, 10 μg of dnaB protein were added and incubation was continued further. At indicated times, reaction mixtures were filtered through a BioGel A-5m column (700 μl) equilibrated with the buffer containing 0.5 mM ATP. When p[NH]ppA^2 was included in the reaction mixture, 0.5 mM p[NH]ppA instead of ATP was present during gel filtration. The ^{3}H radioactivity associated with DNA was determined. SSB, single-strand-binding protein.

TABLE IV
PROPERTIES OF TWO PRIMING SYSTEMS EMPLOYING dnaB PROTEIN AND PRIMASE

Property	General priming	Specific priming[a]
Requirement for prepriming proteins (n, n', n", i, dnaC)	No	Yes
Single-strand-binding protein	No	Yes
K_m for rNTPs	100 μM	10 μM
K_m for dNTPs	20 μM	2 μM
Poly(dT) as template	Yes	No
General sequence recognition by dnaB protein and primase	Yes	Yes
Specific sequence recognition by protein n'	No	Yes
Stability of bound dnaB protein	Unstable	Stable

[a] K_m values for specific priming are from reference 22.

specific priming system can be compared to that in the general priming system (Fig. 7). The (dnaB protein) · DNA complex formed with primase and ATP in the nonspecific system is unstable and is rapidly exchanged with uncomplexed dnaB protein. In contrast, dnaB protein bound with p[NH]ppA is stable and is not displaced. Similarly, dnaB protein bound to SSB protein-coated DNA by prepriming proteins is also remarkably stable and practically nonexchangeable; even the addition of primase and the four rNTPs does not bring about the dissociation of dnaB protein. These results are consistent with the interpretation that dnaB protein in the prepriming replication intermediate catalyzes the synthesis of many primers processively and resists the inhibitory action of SSB protein if primase is added. It would thus seem that some of the prepriming proteins must be linked to dnaB protein to confer upon it these remarkable properties. The processive features of the specific priming systems contrast with the distributive mechanism of the general priming system (Table IV).

VIII. Origin of Complementary Strand Replication

If SSB protein-coated φX DNA has a unique origin for complementary strand replication, then the dnaB protein, as a replication promoter, must migrate along the DNA from the origin, for primase to form multiple transcripts. Thus, the continuous and rapid movement of the replication promoter may obscure the true origin of replication, as primers arise at many sites. To restrict the mobility of the priming system, φX DNA was digested by HaeIII or HhaI endonuclease, known to cleave ssDNA at specific sites (31). Among the fragments produced, the HaeIII fragment Z1 and the HhaI fragment 6 each contain the recognition locus for protein n' (24, 32) (Fig. 8).

Using these fragments as templates, DNA synthesis was carried out in the presence of excess SSB protein. G4 DNA served as a control. As shown in Fig. 9, HhaI fragment 1 of G4 DNA, containing the origin of G4 complementary strand replication (33), specifically directed primase-dependent DNA synthesis. These results show that a circular DNA structure is not essential for recognition by primase.[1]

In φX, the 1353-nucleotide HaeIII Z1 fragment and the 300-nucleotide HhaI fragment 6 specifically directed DNA synthesis. HhaI fragment 3 + 6, generated by partial HhaI digestion, was also active as a template. Each of these fragments contains the 55-nucleotide

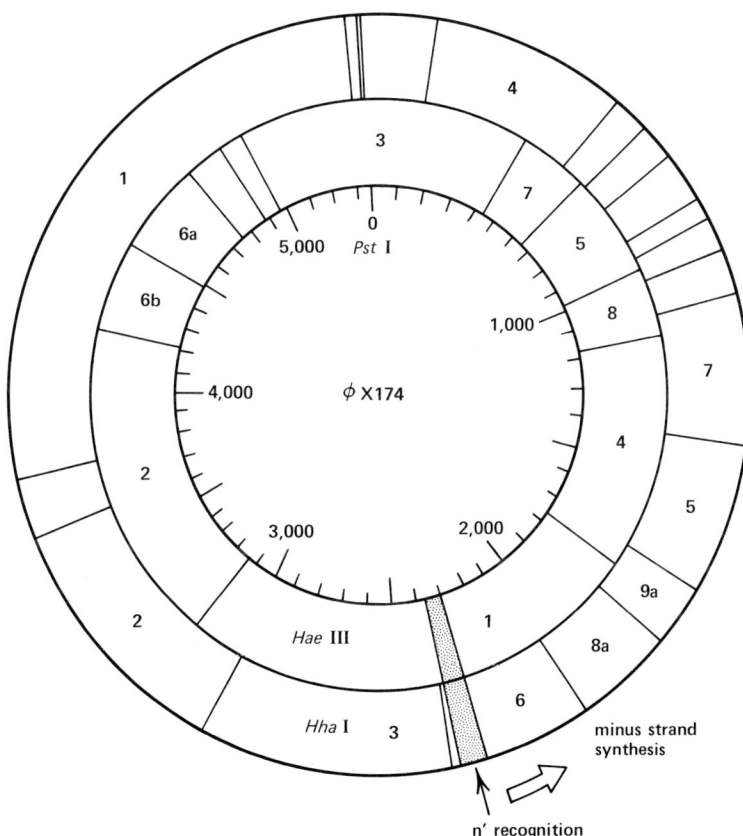

FIG. 8. Restriction cleavage map of φX174 DNA by HaeIII and HhaI and location of the protein n' recognition locus (25, 32).

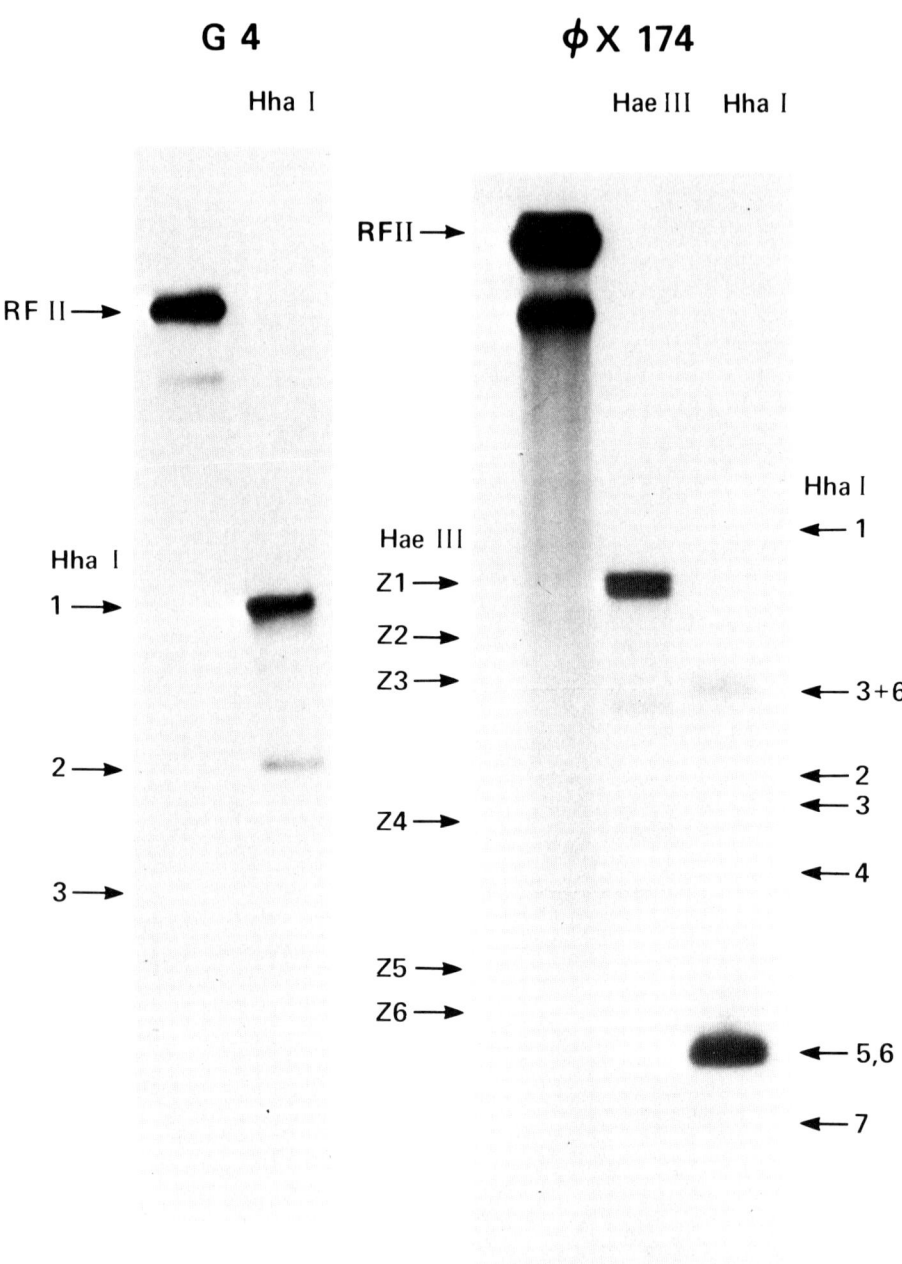

FIG. 9. Electrophoresis of DNA products. DNA synthesis was carried out with phage DNAs or a mixture of phage DNA fragments digested by *Hha*I or *Hae*III as templates. For the G4 system, single-strand-binding protein (SSB), primase, and DNA

locus recognized by protein n'. The other fragments are essentially inactive in the presence of a sufficient amount of SSB protein. These results demonstrate that DNA synthesis is initiated solely at or near the protein n' recognition site. Thus, like G4 and M13, φX DNA has a unique origin for complementary strand replication and forms primers at many regions, because the mobile replication promoter travels processively along the φX DNA.

IX. Polarity of the Mobile Replication Promoter

Polarity of migration of the mobile replication promoter was determined relative to the protein n' recognition locus. According to one model (Fig. 10A), the promoter migrates in an anti-elongation direction, a direction opposite to that of primer and DNA synthesis. This model would fit a role for the mobile replication promoter on the lagging strand at the replication fork entailing synchronous and linked movement of the fork and the promoter. Another model (Fig. 10B) assumes migration of the mobile replication promoter in the same direc-

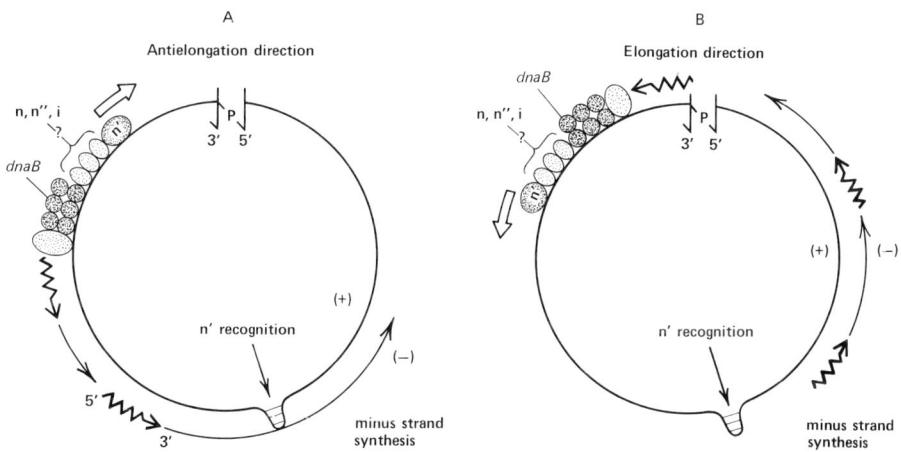

FIG. 10. Two possible models for polarity of the mobile replication promoter.

polymerase III holoenzyme were included. For the φX system, reaction mixtures were further supplemented with proteins n, n', n", i, dnaC, and dnaB. [α-^{32}P]dNTPs were used as substrates. Incubation was at 30°C for 30 minutes. DNA products were fractionated by gel electrophoresis in 1.5% agarose and located by autoradiography. Arrows indicate locations of RF-II and ssDNA fragments generated by HaeIII or HhaI.

tion as that of priming and DNA chain elongation, and requires that the promoter travel ahead of DNA polymerase III holoenzyme.

In a test of these two models, the results were consistent with the anti-elongation direction model (Table V). First, the DNAs synthesized on the Z1 fragment, while heterogeneous, were all longer than 600 nucleotides and were most abundant near the full 1300-nucleotide size of the fragment. The DNA product synthesized on the HhaI fragment 6 was homogeneous in size and always longer than 250 nucleotides. Knowing the relative location of the protein n' recognition locus on these fragments, and assuming movement of the priming system in the direction of RNA and DNA chain elongation, we would have expected to find DNA products shorter than 500 nucleotides on the Z1 fragment and shorter than 250 nucleotides on the HhaI fragment 6. Second, DNA products synthesized on the Z1 fragment hybridized not only to HhaI fragment 6 but also to fragment 3. Similarly, products on HhaI fragment 3 + 6 hybridized with fragment 3. (Note that HhaI fragment 3 is located in the anti-elongation direction relative to the n' recognition site.) Furthermore, by synchronizing initiation of priming through a preincubation with p[NH]ppA and by measuring initiation and abundancy of primers on an intact ϕX circle, a distribution of primers in keeping with the anti-elongation-direction model was obtained (K. Arai, unpublished).

TABLE V
EVIDENCE FOR ANTI-ELONGATION-DIRECTION POLARITY OF THE MOBILE REPLICATION PROMOTER

1. Size of DNA products synthesized on fragments

```
                                   Origin (2301-2355)
              3128         Z1           ↓              1775
   3'  HaeIII  ↓                        □               ↓    5'

       HhaI     ↑                       ↑               ↑
               2977       #3           2363    #6      2063
```

[The HaeIII and HhaI restriction endonuclease cleavage sites at arrows are taken from Sanger et al. (32).]
  a. HaeIII Z1 products are 600–1300 residues
  b. HhaI #6 products are >250 residues
2. Hybridization of DNA products synthesized on fragments
  a. HaeIII Z1 products to HhaI #3
  b. HhaI #3 + 6 products to HhaI #3
3. Synchronized primer formation on ϕX circles
   HaeIII: Z1 >Z2 > Z3 > Z4 is expected for 5' → 3' migration

X. Role of Protein n' in the Mobile Replication Promoter

Hypothetically, assembly of prepriming proteins into the mobile replication promoter takes place at or near the hairpin region recognized by protein n'. Since p[NH]ppA or [S]pppA also promote the formation of the prepriming replication intermediate (K. Arai, unpublished), all the prepriming proteins probably participate in the initial assembly process around the protein n' recognition site. The possibility of the melting of the hairpin by protein n' or the mobile replication promoter needs to be clarified. A unique feature of ϕX complementary strand replication is that its origin need not be the site of formation of a specific primer, since the mobile replication promoter moves rapidly from this site in the 5' to 3' direction of the template strand. Based on an experiment not described here, synthesis of multiple primers on SSB protein-coated ϕX DNA requires the energy of ATP or dATP hydrolysis by protein n'. Through the distinctive dATPase action of protein n' (13, 23) and rNTPase action of dnaB protein (34, K. Arai, unpublished), we have concluded that protein n' plays a key role in driving dnaB protein along DNA after specific sequence recognition at the origin has taken place.

Possible structure–function relationships of the mobile replication promoter are presented in Fig. 6b. In contrast to general priming, the specific priming system employs two functionally different ATPases. The role of dnaB protein as an "engineer" of DNA structure is the same as in general priming. Protein n' acts as an "engine" in utilizing ATP or dATP as fuel for processive migration. Persistence of strong dATPase activity in the replication intermediate is consistent with this model. The stoichiometry of protein n' in prepriming replication intermediate is about one molecule per circle (R. Low, unpublished). Although protein i (18) and dnaC protein (J. Kobori, unpublished) may not be included in the complex, some of the prepriming proteins may mediate the interaction between protein n' and dnaB protein.

Under coupled conditions, primers synthesized by the mobile replication promoter are extended by the holoenzyme of DNA polymerase III to produce RF with a short gap (RF-II). DNA polymerase I excises primers and fills the gaps; DNA ligase seals the ends to produce a covalently complete, but relaxed, RF-I. Finally, DNA gyrase introduces supercoils. We have now completely reconstituted this sequence of events with purified proteins (35). The supercoiled RF-I synthesized by this SS → RF system is as effective a substrate for the next stage of replication (RF → SS) as supercoiled RF isolated from infected cells.

XI. Association of rep Protein and Holoenzyme with (Gene-A Protein) · (RF-II) Complex in a Looped, Rolling-Circle Intermediate

The crude RF replication system *in vitro* (36, 37) can be divided into two stages (38, 39). In the first stage, only viral (+) strand is synthesized (the RF → SS system). Replication of viral strand is initiated

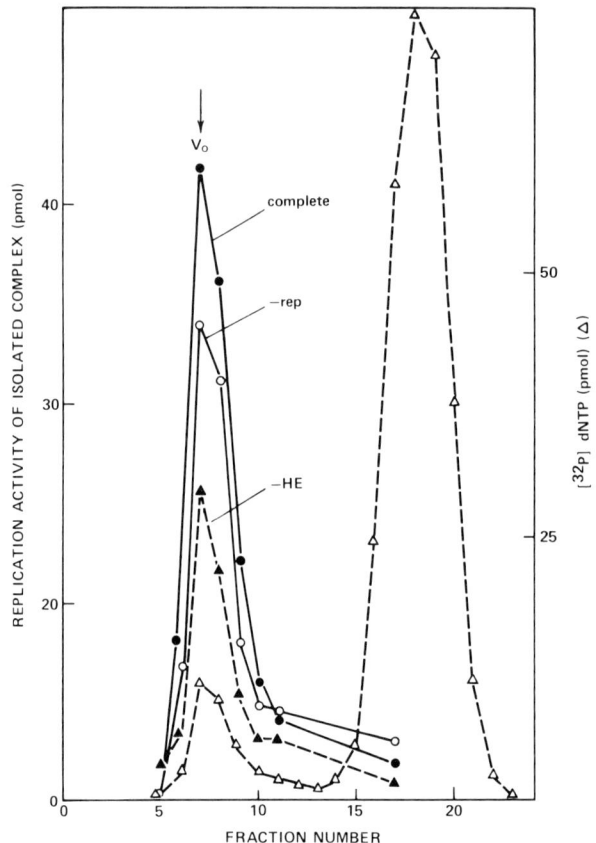

FIG. 11. Isolation of replication intermediate of RF → SS by gel filtration. DNA synthesis was in two steps in a reaction mixture (25 μl) containing φX RF-I, gene-A protein, rep protein, SSB, DNA polymerase III holoenzyme (HE), ATP, and four [α-^{32}P]dNTPs. After incubation at 30°C for 3 minutes, the mixture was passed through Bio-Gel A-5m (700-μl column) to separate the DNA · protein complex from uncomplexed proteins. The DNA · protein complex was then supplemented with single-strand-binding protein (SSB), four [^{3}H]NTPs, rep protein, and DNA polymerase III holoenzyme as indicated. Incubation was for 15 minutes at 30°C. The symbols, except for open triangles, represent replication activity.

by the introduction of a nick by gene-A protein at the origin of RF replication located at a diester bond nearly two-thirds of the distance around the circle from the protein n' recognition site (40–43). Gene-A protein binds covalently to the 5' phosphate end of the nick (44). The rep protein acts as a "helicase," using the energy of ATP hydrolysis (45, 46) to enable replication to proceed continuously in a looped, rolling-circle intermediate (42). Stable association of one molecule of rep protein with the (gene-A protein) · (RF-II) complex has been observed during the unwinding, whether coupled (Fig. 11) to DNA replication or uncoupled from it (data not shown). These results suggest a processive mechanism for rep protein action at the replication fork, in contrast to the distributive mechanism of rep ATPase reaction on ssDNA (N. Arai, K. Arai and A. Kornberg, unpublished). Interaction of rep protein with gene-A protein may prevent the dissociation of rep protein from the replication fork.

When the replication fork has come full circle to the origin, gene-A protein cleaves it again and then ligates to produce a unit-length viral circle. In a subsequent stage, the viral circle can serve as the template for conversion to RF using the same enzyme system described in the SS → RF reaction (38).

XII. Reconstitution of the RF → RF System from Purified Proteins

Earlier attempts to reconstitute the RF → RF system by combining the RF → SS and the SS → RF systems were unsuccessful because of inhibitors in the cruder enzyme preparations. After extensive purification, inhibitors were eliminated to an extent sufficient to enable an RF → RF replication system to be reconstituted from purified proteins (47) (Fig. 12). The products of the RF → SS system are exclusively single-stranded circles; five or more circles are produced per input RF-I after 30 minutes. In the coupled RF → RF system, by contrast, the major product is RF-II. Virtually no single-stranded circles accumulated and about four RF-II molecules were produced per input RF-I. In addition, large complex molecules that moved slowly in gel electrophoresis appeared late in the reaction. These products included ϕX unit-length duplex circles with largely duplex, multigenome-length tails. Synthesis of RF-II and the slow-moving molecules was strictly dependent on each component of SS → RF system. Omission of any one of the prepriming and priming proteins resulted in the synthesis of only viral strands and complete uncoupling from complementary strand replication.

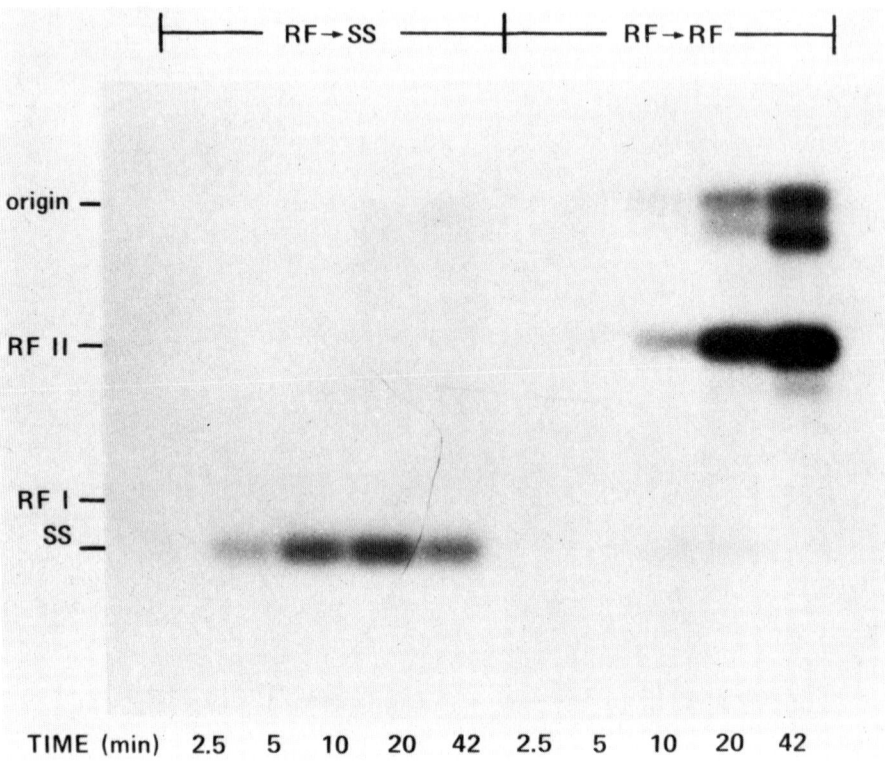

FIG. 12. Separation of DNA products by agarose gel electrophoresis. DNA syntheses (RF → RF; RF → SS) were performed at 30°C. At indicated times, aliquots were treated with sodium dodecyl sulfate and subjected to agarose gel electrophoresis (0.7%). The products were located by autoradiography (47).

Nicking of RF-I by gene-A protein to form the covalent complex is the rate-limiting step in the RF → SS reaction. The (gene-A protein) · (RF-II) complex also proved to be a superior template in the coupled RF → RF replication. More than 10 RF-II molecules were produced per input complex (Fig. 13).

XIII. RNA Primers Are Distributed Only on the Lagging, Complementary Strand

Analysis of RF-II product showed that RNA primers are attached only to the complementary strand and at many regions of the ϕX circle (48). Despite observations of RF multiplication *in vivo* that have led others to propose that both strands are synthesized discontinuously

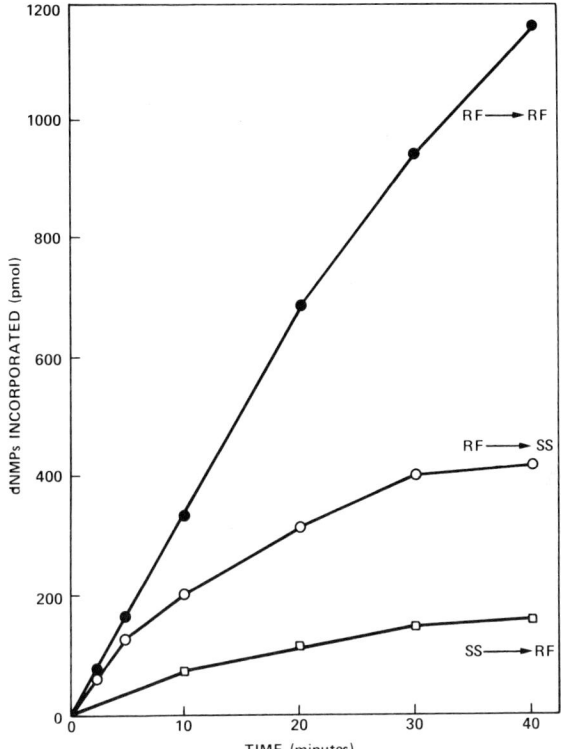

FIG. 13. Rate of DNA synthesis with (gene-A protein) · (RF-II) complex as template. DNA synthesis was with 100 pmol (as nucleotide) of (gene-A protein) · (RF-II) complex for RF → RF or RF → SS and 250 pmol (as nucleotide) of φX DNA for stage I (SS → RF).

(49, 50), the evidence from purified *in vitro* system suggests that only the complementary strand is synthesized discontinuously and that synthesis of viral strand is largely or completely continuous. RNA primers can be excised by DNA polymerase I, which, combined with its gap-filling activity, sets the stage for conversion of RF-II to RF-I by DNA ligase. Participation of more than 20 polypeptides reconstitutes replication of RF → RF in a single reaction mixture. Further work is needed to elucidate precisely how the complementary strand is initiated in RF → RF replication.

A possible structure of the replication fork of the *E. coli* chromosome emerges from these enzyme studies of φX DNA replication (Fig. 14). The leading strand is synthesized by a relatively continuous mechanism; synthesis of the lagging strand is discontinuous. A mobile

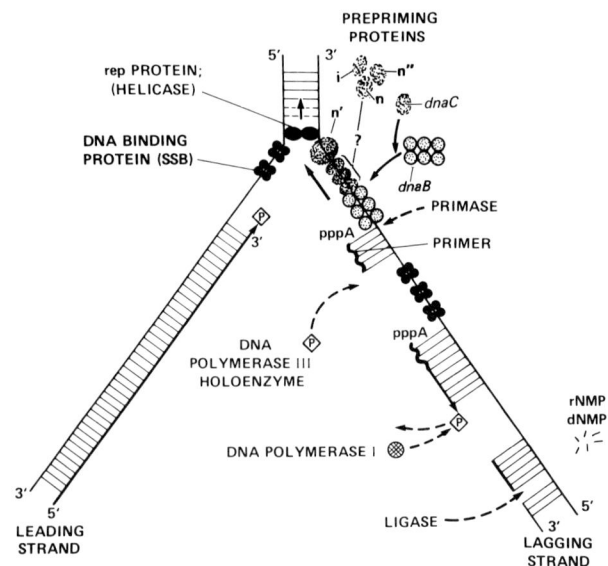

FIG. 14. A scheme for the structure of the replication fork of the *Escherichia coli* chromosome.

replication promoter, assembled at or near the origin of replication, migrates processively, coupled to movement of the replication fork. Multiple, short primers synthesized by primase at numerous, preferred regions are elongated into Okazaki fragments by DNA polymerase III holoenzyme. For efficient coupling of the lagging and leading strands, there may be a mechanism that adjusts the movement of the mobile replication promoter to the rate of unwinding of the duplex at the replication fork. An interesting comparison exists in phage T7 replication. The T7-encoded gene-4 protein is a helicase (51) as well as a primase (52). In phage T4 replication, the T4-encoded gene-41 and -61 proteins appear to be analogous to dnaB protein and primase in facilitating the propulsion of the replication fork (53). Further enzymic studies with various DNA templates, such as plasmids containing the origin of *E. coli* replication (54, 55), may reveal novel features of replication of the *E. coli* chromosome.

References

1. D. T. Denhardt, *CRC Crit. Rev. Microbiol.* **4**, 1611 (1975).
2. E. S. Tessman, *JMB* **17**, 218 (1966).
3. D. T. Denhardt, M. Iwaya and L. L. Larison, *Virology* **49**, 486 (1972).
4. H. Fujisawa and M. Hayashi, *J. Virol.* **19**, 416 (1976).
5. E. S. Tessman and P. K. Peterson, *J. Virol.* **20**, 400 (1976).

6. R. McMacken, J.-P. Bouché, S. L. Rowen, J. H. Weiner, K. Ueda, L. Thelander, C. McHenry and A. Kornberg, in "Nucleic Acid–Protein Recognition (H. J. Vogel, ed.), p. 15. Academic Press, New York, 1977.
7. C. McHenry and A. Kornberg, JBC **252**, 6478 (1977).
8. R. R. Meyer, J. Shlomai, J. Kobori, D. L. Bates, L. Rowen, R. McMacken, K. Ueda, and A. Kornberg, CSHSQB **43**, 289 (1978).
9. A. Kornberg, "DNA Replication." Freeman, San Francisco, California, 1980.
10. K. Geider and A. Kornberg, JBC **249**, 3999 (1974).
11. K. Zechel, J.-P. Bouché and A. Kornberg, JBC **250**, 4684 (1975).
12. J.-P. Bouché, K. Zechel and A. Kornberg, JBC **250**, 5995 (1975).
13. J. Shlomai and A. Kornberg, JBC **255**, 6794 (1980).
14. K. Ueda, R. McMacken and A. Kornberg, JBC **253**, 261 (1978).
15. L. J. Reha-Krantz and J. Hurwitz, JBC **253**, 4043 (1978).
16. S. Wickner and J. Hurwitz, PNAS **72**, 921 (1975).
17. J. H. Weiner, R. McMacken and A. Kornberg, PNAS **73**, 752 (1976).
18. R. McMacken, K. Ueda and A. Kornberg, PNAS **74**, 4190 (1977).
19. B. E. Uhlin, S. Molin, P. Gustafsson and K. Nordström, Gene **6**, 91 (1979).
20. K. Geider, E. Beck and H. Schaller, PNAS **75**, 645 (1978).
21. J.-P. Bouché, L. Rowen and A. Kornberg, JBC **253**, 765 (1978).
22. R. McMacken and A. Kornberg, JBC **253**, 3313 (1978).
23. S. Wickner and J. Hurwitz, PNAS **72**, 3342 (1975).
24. J. C. Fiddes, B. G. Barrell and G. N. Godson, PNAS **75**, 1081 (1978).
25. J. Shlomai and A. Kornberg, PNAS **77**, 799 (1980).
26. R. Schekman, A. Weiner and A. Kornberg, Science **186**, 987 (1974).
27. K. Arai and A. Kornberg, PNAS **76**, 4308 (1979).
28. R. H. Rice and G. E. Means, JBC **246**, 831 (1971).
29. Y. Kaziro, BBA **505**, 95 (1978).
30. A. Sugino, N. P. Higgins, P. O. Brown, C. L. Peebles and N. R. Cozzarelli, PNAS **75**, 4838 (1978).
31. R. W. Blakesley, J. B. Dodgson, I. F. Nes and R. D. Wells, JBC **252**, 7300 (1977).
32. F. Sanger, A. R. Coulson, T. Friedmann, G. M. Air, B. G. Barrell, N. L. Brown, J. C. Fiddes, C. A. Hutchison, P. M. Slocombe and M. Smith, JMB **125**, 225 (1978).
33. G. N. Godson, B. G. Barrell, R. Staden and J. C. Fiddes, Nature **276**, 236 (1978).
34. L. J. Reha-Krantz and J. Hurwitz, JBC **253**, 4051 (1978).
35. J. Shlomai, L. Polder, K. Arai and A. Kornberg, unpublished.
36. S. Eisenberg, J. F. Scott and A. Kornberg, PNAS **73**, 1594 (1976).
37. C. Sumida-Yasumoto, A. Yudelevich and J. Hurwitz, PNAS **73**, 1887 (1976).
38. S. Eisenberg, J. F. Scott and A. Kornberg, PNAS **73**, 3151 (1976).
39. S. Eisenberg, J. F. Scott and A. Kornberg, CSHSQB **43**, 295 (1978).
40. T. J. Henry and R. Knippers, PNAS **71**, 1549 (1974).
41. J. Ikeda, A. Yudelevich and J. Hurwitz, PNAS **73**, 2669 (1976).
42. S. Eisenberg, J. Griffith and A. Kornberg, PNAS **74**, 3198 (1977).
43. S. A. Langeveld, A. D. M. Van Mansfeld, P. D. Bass, H. S. Jansz, G. A. Van Arkel and P. J. Weisbeek, Nature **271**, 417 (1978).
44. A. Eisenberg and A. Kornberg, JBC **254**, 5328 (1979).
45. J. F. Scott, S. Eisenberg, L. L. Bertsch and A. Kornberg, PNAS **74**, 193 (1977).
46. A. Kornberg, J. F. Scott and L. L. Bertsch, JBC **253**, 3298 (1978).
47. K. Arai, N. Arai, J. Shlomai and A. Kornberg, PNAS (in press).
48. N. Arai, L. Polder, K. Arai and A. Kornberg, unpublished.
49. Y. Machida, T. Okazaki and R. Okazaki, PNAS **74**, 2776 (1977).

50. P. D. Bass, W. R. Teerstra and H. S. Jansz, *JMB* **125,** 167 (1978).
51. R. Kolodner and C. C. Richardson *PNAS* **74,** 1525 (1977).
52. E. Scherzinger, E. Lanka, G. Morelli, D. Seiffert and A. Yuki, *EJB* **72,** 543 (1977).
53. C. C. Liu, R. L. Burke, U. Hibner, J. Barry and B. Alberts, *CSHSQB* **43,** 469 (1978).
54. S. Yasuda and Y. Hirota, *PNAS* **74,** 5458 (1977).
55. K. Von Meyenburg, F. G. Hansen, E. Riise, H. E. N. Bergmans, M. Meijer and W. Messer, *CSHSQB* **43,** 121 (1978).

The DNA Replication Origin (*ori*) of *Escherichia coli*: Structure and Function of the *ori*-Containing DNA Fragment

YUKINORI HIROTA
MASAO YAMADA
AKIKO NISHIMURA
ATSUHIRO OKA
KAZUNORI SUGIMOTO
KIYOZO ASADA AND
MITSURU TAKANAMI

National Institute of Genetics
Yata 1,111 Mishima-shi, Japan, and
Institute for Chemical Research of Kyoto University
Uji-shi, Japan

The cell division cycle of *Escherichia coli* involves the regulation of many cellular activities, including initiation of DNA replication, formation of bacterial membrane, and equipartition of DNA copies. The ability of the cell to maintain its genetic content through successive generations and through changes in the environment that alter growth rate indicates that these processes are tightly regulated in concert.

A number of models for the control of initiation of DNA replication and cell division have been proposed (*1*–*6*). It is generally agreed that an essential factor in DNA replication is regulatory and that it operates by changing the frequency of initiation at the replication origin. It has been speculated (*1*) that initiation of DNA replication takes place in concert in the initiation complex, which is a part of the membrane structure. Daughter chromosomes could be segregated by the growth of the membrane between the points of attachment of the DNA replication origin to the membrane in the complex. This complex recognizes the nucleotide sequence of the origin. Thus the origin could control both DNA replication and cell division. It has been shown that the bidirectional DNA replication is triggered at the replication origin located at 82.8 minutes (*7*–*9*) on the chromosome. However, detailed elucidation of the reaction at the origin has been hindered by the large bulk of chromosomal DNA and membrane materials compared with

Progress in Nucleic Acid Research
and Molecular Biology, Vol. 26

ISBN 0-12-540026-8

the specific site of the replication origin and the site of its attachment to the membrane.

Recently, a major breakthrough for this problem came in the successful isolation of the replication origin of *E. coli* chromosome by recombinant DNA techniques. Fragments of *E. coli* chromosomal DNA, obtained by digestion with *Eco* RI, were ligated to a selectable, non-self-replicating DNA fragment (b1a) coding for ampicillin resistance, and the mixture was used to transform *E. coli* cells for the selective marker. Using this procedure, cells carrying the *E. coli* origin region were isolated first by us (*10*), then by others (*11*, *12*). Analyses of this DNA fragment by two independent groups were in good agreement (*13–16*). This approach facilitated the further reduction of the size of the origin region (*17*). A similar isolation of the origin fragment from *Salmonella typhimurium* has been reported (*18*).

In this presentation we report our extensive investigation of the structure and function of the replication origin of *E. coli* DNA based on the nucleotide sequence of the segment, and we propose a structural model of the replication origin.

I. General Survey of the Nucleotide Sequence of the Origin-Containing DNA Fragment of *E. coli*

The cloning and sequencing of the *E. coli* DNA replication origin are now well documented (*10*, *13–16*). A nucleotide segment of 2296 base-pairs containing the DNA replication origin is shown in Fig. 1.

The minimum extent of the DNA segment carrying the information for autonomous replication is reduced to 245 base-pairs [positions 23(G)–267(C)] (*17*), which are boxed in Fig. 1. The nucleotide sequences of the origins of lambda-type phages are within the coding

---

FIG. 1. Nucleotide sequence of the *ori* region between −100 site and +2196 site (*Bam*HI site 4). Nucleotides are numbered from the *Bam*HI site 2 in the left-to-right direction (5′ to 3′). *Bam*HI sites 2 and 3, *Bgl*II sites 1 and 2, *Hin*dIII site 1, and *Ava*II site 1 are at positions 1, 92, 22, 38, 244, 156, respectively. The maximum size of *ori* deduced from the result (*17*) is boxed. The GATC sequence and the recognition sequence of *Dpn*I and *Dpn*II are underlined. The frequencies of appearance of the GATC sequence within each 100 nucleotides from −100 to +2196 are listed at the left of the entire sequence. Amino-acid sequences of the postulated structural gene of X-protein and of the *asn* gene are deduced from the genetic code and are described under the nucleotide sequences of the genes. Nucleotide sequences of the Pribnow-box-like promoter sequence of *asn* gene and hypothetical *proX* gene are assigned and boxed. The ribosome-binding-site-like sequences upstream of the reading frames of *proX* and *asn* are underlined (wavy line).

region of an essential gene for replication; however, the *E. coli* origin appears to be in a nontranslated area. Numerous direct and inverted repeats are seen within the 245 base-pair segment. Assuming that the double strands in this area are opened during the initiation event, it is possible to deduce potential secondary structures (discussed in Section VII) from one of the separated strands. We have previously noted that two coding frames are located in the vicinity of the *ori* segment (*14*). We now predict that one of the open frames, starting from ATG at position 734(T) and ending at TAA at position 291(T), corresponds to a structural gene. The prediction is supported by the fact that when the *E. coli* sequence is compared with the *S. typhimurium* sequence (*18*), replacement of bases at every third position is often seen in the coding frame. The predicted gene, named *proX*, accompanies a potential ribosome-binding sequence AGGTG at positions 746(T)–742(C). A possible candidate for the Pribnow-box-like promoter sequence is the TACAATA at positions 768(A) to 762(T). The hypothetical gene product was named X-protein. From the nucleotide sequence, the amino-acid sequence of the X-protein was deduced from the genetic code, and the result is described under the nucleotide sequence of the postulated structural gene, *proX* (Fig. 1).

Another promoter-structural gene (*asn*), coding for asparagine synthetase (*19, 20*), was assigned previously (*13*); it too is shown in Fig. 1. The initiation codon (ATG) of the structural gene of *asn* begins from position 1434(A) and extends to the outside of the sequenced region [position 2196(C)]. From the sequence of the assigned structural gene, the amino-acid sequence of the asparagine synthetase was deduced; the sequence is described under the nucleotide sequence of the structural gene of *asn* in Fig. 1. A Pribnow-box-like promoter sequence is located upstream from the structural gene of *asn* (*13*) [TTTAATG at positions 1331(T) to 1337(G)], and a potential ribosome-binding sequence AGGAG appears at positions 1421(A) to 1425(G).

The content of adenine plus thymine along the entire 2296 base-pair fragment is shown in Fig. 2. Six major (A + T)-rich peaks are found within the fragment. The 245 base-pair region at the origin contained 57% to 60% A + T (peak A). The *proX* region contained a peak of 55% A + T (peak B), the upstream region of *proX* one of 57% (peak C), the upstream region of *asn* gene one of 71% (peak D), and the structural gene of *asn* the relatively low 38% to 47% (peaks E and F). The (A + T)-content of the origin region is invariably high. This property might provide for the more ready separation of the strands during

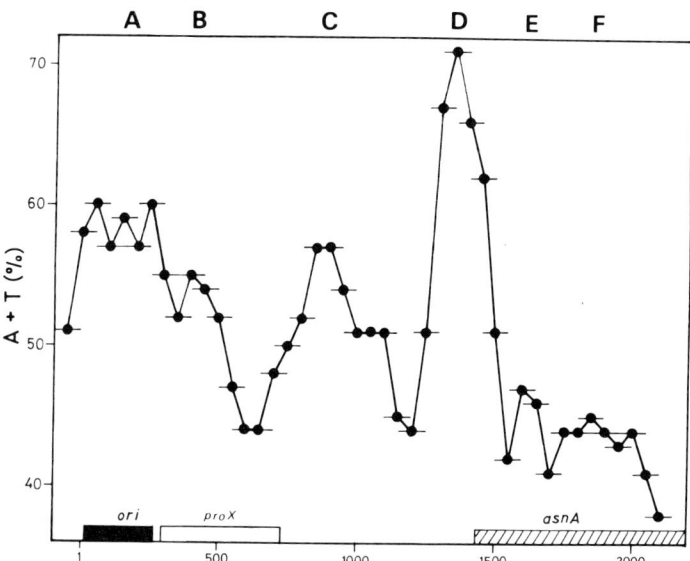

FIG. 2. The (A + T) contents along the entire 2296 base-pair fragment. The average (A + T) content in every 100 base-pairs is plotted at 50 base-pair intervals.

initiation. However, the region richest in A + T (peak D in Fig. 2) is proved to be nonessential for the *ori* functions. The positions of peaks C and D appear to correspond to the promoter regions for genes X and *asn*, respectively. High (A + T) ratios are often found at the promoter region. This property might provide for the ready separation of strands during transcription.

II. Minimal Size of the Origin Region and Its Location

The *Eco* RI fragment of 6×10^6 molecular weight was recloned on the bacteriophage λgtλC *(21)*, and the hybrid was named λgt *ori*. It existed in the cell, either on *recA*⁻ or λ-lysogen, in an extrachromosomal state *(13)*. Extrachromosomal replication of the prophage of λgt *ori* does not depend upon λDNA replication functions. Derivatives of the λgt *ori* having two nonsense mutations in the *susO29*, and *susP3* genes, both required for vegetative DNA replication, were prepared. A λ mutant, λgt *cI857 susO29 susP3 ori*⁺ *asnA*⁺, thus constructed, was used for a detailed study of DNA replication. It existed as a prophage in a host cell of *E. coli* K12, *asnA*⁻ *recA*⁻ *su*⁻ in an extrachromosomal state (H. Yamagishi and A. Murakami, unpublished) as did the λgt *ori*⁺

prophage or ori^+ DNA cloned on the nonreplicating DNA, pSY221, in the same host (10, 13). Furthermore, a subfragment of the Eco RI fragment was recloned on pBR322 (22), and the hybrid DNA thus constructed was named pTSO125. It replicated in a $polA^-$ host (14). The implication of the phenotype of λgt ori and pTSO125 is that the origin DNA on the two DNA species carries information for the DNA replication origin of E. coli irrespective of the cloning vectors used. The growth of ColE-ori DNA (such as pTSO125) on a $polA^-$ host, or the extrachromosomal manifestation of λgt ori DNA on the $recA^-$ host, was used to demonstrate a functioning ori^+ gene.

Using these phenotypes, the precise location and size of the E. coli replication origin at the nucleotide level has been determined. A series of pBR322-ori hybrid plasmids derived from pTSO125 were used to introduce deletions of various sizes. It was found that the left-hand end of the origin is between positions 24(A) and 37(A), and the right-hand end is between positions 267(C) and 268(A). The origin phenotypes, either active or defective, were clearly resolved at the both sides of these boundaries. From the result, it is concluded that the maximum size of the ori segment is 245 base-pairs (17).

We extended this observation further. A series of $recA^+$ bacterial strains carrying defective mutants of the pBR322-ori hybrid plasmid were constructed. These plasmid DNAs carried deletions (Δ) or insertions (Ω) at these restriction sites: pTSO202 (BamHI-3 Ω4 base-pairs), pTSO216 (BamHI-3 Δ4 base-pairs), pTSO190 (HindIII-1 Ω4 base-pairs), pTSO212 (HindIII-1 Δ7 base-pairs), and pTSO157 (BglII Δ16 base-pairs), were introduced into C600 (F^- thr^- leu^- thi^-). From the strains, cleared lysates were prepared, the extracts were used to transform a $polA1^-$ strain (JE269F^-) to ampicillin resistance, and the transformed resistant colonies were isolated. Plasmid DNAs were prepared from the resistants. These plasmid DNAs were able to replicate on the $polA1^-$ host.

It was thought that if the nucleotide sequence deleted in the mutant plasmid were essential for the Ori^+ function, then the recovered Ori^+ plasmids must regain the sequence. The Ori^+ plasmid DNAs

---

FIG. 3. Deletion map of the mutants isolated from λgt ori. The names and genotypes of the mutants are shown at the left, and deleted regions are shown in white. Restriction sites in the 6×10^6 dalton EcoRI fragment containing the ori and asn genes are shown at the top. Fragments are named alphabetically in the order of decreasing size. The locations of ori and asn genes are indicated at the bottom, with corresponding base-pairs numbered from the BamHI site at the junction between the BamHI-F fragment and G in the left-to-right direction.

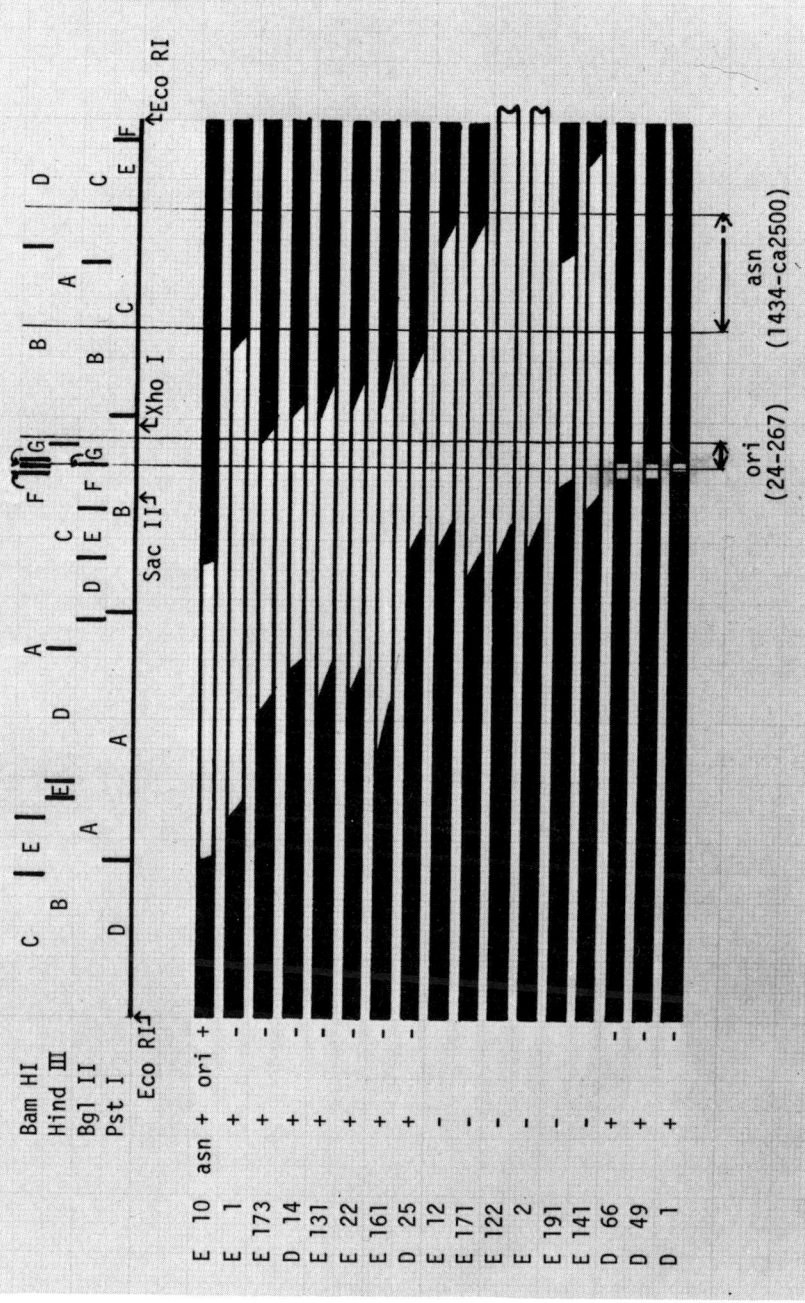

were isolated, and restriction sites of the plasmid DNAs were examined. All the plasmids were found to carry the wild-type ori^+ DNA. This proves that the integration of the ori^- plasmid into the replication origin region of E. coli during its residence in the E. coli C600 host took place because of the nucleotide sequence homology. The integrated plasmid could be excised in a way to replace the defective nucleotide with that of the wild type. Thus, the indispensable role of these nucleotide sequences in the mutants is established.

Furthermore, during extensive analyses on the structural and functional relationship of the E. coli origin carried out using mutants of λgt ori with deletions at the inserted origin, a fragment indispensable for the extrachromosomal replication of λ was isolated and characterized. Figure 3 summarizes the results of such analyses.

All 17 DNAs extracted from the mutants of λgt ori were analyzed by restriction endonuclease digestion of BamHI, BglII, HindIII, PstI, SacII, and XhoI, and the detailed restriction maps of the mutant DNAs were established. Without exception, the mutant lacking the nucleotide segment of positions 24(A) to 267(C), either entirely or in part, had phenotypes defective for the origin function. Thus deletion analyses on the replication function of λgt ori agree with the results of the cloning study using pTSO125 DNA.

III. Is the Cloned Origin Fragment the Replication Origin of E. coli?

We document here the following supporting observations that our cloned origin segment is the replication origin of the E. coli chromosome. The first is the previous observation that the HindIII restriction map of the 6×10^6-dalton EcoRI fragment (10) agrees with that of the HindIII digest of the EcoRI fragment of the origin of the E. coli chromosome as determined by pulse-labeling (23). The second supporting evidence is that the genetic locus of the replication origin of E. coli (7–9) coincides with the site of integration of the cloned origin segments (10). The mapping location of the oriJ fragment (24) was at the terminus of the DNA replication. The oriJ fragment was shown to be the DNA replication origin of a defective prophage. Recently, Von Mayenburg and Hansen (25) presented genetic evidence that the integration of the cloned origin segment could rescue mutant bacteria in which the replication origin was deleted. Taking all this together, we may conclude that the 245 base-pair segment carries the essential origin region of the E. coli chromosome, at least in part.

IV. Sites of the DNA–RNA Junctions within the 245 Base-Pair Fragment and the Directionality of DNA Replication

The nucleotide sequence of the *E. coli* regions shows the absence of a bidirectional sequence within the origin, which might have been expected inasmuch as replication proceeds bidirectionally. The implication is that bidirectionality requires an asymmetric starting process. It is conceivable that the primer RNA synthesis occurs at the origin, and synthesis of the displaced (lagging) strand then occurs discontinuously. When the lagging strand reaches the vicinity of the origin, its 3'-OH end can serve as a primer for leading strand synthesis in the other direction. The DNA–RNA junctions found within the fragment are located at positions 71(G) and 101(A) (26). Both junctions are on one of the two strands. The two positions are on the same strand in the same directionality. These observations strongly suggest an asymmetric start process of the *E. coli* replication origin.

V. The GATC Sequence

The GATC sequence, the recognition sequence of restriction enzymes *Dpn*I or *Dpn*II, identical with that of *E. coli* K12 *dam* methylase (27), is frequently found at the *ori*-containing region (13). This sequence is most frequently found within the 245 base-pair fragment, in the same polarity (Fig. 1). Mutants of *E. coli* defective in the *dam* methylase grow normally (28). A high occurrence of the GATC sequence at the *ori*-containing region in *S. typhimurium* is also reported (18). The *ori* fragments from the both bacteria replicate autonomously in the both bacterial species (29). An explanation for the multiplicity of GATC sequences in the origin region might be that they serve as a part of the recognition site for DNA initiation proteins to form the origin-initiation complex. The high content of the GATC sequences at the origin region might be a strong selective force for the preferential and the effective replication of the chromosome.

VI. Diversity of DNA Replication Origins of the Replicons in *E. coli*, and Homology of the *E. coli* and *S. typhimurium* Origins

The sequences of the origin regions of the lambda-type phages, plasmids, and that of the *E. coli* show substantial differences (30–35); i.e. the nucleotide sequence of the lambda-type phage origins are within the coding region of a gene essential for replication, but the *E. coli* origin appears to be in a nontranslated area. It seems unlikely that the replication of all the replicons in *E. coli* are under similar control. On the other hand, the DNA replication origins of different species of enteric bacteria, such as *E. coli* and *S. typhimurium*, have a striking nucleotide sequence homology (*18*). This homology strongly suggests the presence of a common mechanism for control of DNA initiation and cell division among the enteric bacteria. These observations leave little doubt as to the need to examine in detail the characteristic features of the nucleotide sequences that regulate DNA replication and cell division.

VII. Potentials for Secondary Structure

Based upon the extensive symmetry found in the *E. coli* origin sequence, secondary structures (*13–17*) that may stabilize hairpin or more elaborate shapes, such as cloverleaf structures (*18*) or branched forms (*15*), have been reported. The minimal number of base-pairs for the Ori⁺ function is reduced to 245 from the previous 422 (*17*). As a consequence, the number of potential secondary structures that may be formed is considerably reduced. The secondary structure to form

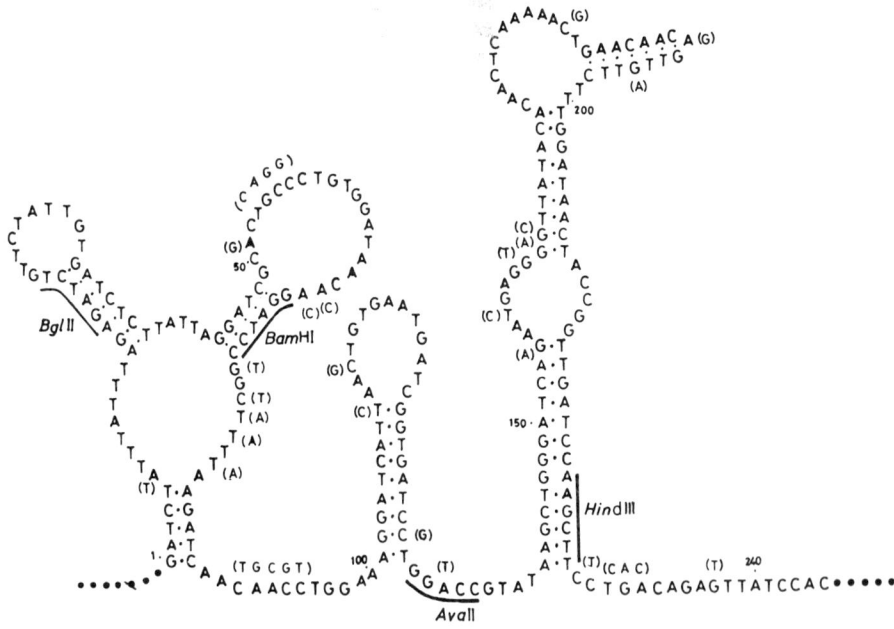

FIG. 4. A possible secondary structure for the *Escherichia coli* ori region. The sequence goes from position 23 to 267 in Fig. 1, and the nucleotides of the *Salmonella typhimurium* ori differing from those of *E. coli* ori are shown in parentheses.

stable base-pairs within the 245 base-pair fragment differs from that of the larger fragment (15), and the branched structures proposed by Messer et al. cannot be formed from the smaller fragment as stable structures. A cloverleaf structure of the *Salmonella* origin proposed by Zyskind and Smith (18) could be constructed from the 245 base-pair fragment of the *E. coli* origin, and many other structures could be constructed. One structure of this kind is illustrated in Fig. 4. This structure had the smallest free energy, and it is common to the origins of both *E. coli* and *S. typhimurium*.

The primary structural homology between *E. coli* and *S. typhimurium* may suggest the presence of common secondary structure, and similar secondary structures could be constructed at several origins in phages and plasmids of *E. coli* even where primary structure differs. However the following fact should be taken into account. The formulation of the secondary structure of a segment from its primary nucleotide sequence is based solely on the thermodynamic stability of base-pair formation and it has little theoretical bearing on the specific determination of DNA initiation. Thus, these structures remain gratuitous until supported by rigorous experimental tests.

VIII. Functional Analysis of the Nucleotide Sequence of the Origin Segment

In order to establish the structural and functional relationship of the DNA replication origin of *E. coli*, a series of plasmids having short sequences inserted or deleted in the vicinity of restriction sites, either *Bgl*II-2, *Bam*HI-3, *Ava*II-1, or *Hin*dIII-1, were constructed *in vitro*. The correlation between the phenotypes of the mutants and the nucleotide sequences were analyzed. The results are summarized in Fig. 5.

The majority of the mutants carrying insertion or deletion mutation were Ori$^-$ (17) (Figs. 5 and 6). Among the mutants examined, two exceptional ones having the Ori$^+$ phenotype, yet also having the modified sequences, were found. One Ori$^+$ mutant (pTSO225) carried an inserted GATC sequence at the *Bgl*II site between positions 42(C) and 43(T). However, the mutant carrying the inserted GATC sequence at the *Bam*HI-3 site between positions 96(C) and 97(C) showed the Ori$^-$ phenotype (Fig. 5). The other Ori$^+$ mutant plasmid (pTSO193) carried the deletion of 12 base-pairs at the *Hin*dIII-1 site between 244(A) and 257(G). However, deletions of either five, seven, or eight base-pairs at the same *Hin*dIII-1 site yielded the Ori$^-$ phenotype.

```
                    GATC (+4, Ori⁺)                                    (+4, Ori⁻) GATC
          BglII  ↓                                              BamHI  ↓      100
   AAGATCTATTTATTTAGAGATCTGTTCTATTGTGATCTCTTATTAGGATCGCACTGCCCTGTGGATAACAAGGATCCGGC
   ΔΔΔΔΔΔΔΔΔΔΔΔΔΔΔΔ(-16 Ori⁻)                                  (-4, Ori⁻) ΔΔΔΔ
                                                               (-15, Ori⁻) ΔΔΔΔΔΔΔΔ

                                    GAC (+3, Ori⁻)
                                     AC (+2, Ori⁻)
                               AvaII  ↓                                          200
   TTTTAAGATCAACAACCTGGAAAGGATCATTAACTGTGAATGATCGGTGATCCTGGACCGTATAAGCTGGGATCAGAATGAGGGGTTATACACAACTCAA
   ΔΔΔΔΔΔΔ                                    ΔΔΔ     (-3, Ori⁻)
                                              ΔΔΔΔ    (-4, Ori⁻)
                                              ΔΔΔΔ    (-5, Ori⁻)
                                              ΔΔΔΔΔΔ  (-6, Ori⁻)
                                              ΔΔΔΔΔΔΔ (-7, Ori⁻)
                         (+4, Ori⁻) AGCT      ΔΔΔΔΔΔΔΔ(-8, Ori⁻)
                               HindIII ↓
   AAACTGAACAACAGTTGTTCTTTGGATAACTACCGGTTGATCCAAGCTTCCTGACAGAGTTATCCACAG
                                              ΔΔΔΔΔ    (-5, Ori⁻)
                                              ΔΔΔΔΔΔΔ  (-7, Ori⁻)
                                              ΔΔΔΔΔΔΔΔ (-8, Ori⁻)
                                              ΔΔΔΔΔΔΔΔΔΔΔΔ(-12, Ori⁺)
```

FIG. 5. Modification of the *ori* sequence. The sequence essential for Ori function is indicated in the box, and the sites and nucleotides of insertion and deletion are shown above and below the *ori* sequence, respectively.

A simple interpretation of these results is that the nucleotide sequences essential for the Ori⁺ phenotype were conserved in those Ori⁺ mutants at the fixed positions of the *ori* segment. As shown in Fig. 6b, an Ori⁺ derivative containing a four base-pair insertion mutation at the *Bgl*II-2 site (pTSO225) created the same sequence at positions 22(A) to 32(A) as the wild-type sequence, owing to the presence of a sequence A · AT · TATTTA (Fig. 6b). The other Ori⁺ derivative of pTSO193 containing, the 12 base-pair deletion mutation created the same sequence, ACAG······CAC, at positions 254(A) to 267(C) as in *E. coli* or *S. typhimurium*.

No such homology is recognized in the other Ori⁻ derivatives. These results led us to the conclusions that (*a*) the region around the *Hin*dIII-1 site plays a role in the maintenance of the appropriate distance between two sequences at the both sides of *Hin*dIII-1 site; (*b*) the presence of the primary sequences A · AT · TATTTA and

---

FIG. 6. Nucleotide sequence of insertion and deletion mutants. The sequences of derivatives with insertion or deletion at *Hin*dIII site 1 (a), the sequence of a derivative with insertion at *Bgl*II site 2 (b), the sequence of a derivative with insertion or deletion at the left end (c) or the right end (d) of *Ava*II, are aligned for comparison with the original *ori* sequence. A dot above a residue indicates the same residue as in *Escherichia coli ori*⁺. *Hin*dIII site 1 (a) and *Bgl*II site 2 (b) have a line above. The two arrows indicate the left (c) and right (d) ends of the maximum size of *ori* determined. The sequence of *Salmonella typhimurium ori* reported by Zyskind and Smith (1980) is also presented for comparison. bp = base-pair.

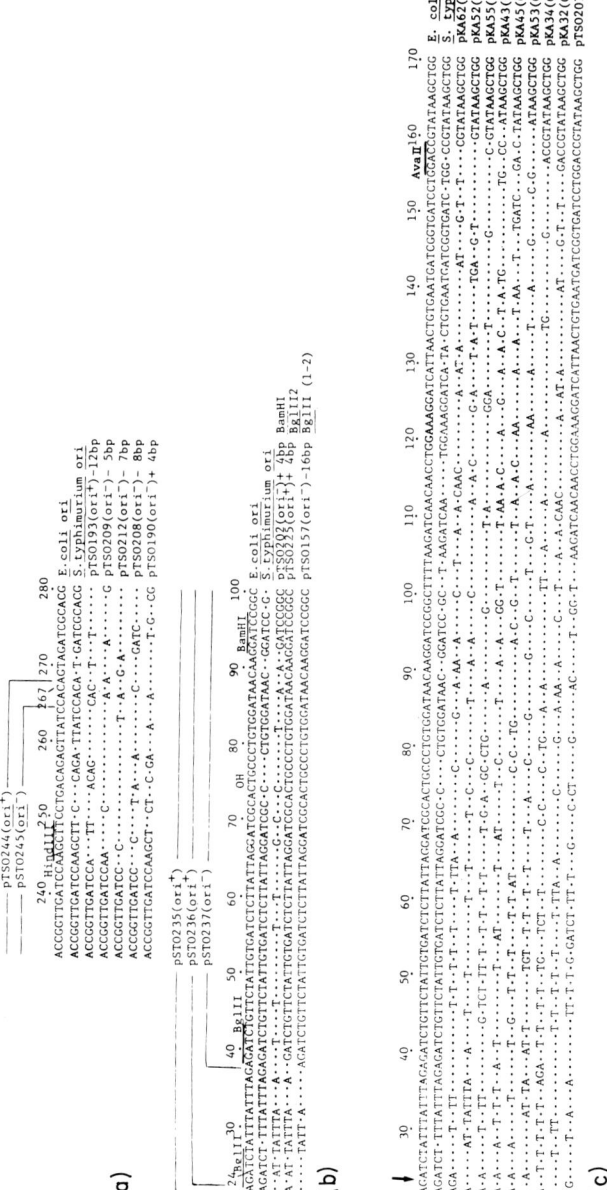

ACAG·······CAC at the fixed positions in the *ori* segment is indispensable for the Ori⁺ function.

Based on these results, we propose a "linear-frame model" for the DNA replication origin. The specific nucleotide sequences that determine the sites where DNA initiation proteins function are separated by nucleotides at a precise distance. In this model, the two kinds of nucleotide segments play the following roles: *class A, recognition sequence;* the sequences that determine the recognition sites for the initiation proteins; *class B, distance sequence,* the sequences that determine the precise distance between the recognition sequences.

Based on the model, certain predictions can be made, as follows. Mutations that occur in the class A sequence, either base substitution, deletion, or insertion, produce Ori⁻ phenotypes due to the defect in the recognition sequence. Base-substitution mutations occurring in the class B sequence, however, do not produce the Ori⁻ phenotype. However, either insertion or deletion mutations introduced within the class B sequences produce the Ori⁻ phenotype.

These predictions are supported by the following observations. A deletion or a insertion introduced at the inner part of the 245 base-pair *ori* segment causes the Ori⁻ phenotype. However, an insertion or deletion mutation introduced at the regions near the termini of the origin fragment produces the Ori⁺ phenotype of the mutant, owing to the accidental appearance of a class A sequence at the right position. As predicted, all the insertions or deletions introduced into the *Ava*II site result in the Ori⁻ phenotype (Fig. 5). Phenotypes of mutants having an inserted GATC at the *Bgl*II site an Ori⁺, but those having the same insertion at the *Bam*HI-3 site are Ori⁻ (Fig. 5).

The striking differences among the bases at the positions, as the result of insertions or the deletions, are illustrated in Fig. 6. The secondary structural models proposed so far have failed to visualize the striking differences in secondary structure of the mutants. The roles of the sequences within the defined *ori* segment could be allocated by isolation of more base-substitution mutants, and such experiments are in progress.

IX. Concluding Remarks

The merging of the genetic analyses of *E. coli*, rapid nucleotide sequence analysis, and gene cloning technology in *E. coli* promises a major breakthrough in our understanding of the regulation of DNA replication and cell division. The *in vitro* initiation of replication at the *E. coli* chromosome origin has not yet been successful, and the absence of an *in vitro* DNA initiation system does not allow us to dissect the detailed reactions involving DNA initiation reactions.

As generally believed (e.g., *36*), there is little doubt that advances in the techniques for gene manipulation and *in vitro* mutagenesis aided by classical bacterial genetics, will greatly aid future work directed at unraveling the process of replication control and cell division. Much remains to be done, but the prospects for major advances regarding this problem are both good and exciting.

ACKNOWLEDGMENTS

We thank H. Iketani for excellent technical assistance. This work was supported by research grants from the Ministry of Education of Japan.

REFERENCES

1. F. Jacob, S. Brenner and F. Cuzin. *CSHSQB* **28**, 329 (1964).
2. K. G. Lark, *Bacteriol. Rev.* **30**, 3 (1966).
3. R. H. Pritchard, *in* "CIBA Foundation Symposium on Bacterial Episomes and Plasmids" (G. E. W. Wolstenholme and M. O'Connor, eds.), p. 65. Churchill, London, 1969.
4. N. Sueoka and W. G. Quinn. *CSHSQB* **33**, 695 (1969).
5. C. Helmstetter, S. Cooper, O. Pierucci and E. Revelas, *CSHSQB* **33**, 809 (1969).
6. Y. Hirota, A. Ryter, M. Ricard and U. Schwarz, *in* "Mechanism and Regulation of DNA Replication" (A. R. Kolber and M. Kohiyama, eds.), p. 407. Plenum, New York, 1974.
7. L. G. Caro and C. M. Berg, *CSHSQB* **33**, 559 (1968).
8. R. E. Bird, J. Louarn, J. Martuscelli and L. G. Caro, *JMB* **70**, 549 (1972).
9. J. Louarn, M. Funderburgh and R. E. Bird, *J. Bact.* **120**, (1974).
10. S. Yasuda and Y. Hirota, *PNAS* **74**, 5458 (1977).
11. W. Messer, H. E. N. Bergmans, M. Meijer and J. E. Womack, *Mol. Gen. Genet.* **162**, 269 (1978).
12. T. Miki, S. Hiraga, T. Nagata and T. Yura, *PNAS* **75**, 5099 (1978).
13. Y. Hirota, S. Yasuda, M. Yamada, A. Nishimura, K. Sugimoto, H. Sugisaki, A. Oka and M. Takanami, *CSHSQB* **43**, 129 (1978).
14. K. Sugimoto, A. Oka, H. Sugisaki, M. Takanami, A. Nishimura, S. Yasuda and Y. Hirota, *PNAS* **76**, 575 (1979).
15. W. Messer, M. Meijer, H. E. N. Bergmans, F. G. Hansen, K. Meyerberg, E. Beck and H. Schaller, *CSHSQB* **43**, 139 (1978).
16. M. Meijer, E. Beck, F. G. Hansen, H. E. N. Bergmans, W. Messer, K. Meyenburg and H. Schaller, *PNAS* **76**, 580 (1979).
17. A. Oka, K. Sugimoto, M. Takanami and Y. Hirota, *Mol. Gen. Genet.* **178**, 9 (1980).
18. J. W. Zyskind and D. W. Smith, *PNAS* **77**, 2460 (1980).
19. B. J. Bachmann and K. Brooks Low, *Microbiol. Rev.* **44**, (1980).
20. K. Von Meyenburg, F. G. Hansen, L. D. Nielsen and E. Rise, *Mol. Gen. Genet.* **160**, 287 (1978).
21. M. Thomas, J. R. Cameron and R. W. Davis, *PNAS* **71**, 4579 (1974).
22. F. Bolivar, R. L. Rodrigues, P. J. Greene, M. C. Betlach, H. L., Heyneker, H. W. Boyer, J. H. Crosa and S. Falkow *Gene* **2**, 95 (1977).
23. R. C. Marsh and A. Worcel, *PNAS* **74**, 2720 (1977).
24. R. Diaz and R. H. Pritchard, *Nature* **275**, 561 (1978).

25. K. Von Meyenburg and F. G. Hansen, *ICN-UCLA Symp. Mol. Cell. Biol.* 1980, 19th (in press)
26. T. Okazaki, S. Hirose, T. Ogawa, A. Fujiyama and Y. Kohara, *ICN-UCLA Symp. Mol. Cell. Biol.* 1980, 19th (in press)
27. G. E. Geier and P. Modrich, *JBC* **254**, 1408 (1979).
28. M. G. Marinus and N. R. Morris, *JMB* **85**, 309 (1974).
29. K. Joh and S. Hiraga, *J. Bact.* **138**, 297 (1979).
30. J. Tomizawa, H. Ohmori and R. E. Bird, *PNAS* **74**, 1865 (1977).
31. J. C. Fiddes, B. G. Barrell and G. N. Godson, *PNAS* **75**, 1081 (1978).
32. J. Sims and D. Dressler, *PNAS* **75**, 3094 (1978).
33. K. Denniston-Thompson, D. D. Moore, K. E. Kruger, M. E. Furth and F. R. Blattner, *Science* **198**, 1051 (1977).
34. G. Hobom, R. Grosschedl, M. Lusky, G. Scherer, E. Schwarz and H. Kössel, *CSHSQB* **43**, 165 (1979).
35. R. Grosschedl and G. Hobom, *Nature* **277**, 621 (1979).
36. R. Kolter and D. R. Helinski, *Annu. Rev. Genet.* **13**, 355 (1979).

Replication of Linear Duplex DNA *in Vitro* with Bacteriophage T5 DNA Polymerase

R. K. FUJIMURA
S. K. DAS
D. P. ALLISON AND
B. C. ROOP

Biology Division, Oak Ridge National Laboratory, and The University of Tennessee—Oak Ridge Graduate School of Biomedical Sciences
Oak Ridge, Tennessee

I. Properties of Replicative DNA Polymerases

Bacteriophage T5 DNA polymerase is one of the first DNA polymerases shown to be essential for replication of DNA (1). Yet not much work was done with it until we started to characterize it. Our findings are summarized in Table I, where its properties are compared with those of other replicative DNA polymerases (2). After purification to apparent homogeneity, only one protein band was detectable on electrophoresis with dodecyl sulfate, with a mobility corresponding to a molecular weight of 96,000. Therefore, the enzyme is a single polypeptide. It has an associated $3' \rightarrow 5'$ exonuclease.

T5 DNA polymerase has at least two unique properties. It is capable of unwinding a strand from a nick in a circular duplex DNA as it elongates a strand from the 3'-OH end as a primer. The only other polymerase that is definitely known to do this is *Escherichia coli* DNA polymerase I (EC 2.7.7.7) and its fragment without the $5' \rightarrow 3'$ exonuclease ("Klenow fragment") (3). Also, T5 DNA polymerase translocates along the template quite efficiently. It is by far the most processive enzyme we have studied. Its associated $3' \rightarrow 5'$ exonuclease is also processive with single- or double-stranded homopolymers. Substrates 20–200 nucleotides long are hydrolyzed processively until only about 5 nucleotides long (4). Other replicative DNA polymerases studied, such as T4 DNA, require many auxiliary proteins to become processive (2, 5). Recently, processiveness of the α polymerase was also shown to be determined by an auxiliary protein copurified with a

TABLE I
PROPERTIES OF REPLICATIVE DNA POLYMERASES[a]

Property	T5	Escherichia coli polIII	T4	T7	α	Bacillus subtilis polIII
Molecular properties						
Molecular weight	96,000	140,000	114,000	87,000	149,000	166,000
Number of subunits	1	≥4?	1	2	≥2	1
Homogeneous	Yes	No	Yes	Yes	Yes?	No
Repl⁻ mutant?	Yes	Yes	Yes	Yes	No	Yes
Template primer						
Primed single strand	+	±	+	+	+	−
Nicked circular duplex	+	−	−	−	−	?
Gapped duplex	+	+	+	+	+	+
De novo synthesis	−	−	−	−	−	−
Functions						
Polymerization: 5' → 3'	+	+	+	+	+	+
Exonuclease: 3' → 5'	+	+	+	+	−(?)	+
dNTP → dNMP turnover	+	+	+	+	−	?
Exonuclease: 5' → 3'	−	+	−	−	−	−
Strand displacement	+	−	−	−	−	−
Processiveness: nucleotide added/E., 37°C	160	?	12	?	8,11	7?

[a] From Fujimura and Das (2), where references to the original papers appear.

certain fraction of the polymerase preparation (6). Therefore, the T5 DNA polymerase system may be the simplest system to study many of the functions associated with replication of duplex DNA.

The two most active areas of research on the mechanisms of DNA replication concern the fidelity of replication, and initiation of replication of duplex DNA. Here we would like to present two sets of experiments that may contribute to understanding these mechanisms.

II. Mechanisms of Base Selection and Editing

One of the popular hypotheses on fidelity of replication of DNA is that the $3' \rightarrow 5'$ exonuclease associated with the polymerase eliminates a wrong base as it is being incorporated. Several investigators have shown that prokaryotic DNA polymerases have error rates of only one in 10^5 to 10^6 base pairs (7). A recent report shows that proteins in T4 DNA replication complex improve the "editing" step, thus making such a system as accurate as *in vivo* (5). Galas and Branscomb proposed that hydrogen-bonding strength between base pairs is sufficient to account for accuracy of DNA polymerase (8): the polymerase bound at a primer end prefers to condense a nucleotide that forms the strongest hydrogen bond with the adjacent nucleotide in the template, and its associated $3' \rightarrow 5'$ exonuclease tends to hydrolyze a nucleotide that is "mismatched" in the Watson–Crick sense. These two processes combine to give a polymerase a high degree of accuracy. Certain experiments *in vivo* and *in vitro* with an antimutator strain of T4 DNA polymerase support this two-step mechanism (9).

All the replicative DNA polymerases so far studied are processive in incorporation of nucleotides for at least a short stretch (10). A study with T4 replication complex suggests that, with such a complex, processiveness becomes greater (11). One question that has been ignored is whether it is possible for a polymerase going in one direction processively to reverse itself and go in the opposite direction. T5 DNA polymerase is uniquely suited to answer this question. It preferentially binds to primer ends if primers are present in excess, and essentially 100% of it functions as a polymerase in the presence of the four common triphosphates. This is shown by the data presented in Fig. 1 (12). In the presence of poly(dA) without primers, essentially all the polymerases were bound to the end of poly(dA) and acted as exonucleases. As primers were added, exonuclease activity decreased, and polymerase activity increased concomitantly if dTTP was present. The sum of the two activities remained constant. This suggests that polymerases bind preferentially at the ends of polynucleotides inde-

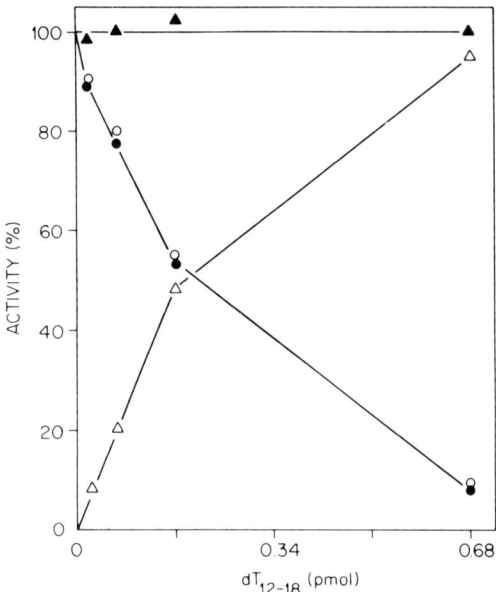

FIG. 1. Relationship between polymerase and exonuclease activities of T5 DNA polymerase. In a reaction volume of 300 μl, 130 pmol (determined by release of 5'-terminal nucleotides) of [^{3}H]poly(dA) and 0.24 pmol of enzyme were used with various amounts of $(dT)_{12-18}$. The mixture was incubated at 37°C for 1 minute. Exonuclease activity was assayed by spotting on DE-81 paper strips. Nucleotides were eluted with 0.3 M NH$_4$ formate by gentle shaking and then counted. The polymerase assay was carried out by acid precipitation of the products on glass fiber filters. Exonuclease activity under the standard conditions (○) and in the presence of 200 μM dTTP (●); △, fraction of enzyme molecules functioning as polymerase; ▲, the sum of enzyme functioning as exonuclease and polymerase. Reprinted from Das (12) with permission of the publisher.

pendently of dNTPs. When the primer concentration became excessive, essentially all the enzyme activities were polymerase activities, suggesting that, at high enough concentration of primers, all were bound to primer ends.

On the basis of these observations, S. Das devised an experiment that showed that a polymerase can switch direction, depending on

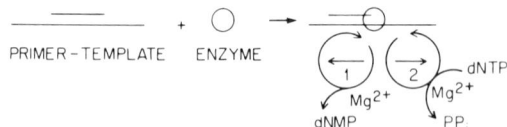

FIG. 2. Schematic diagram of exonucleolytic (arrow 1) and polymerization (arrow 2) reactions.

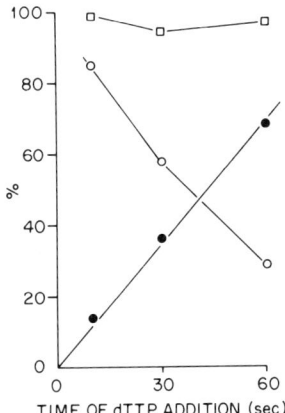

FIG. 3. Switching of direction of catalysis by the addition of dTTP (33 μM) (13). For this experiment dA$_{600}$ · (dT$_8$-[^{3}H]dT$_5$) was used as primer-template. Enzyme/primer ratio was 20. For each time point, 1.0 pmol of substrate was used. Denatured calf thymus DNA was added in excess (20 μg) after the enzyme and poly(dA) · oligo(dT) were mixed, but prior to incubation; Mg^{2+} was added to start hydrolysis, and at indicated times dTTP was added. Percentage of primers hydrolyzed, percentage polymerized, and the sum of two processes are shown by ●, ○, and □, respectively. The temperature of incubation was 25°C.

conditions (13). The enzyme was bound to the primer end of poly(dA) · oligo(dT) in the absence of Mg^{2+}, as shown schematically in Fig. 2. Then excess denatured calf thymus DNA was added to bind excess enzymes and make it unlikely for the enzyme bound to oligo(dT) to rebind if it were released. The incubation was started as Mg^{2+} was added, which makes the enzyme act as an exonuclease hydrolyzing the primer ends and releasing dTMP from oligo(dT). Then the system was tested to see whether the bound enzyme would act as polymerase when dTTP was added as the precursor. Oligo(dT) used as a primer is acid-soluble under these conditions, but binds to DEAE paper. The dTMP freed by hydrolysis did not bind to the DEAE paper in 0.3 M NH$_4$-formate; thus this was the assay for exonuclease. The polymerization made the primer acid-insoluble. In the experiment shown in Fig. 3, dA$_{600}$ · (dT$_8$-[^{3}H]dT$_5$) was used, and unlabeled dTTP was added at various times after the hydrolysis had started. As soon as dTTP was added, hydrolysis stopped and essentially all the remaining primers became acid-insoluble, indicating elongation. This suggests that the enzyme in the exonuclease mode was able to reverse its direction of action and act in the polymerase mode.

There is no reason not to believe that such reversal can occur; that is, that T5 polymerase acting in the polymerization mode can change its direction and act in the exonucleolytic mode.

III. Replication of Linear Duplex DNA

Another major question being asked is how initiation of replication of linear duplex DNA occurs. The currently popular model says that there is a unique origin. The primer may be a 3'-OH terminus created by breakage of phosphodiester bond in one of the strands, or it could be a primer synthesized de novo situated at the origin (14). On the leading strand, the synthesis is probably continuous from the origin to the end of replicon. On the lagging strand the new primers are synthesized as single-stranded regions are created, and synthesis goes discontinuously in the 5'-to-3' direction toward the origin.

According to electron microscopic analysis, initiation of T5 DNA replication occurs at unique but multiple sites, most frequently from near the middle (15). In vitro, the purified T5 DNA polymerase synthesizes DNA with nicked PM2 DNA as template quite extensively by the rolling circle model of synthesis (Fig. 4), but linear duplex DNAs, even when nicked, are not very good primer-templates (16, 17). With the latter, synthesis stops within a short time with many branched structures (Fig. 5). Thus we have searched for a system that will replicate T5 DNA efficiently with some specificity.

In crude extracts prepared by lysozyme–EDTA treatment (18), T5 DNA synthesis is dependent on exogenous DNA polymerase (Fig. 6). There is essentially no synthesis with noninfected extracts or by purified polymerase alone. Thus synthesis is dependent on T5 DNA polymerase and some other phage-induced factor(s). When extracts of infected cells were prepared as above from polA$^-$, endo I$^-$ cells (BT 1000) and fractionated by means of metrizamide gradient centrifugation, two fractions that activated exogenous T5 DNA into primer-template were obtained (Fig. 7). The faster sedimenting fraction contained most of the endogenous T5 DNA polymerase and endonucleases, as well as DNA and DNA polymerase, and M13 synthesis resistant to rifampicin. The slower sedimenting fraction (fraction II) was almost free of endogenous polymerase and endonucleases, but synthesized T5 DNA as well or better than the faster sedimenting fraction when exogenous polymerase was added. When extracts from noninfected cells were fractionated and analyzed in a similar manner, the activation factor was present in about the same position in the gradient as fraction II, and it was essentially free of en-

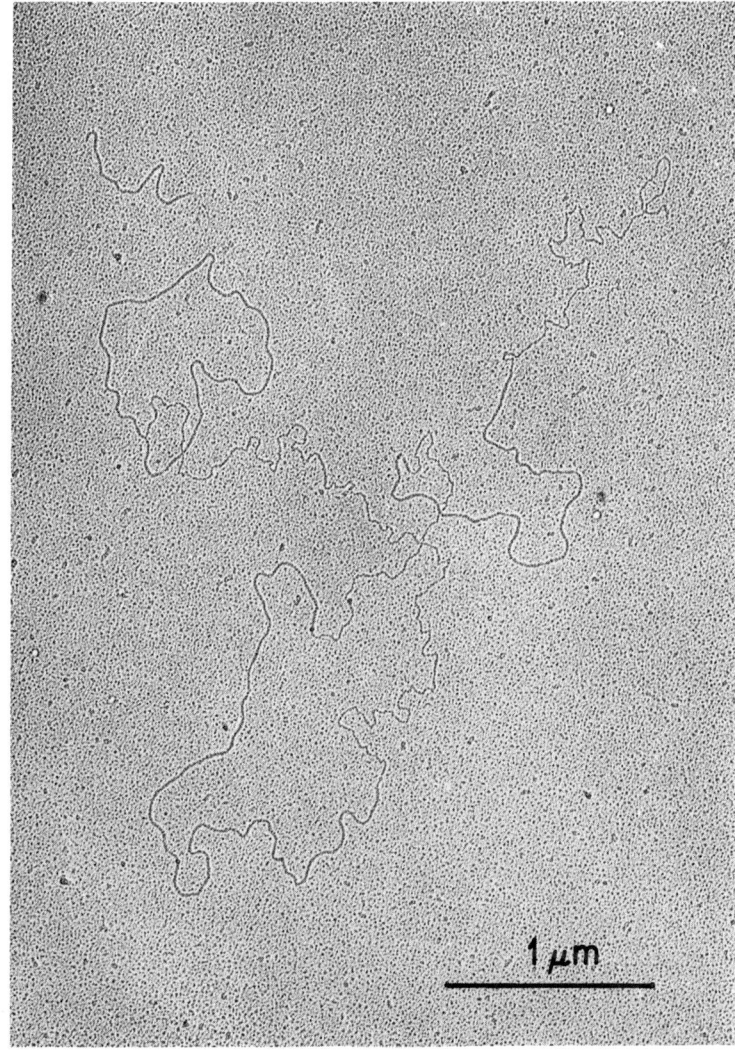

FIG. 4. Electron micrograph of nicked PM2 DNA after an extensive synthesis by T5 DNA polymerase. From Fujimura and Allison (*17*), with permission of the publisher.

donucleases when assayed with PM2 DNA (*19*) (Fig. 7). With the same amount of proteins from the fraction II region of infected and noninfected cell extracts, there was less synthesis with the fraction from noninfected cells (Fig. 8). There was definite synthesis regardless of the levels of endonucleases, and there was hardly any synthesis

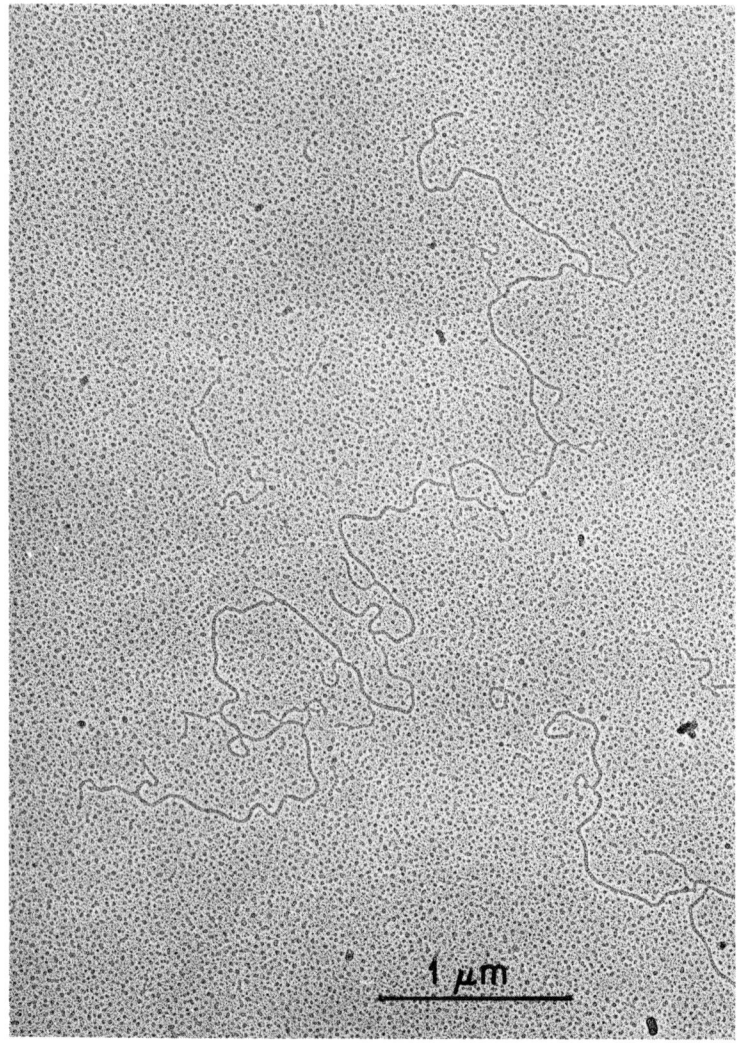

FIG. 5. Electron micrograph of nicked T7 DNA after the limit of synthesis by T5 DNA polymerase. From Fujimura and Allison (17), with permission of the publisher.

with PM2 DNA (Fig. 9), which suggests that primer-templates formed by nicks were insignificant.

The fraction containing the activation factor from noninfected cells was extracted on a larger scale by sequential fractionation by polyethylene glycol and $(NH_4)_2SO_4$, yielding material of specificity similar to

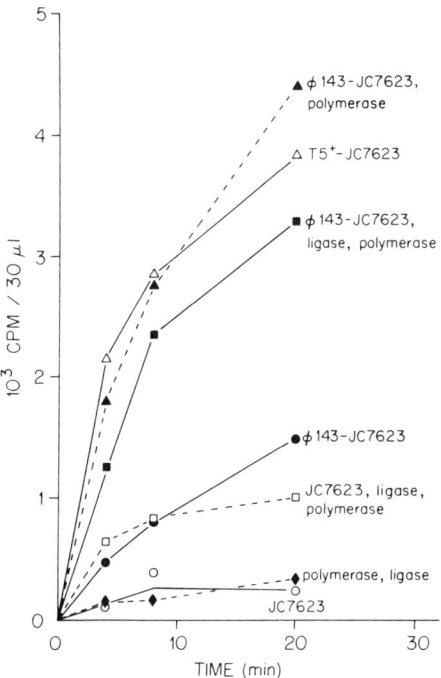

FIG. 6. JC7623 rec BC⁻ extracts with exogenous T5⁺ DNA polymerase and T5 DNA. Extracts were prepared by Staudenbauer's method (18) with some modification. Content of endogenous polymerase in T5⁺-JC7623 was determined with denatured calf thymus DNA as primer-template. An equivalent amount of exogenous polymerase was added to φ143-infected JC7623. (φ143 is a T5 mutant in the polymerase and 5′ → 3′ exonuclease gene.) The amount of extract added to each sample was 16 µg of protein per 100 µl of reaction mixture. A 30-µl aliquot was taken out at various times.

that from metrizamide gradient centrifugation. When the fraction was added in a small amount to the polymerase reaction mixture consisting of T5 DNA and T5 DNA polymerase, synthesis occurred linearly for at least 60 minutes (Fig. 10). The system requires Mg^{2+} and four dNTPs. It has no obvious requirement for rNTPs. D5 protein, a DNA binding protein from T5-infected cells (20), stimulated the synthesis more. The synthesis was stimulated by the fraction at 37°C, but not at 25°C. The fraction did not nick nor relax PM2 DNA significantly. The T5 DNA polymerase was preferred over Klenow fragments of *E. coli* DNA polymerase I. T4 DNA polymerase, which does not synthesize from nicks, caused no synthesis.

The template specificity of the system is summarized in Table II. The synthesis was not inhibited by ligase. The system did not synthe-

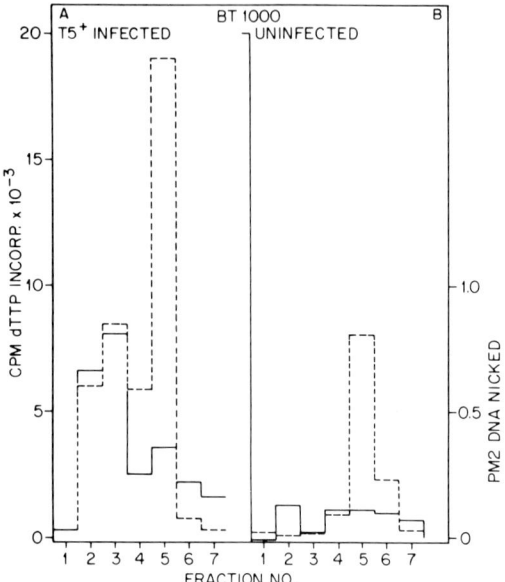

FIG. 7. Metrizamide gradient centrifugation of extracts. Extracts were prepared from (A) T5+ infected and (B) noninfected BT $1000_{polA^-, endo\,I^-, thy^-}$ by the method of Staudenbauer. Three hundred microliters of each extract were layered on a gradient made with 16% w/v and 40% w/v metrizamide solution in 25 mM Hepes, pH 8.0, 5 mM EGTA, and then centrifuged in a SW65 rotor at 50,000 rpm for 16 hours at 0°C. Samples were collected in seven fractions. Assay for DNA synthesis was carried out with 10 μl of fractions in the standard conditions with exogenous T5 DNA and T5 DNA polymerase. The standard conditions are, in 30 μl: 67 mM TrisCl pH 7.6, 10 mM $MgCl_2$, 10 mM dithiothreitol, 100 μM each of four deoxynucleoside triphosphates with one of them labeled with 100 cpm per picomole of ^{32}P- or ^3H-labeled nucleoside triphosphates, 1 μg of T5 DNA, 0.5 pmol of T5 DNA polymerase, and variable amounts of stimulating factor. Incubation was for 30 minutes at 37°C. Endonuclease assay was carried out with PM2 DNA by the nitrocellulose filter method (19). —, Endonuclease activity with PM2 DNA; ----, T5 HA DNA synthesis with T5 polymerase.

size PM2 DNA or pBr322 DNA. It did not further stimulate the synthesis with nicked PM2 DNA or pBr322 DNA. It did not synthesize from linear duplex DNA with "blunt" (i.e., no single-strand) ends made from pBr322 or PM2 DNA with restriction enzymes. Sucrose density gradient centrifugation of the products showed that, in the native state, the products were associated with intact T5 DNA, but most of the templates were partially broken when denatured; therefore, it was difficult to determine whether synthesis occurred from preexisting 3'-OH termini or de novo. When there is fragmented T5 DNA

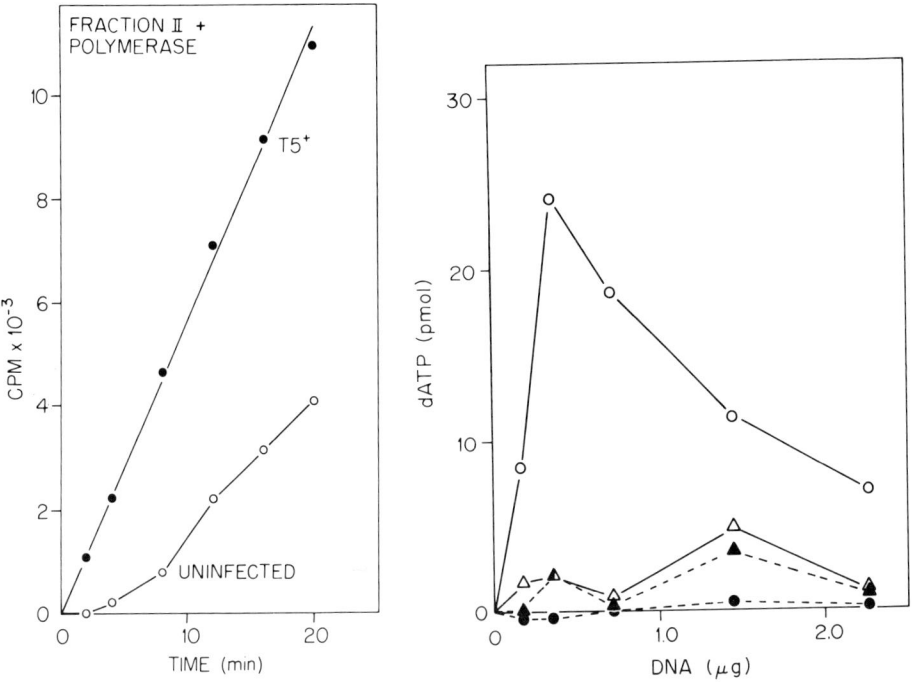

FIG. 8 (left). Comparison of the activation factor from infected and noninfected cells. Assays were carried out at the standard conditions using a 30-µl aliquot at each time point. The optimum amount of each sample from the fraction II (slower) region of metrizamide gradients was used.

FIG. 9 (right). Synthesis of T5 DNA and PM2 DNA. Under the standard assay conditions for T5 DNA synthesis, various amounts of T5 DNA and PM2 DNA were added in the presence of four dNTPs or only dATP. The optimum amount of fraction II from the metrizamide gradient was used; it was incubated at 37°C for 15 minutes. ▲, PM2 DNA, dATP; △, PM2 DNA, four dNTPs; ●, T5 DNA, dATP; ○, T5 DNA, four dNTPs.

present, it is preferentially used as primer-template over intact T5 DNA. This was an additional complication.

These results show that the system consisting of T5 DNA polymerase, D5 protein, and an initiation factor(s) is preferred for T5 DNA synthesis. The initiation factor for T4 DNA replication is a T4-induced topoisomerase (21), but a host gyrase can be substituted for it, albeit less efficiently (22). Topoisomerases with similar properties are found in several eukaryotic cells, and may function in formation of a replication bubble at the origin of replication (23). A host factor in combination with T7 DNA polymerase and gene 4 protein acts like a

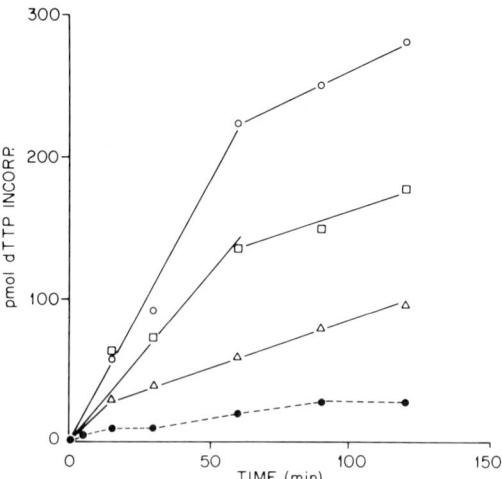

FIG. 10. Time course of T5 DNA synthesis. In 300 µl of standard reaction mixture, 10 µg of T5 DNA (from a mutant free of genetically determined nicks), 1.35 µg of T5 D5 protein, 1.3 µg of the activation factor from BT 1000 extracts, and 9 pmol of T5 DNA polymerase were mixed and incubated at 37°C. Aliquots of 30 µl were taken out at various times for assay of radioactivity in acid-insoluble fractions. ○, D5-DBP, BT 1000 factor, polymerase; □, BT 1000 factor, polymerase; △, D5-DBP, polymerase; ●, polymerase. DBP, DNA-binding protein.

specific endonuclease that may be involved in initiation of T7 DNA replication (24). D5 protein, the most abundant DNA binding protein in T5-infected cells, binds preferentially and cooperatively to duplex DNA; thus it differs from the single-stranded-DNA-binding protein of E. coli or phage T4 and T7 (25). These findings suggest that our system is unique but has some similarity to other systems; further work should contribute to an understanding of the parameters that give specificity to DNA replication systems.

More recently, a multienzyme complex that may be a transcription and replication complex was isolated from phage T5-infected cells, and characterized (26). It consists of six phage-induced proteins and five host proteins, including T5 DNA polymerase, gene D5 protein, and E. coli RNA polymerase. Thus, the faster-sedimenting fraction from metrizamide gradients may contain this complex. It was bound to T5 DNA, and the DNA came preferentially from the midsection of the DNA, where the primary origin of replication appears to lie (15). This is consistent with the results of our preliminary experiments showing that the DNA synthesized in the initial few seconds in vivo and in vitro appears to arise at the midsection of the DNA (27).

TABLE II
Template Specificity of Stimulation of DNA Synthesis by Factors

Sample	Synthesis by polymerase[a]	Stimulation[b] by BT 1000 factors	Stimulation[b] by BT 1000 factors plus T5 DBP[c]
T5 HA DNA (nickless)			
Intact	−	4	10
Ligase treated	−	13	25
EcoRI treated	−	4	5
EcoRI → ligase	−	2	3
SmaI treated	+	1	1
SmaI → ligase	−	3	3
PM2 DNA			
Intact circle	−	1	1
Nicked circle	+	1	1
Nicked → ligase	−	1	1
Linear (HpaII)	−	≤1	1
HpaII → ligase	−	≤1	1
pBR 322			
Intact circle	−	1	1
Nicked circle	+	1	1
Nicked → ligase	−	1	1
Linear (EcoRI)	−	1	1
EcoRI → ligase	−	1	1

[a] When the synthesis occurs for at least 30 minutes at 37°C, then it is positive.

[b] Amount of stimulation was expressed as the ratio of amount synthesized in 30 minutes in the presence of additional factors/polymerase alone.

[c] DBP, DNA-binding protein.

Acknowledgments

Research was sponsored by the Office of Health and Environmental Research, U. S. Department of Energy, under contract W-7405-eng-26 with the Union Carbide Corporation. S. K. Das was supported by a research assistantship granted by the University of Tennessee.

References

1. A. DeWaard, A. V. Paul and I. R. Lehman, *PNAS* **54**, 1241 (1965).
2. R. K. Fujimura and S. K. Das, This Series **24**, 87 (1980).
3. H. Klenow, H. Overgaard and S. A. Patkar, *EJB* **22**, 371 (1971).
4. S. K. Das and R. K. Fujimura, *NARes* **8**, 657 (1980).
5. C.-C. Liu, R. L. Burke, U. Hibner, J. Barry and B. Alberts, *CSHSQB* **43**, 469 (1979).
6. R. A. Bambara and J. W. Hockensmith, *J. Supramol. Struct., Suppl.* **4**, 331 (1980).
7. F. Bernardi and J. Ninio, *Biochimie* **60**, 1083 (1978).
8. D. J. Galas and E. W. Branscomb, *JMB* **124**, 653 (1978).
9. M. F. Goodman, R. Hopkins and W. C. Gore, *PNAS* **74**, 4806 (1977).

10. S. K. Das and R. K. Fujimura, *JBC* **254**, 1227 (1979).
11. B. Alberts, J. Barry, M. Bittner, M. Davies, H. Hama-Inaba, C.-C. Liu, D. Mace, L. Moran, C. F. Morris, J. Piperno and N. K. Sinha, in "Nucleic Acid–Protein Recognition" (H. J. Vogel, ed.), p. 31. Academic Press, New York, 1977.
12. S. K. Das, *BBRC* **79**, 247 (1977).
13. S. K. Das and R. K. Fujimura, *JBC* **255**, 7149 (1980).
14. J. Tomizawa and G. Selzer, *ARB* **48**, 999 (1979).
15. G. J. Bourgignon, T. K. Sweeney and H. Delius, *J. Virol.* **18**, 245 (1976).
16. R. K. Fujimura and B. C. Roop, *JBC* **251**, 2168 (1976).
17. R. K. Fujimura and D. P, Allison, *JBC* **251**, 2174 (1976).
18. W. L. Staudenbauer, *Mol. Gen. Genet.* **145**, 273 (1976).
19. F. T. Gates, III, and S. Linn, *JBC* **252**, 2802 (1977).
20. G. Cinnadurai and D. J. McCorquodale, *Nature* **247**, 554 (1974).
21. L. F. Liu, C.-C. Liu and B. M. Alberts, *Nature* **281**, 456 (1979).
22. D. McCarthy, *JMB* **127**, 265 (1979).
23. L. F. Liu, C.-C. Liu and B. M. Alberts, *Cell* **19**, 697 (1980).
24. C. C. Richardson, L. J. Romano, J. E. Kolodner, J. E. LeClerc, F. Tamanoi, M. J. Engler, F. B. Dean and D. S. Richardson, *CSHSQB* **43**, 427 (1979).
25. A. C. Rice, T. A. Ficht, L. A. Holladay and R. W. Moyer, *JBC* **254**, 8042 (1979).
26. T. A. Ficht and R. W. Moyer, *JBC* **255**, 7040 (1980).
27. R. K. Fujimura and B. C. Roop, *FP* **37**, 1411 (1978).

Mechanisms of Catalysis of Human DNA Polymerases α and β

DAVID KORN
PAUL A. FISHER AND
TERESA S.-F. WANG

Laboratory of Experimental
  Oncology
Department of Pathology
Stanford University School of
  Medicine
Stanford, California

The availability of essentially homogeneous preparations of human DNA polymerases α and β (1-3) has enabled us to undertake a detailed examination of their enzymological properties with a variety of DNA substrates of defined structure (4-8) and to begin to identify some of the specific mechanisms and molecular signals that appear to regulate the polymerase–nucleic acid interactions (9, 10). The complete absence of associated or contaminating endo- and exodeoxyribonuclease activities from our purified polymerase fractions [at exclusion limits of 10^{-4} to 10^{-7} of the polymerase activity (6, 8)] has permitted us to pursue these studies without concern for the potentially misleading consequences of even subtle modifications of substrates by trace levels of nuclease contaminants and provides assurance that the observed results are valid reflections of the inherent catalytic properties of the polymerase proteins themselves. From the results summarized in this contribution, there emerges a set of conclusions that substantially clarify many of the specific features involved in the recognition by polymerases α and β of their nucleic acid substrates. The data also offer novel insights and support provocative speculations into the manner in which the information encoded in the molecular structure of these two most dissimilar polymerase proteins (1-3) may contribute uniquely and effectively to their postulated *in vivo* participation in the processes of DNA replication and repair (11).

The decision to undertake a detailed analysis of the mechanisms of catalysis of DNA polymerases α and β was initially generated by our desire to develop a formal enzymological framework within which to evaluate the interactions of the polymerases with candidate replication factors in mechanistically interpretable, model assay systems and was further stimulated by two sets of prior observations: our studies of the patterns of primer-template utilization by these enzymes, and some curious and perplexing results encountered in our examination of the effects of spermidine on the reactivity of DNA polymerase α (6).

I. Primer-Template Utilization by DNA Polymerases α and β

Polymerases α and β can copy a variety (7, 12) of natural heteropolymeric and synthetic homo- or heteropolymeric polynucleotide templates, although they exhibit substantial differences in their relative efficiencies of utilization of different primer templates and in their relative preferences for Mg^{2+} and Mn^{2+} as divalent metal activators (3–8, 13). Both of the polymerases are highly reactive with gapped ("activated") duplex DNA substrates, and we have investigated their specific utilization of this class of substrates in detail (5, 6, 8). For DNA polymerase α, optimum activity is obtained in the presence of Mg^{2+} (substitution of Mn^{2+} reduces activity by ~95%) on gaps that average 30 to 60 nucleotides in length (Fig. 1) (8); K_m^{app} for DNA is ~40 μM (nucleotide) and ~75 nM of usable 3'-hydroxyl primer termini (6, 14). However, there are two curious features of this reaction (5, 6). First, the "average template length" (14, 15), measured by

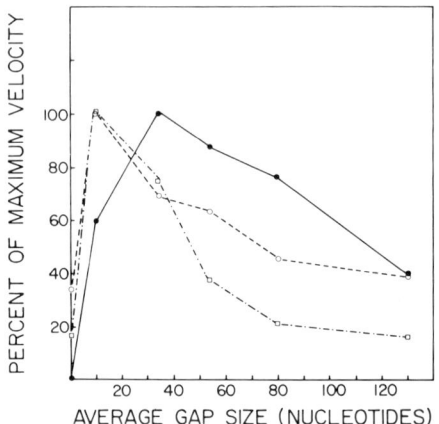

FIG. 1. Relationship of maximal velocity of polymerization to mean gap length in duplex KB DNA. Initial reaction rates were measured with DNA polymerase α (●—●) and with DNA polymerase β in the presence of Mn^{2+} (○---○) or Mg^{2+} (□-·-·-□). Kinetic constants were obtained from Lineweaver–Burk plots generated by the method of least squares. Reprinted from Wang and Korn (8) with permission of American Chemical Society.

polymerase α on gapped duplex DNA substrates, is reproducibly less, by 20–40 nucleotides, than the average physical size of the gaps estimated with phage T4 DNA polymerase or with polymerase β (8). We have concluded (6, 8) from this that polymerase α is unable to fill gaps to completion. Second, enlargement of the gaps in the DNA substrate by resection with *Escherichia coli* exonuclease III leads to a progressive reduction in the rate (Fig. 1) of dNMP incorporation by polymerase α (5, 6), while the extent of dNMP incorporation at a given primer site remains essentially invariant. These results suggest that polymerase α may bind and occlude a substantial length of single-stranded template and be unusually sensitive to localized regions of (template) secondary structure.

In contrast, DNA polymerase β is most reactive on gaps of about

TABLE I
KINETIC PARAMETERS OF THE REACTION OF DNA POLYMERASE β WITH NICKED AND GAPPED DNA

Expt. No.	Primer-template		$E_a{}^a$ (kcal/mol)	$Q_{10}{}^a$	Apparent K_m for DNA		V_{max} (fmol dTMP/min)
					Nucleotide (μM)	3'-OH termini (nM)	
1[a]	Activated DNA		11.5	1.9	—	—	—
	PM2 DNA (1.3)		11.4	1.9	—	—	—
2[b]	PM2 DNA (1.3)		—	—	58	4.3	256
	PM2 DNA (10.0)		—	—	9	5.1	266
	PM2 DNA (21.4)		—	—	4	4.5	220
3[c]	Nicked KB DNA	Mn^{2+}	—	—	7	4.1	16
	Gapped KB DNA		—	—	12.5	7.3	71
	Nicked KB DNA	Mg^{2+}	—	—	108	63.5	60
	Gapped KB DNA		—	—	374	220	483

[a] The values of the thermal constants were obtained from Arrhenius plots. Initial velocities of incorporation were measured in the presence of Mn^{2+} with activated salmon sperm DNA or with duplex circular PM2 DNA that contained an average of 1.3 nicks per molecule at 10°, 20°, 30°, 35°, 40°, and 45°. E_a = activation energy; Q_{10} = ratio of reaction rates at T and $T + 10$.

[b] The kinetic constants were obtained from Lineweaver–Burk plots. Substrate was duplex circular PM2 DNA that contained an average of 1.3, 10.0, or 21.4 nicks per molecule. Reactions were performed at 35°C for 10 minutes in Mn^{2+}.

[c] "Gapped DNA" refers to calf thymus DNA that contained gaps averaging 10 nucleotides in length. "Nicked DNA" refers to calf thymus DNA briefly treated with pancreatic DNase I as described (8). Values of V_{max} are directly comparable to one another within experiments 2 and 3, but not between the two experiments.

ten nucleotides in length, and it appears to be capable of filling gaps completely (Fig. 1) (8). Moreover, with polymerase β, the substitution of Mn^{2+} for Mg^{2+} has a profound effect on the kinetic parameters of the incorporation reaction (3) that appears to be unique to this polymerase. In the presence of Mn^{2+}, the apparent affinity of the enzyme for DNA substrate is increased 10- to 50-fold and the maximal velocity of incorporation is decreased 5- to 10-fold, compared to the values determined in Mg^{2+} (3, 8) (Table I).

Another striking distinction between the two polymerases concerns their ability to replicate nicked duplex DNA substrates. While polymerase α is completely unreactive with such primer-templates, polymerase β can perform a limited synthetic reaction on these substrates to an extent of incorporation of about 15 nucleotides at each nick (Fig. 2) (8). To prove that this reaction does in fact occur at nicked sites, we have exploited the stringent substrate requirements of phage T4 DNA ligase (16). The nicked circular duplex DNA substrate molecules are fully ligatable before the polymerization reaction, and such prior ligation completely abolishes their primer-template capacity. Subsequent to the polymerization reaction, the form-II DNA product molecules can no longer be ligated. Analysis of the effect of temperature on the polymerization reaction with nicked and gapped DNA substrates in Mn^{2+} (8) (Table I) reveals identical values of activation energy (E_a) and Q_{10}, indicating that the frequency of productive interactions of polymerase β with 3'-hydroxyl termini at nicks and gaps is indistinguishable and suggesting that localized destabilization of the 5'-terminated DNA strand at the nick site does not contribute significantly to the rate-determining step(s) of the synthetic reaction. The results of analysis of the polymerization product with the single-strand-specific S1 nuclease suggest (8) that the reaction of polymerase β on these nicked duplex substrates most probably occurs by a strand displacement mechanism limited in its extent by branch migration reactions of the displaced template strand (17). With respect both to the extent of incorporation and the apparent absence of strand switching by the polymerase, the reaction of polymerase β on nicked substrates differs significantly from what had previously been described (4, 5) on a formally similar class of primer-template molecules, D-mtDNA (mitochondrial DNA containing "D-loops") (18). We believe these differences can be most reasonably explained by considering that, with the latter class of substrate molecules, the rate and extent of primary helix unwinding are expected to be facilitated greatly by the removal of negative superhelical turns (19).

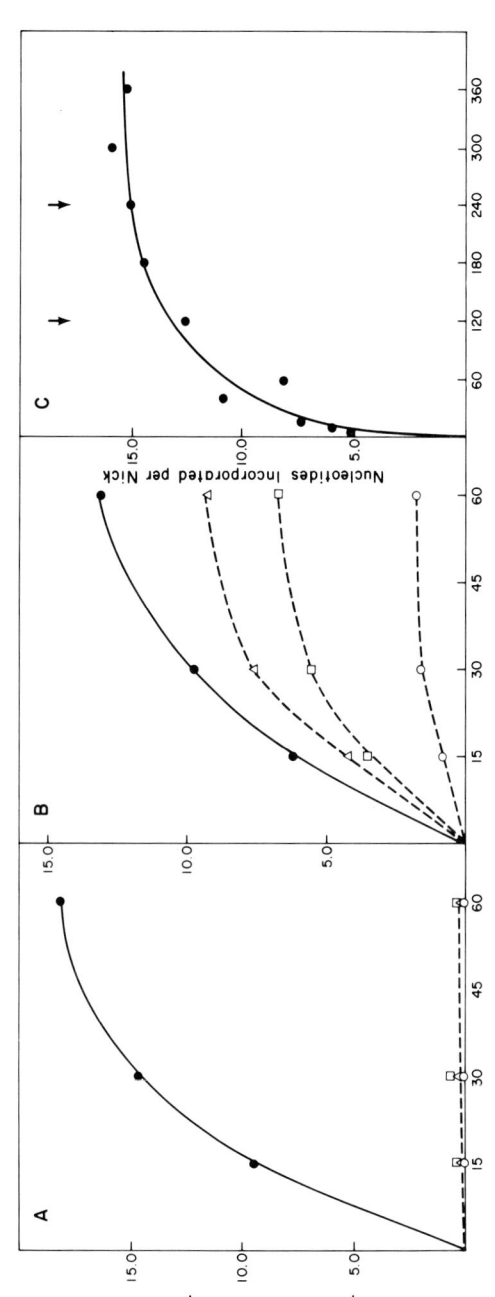

FIG. 2. Utilization of nicked PM2 DNA by DNA polymerases α (A) and β (B and C). Panels A and B: Activated DNA (control) (●); PM2 DNA with averages of 1.3 nicks per molecule (○), of 10 nicks per molecule (□), and of 21 nicks per molecule (△). Panel C: Extent of incorporation by polymerase β on PM2 DNA containing an average of 1.5 nicks per molecule. At 120 minutes (↓) and 240 minutes (↓), fresh polymerase β was added to the reaction in amounts identical to that present at time zero. Reprinted from Wang and Korn (8) with permission of American Chemical Society.

II. Primer-Template Recognition by DNA Polymerase β

DNA polymerase β is a small, basic protein [$M_r \approx 43,000$; $pI = 9.2$ (13)] that binds relatively tightly to columns of denatured DNA-cellulose (at pH 7.2) and has a high binding affinity for single-stranded DNA at neutral pH (20). However, it had not been previously determined to what degree the recognition of single-stranded or duplex regions of the polynucleotide substrate might contribute to the binding of primer-template in the active center of the polymerase protein. Examination of the polymerization reaction with nicked duplex PM2 DNA molecules containing an average of 1.3 to 21 nicks per molecule (8) demonstrated a constancy of K_m^{app} for primer termini and of V_{max} (Table I), and thus suggested that the enzyme has no catalytically significant affinity for duplex DNA.

Further detailed examination (12) of the polymerase β reaction has substantiated these initial interpretations and provided additional clarification of the nature of the polymerase–nucleic acid interaction. Calf thymus DNA was gently treated with pancreatic DNase I to yield a population of molecules containing an average of one 3′-hydroxyl terminus (nick) per 1700 nucleotides. When these molecules were used as substrates in the polymerization reaction, either in Mg^{2+} or in Mn^{2+} (Fig. 3A,B), the values of K_m^{app} were identical to the values of K_i^{app} determined in parallel experiments in which the same DNA molecules were used to inhibit the polymerization reaction on optimally gapped DNA (Fig. 3C,D). (In Mn^{2+}, $K_m^{app} = K_i^{app} = 5$–$10$ nM of 3′-hydroxyl termini.) Moreover, the nicked molecules behaved as inhibitors that were classically competitive with the gapped DNA primer-template. Preparations of calf thymus DNA gently treated with micrococcal nuclease to generate 3′-phosphate-terminated nicks were inactive as polymerase substrates, but behaved similarly to the 3′-hydroxyl-terminated, nicked DNA molecules in their capacity to inhibit the polymerization reaction with gapped DNA (in Mn^{2+}, $K_i^{app} \approx 10$ nM of 3′-phosphate termini). These results indicate that polymerase β has a comparable affinity for 3′ termini, whether they present hydroxyl or phosphate residues.

Intact, relaxed circular duplex (form IV) DNA is without detectable effect on the reactivity of polymerase β with gapped or nicked DNA substrates in the presence either of Mg^{2+} or Mn^{2+} (Fig. 4A,D). Surprisingly, it is also evident in Fig. 4 (panels B, C, E, and F) that the homopolymers $dA_{\overline{100}}$ and $dT_{\overline{100}}$ are similarly inert as tested in these assays. However, single-stranded heteropolymeric DNA (Fig. 5) can readily be demonstrated to be a weak inhibitor of polymerase β, and

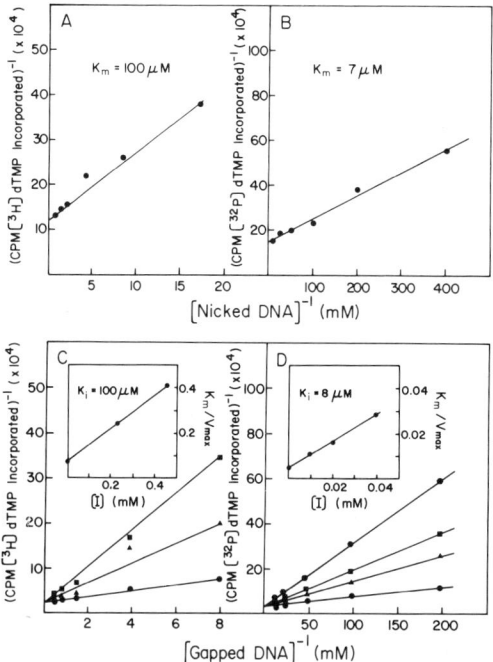

FIG. 3. Interaction of DNA polymerase β with 3'-hydroxyl termini in nicked duplex DNA. The data are presented as Lineweaver–Burk plots. Kinetics of dTMP incorporation in the presence of MG^{2+} (A) and Mn^{2+} (B). Kinetic analysis of the capacity of nicked DNA to inhibit the polymerization reaction on gapped DNA in the presence of Mg^{2+} (C) and Mn^{2+} (D).

by a mechanism kinetically noncompetitive with the DNA substrate. Moreover, the inhibitory capacity of closed-circular φX174 DNA indicates that the mechanism of this noncompetitive inhibition is 3'-terminus-independent, suggesting that the curious inactivity of the synthetic homopolymers (Fig. 4) cannot be attributed to the absence of (potentially) base-pairable 3' termini. In additional experiments (data not shown), we have used synthetic, single-stranded, random heteropolymers, averaging 50 to 80 nucleotides in length (10), to demonstrate both the capacity of polymerase β to catalyze dNMP incorporation by a self-priming snapback mechanism [similar to that described for polymerase α (7)] and the ability of these heteropolymers to inhibit the reaction of polymerase β on gapped DNA. The mechanism of this inhibition is kinetically of the "mixed-type," suggesting that the polymerase is capable of interacting with these polymers by binding

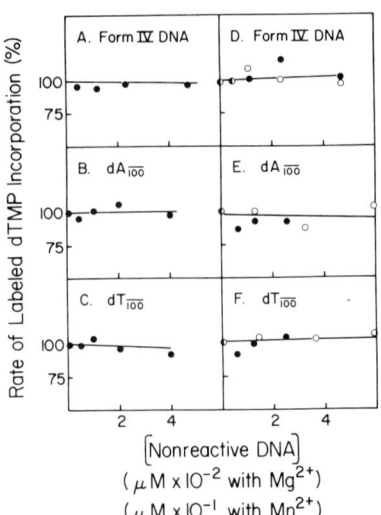

FIG. 4. Effect of intact duplex (form IV) PM2 DNA, dA$_{\overline{100}}$, and dT$_{\overline{100}}$ on the polymerization reaction of DNA polymerase β on gapped DNA (A, B, C) and nicked DNA (D, E, F). The assays in panels D, E, and F were performed both in Mg^{2+} (●) and in Mn^{2+} (○).

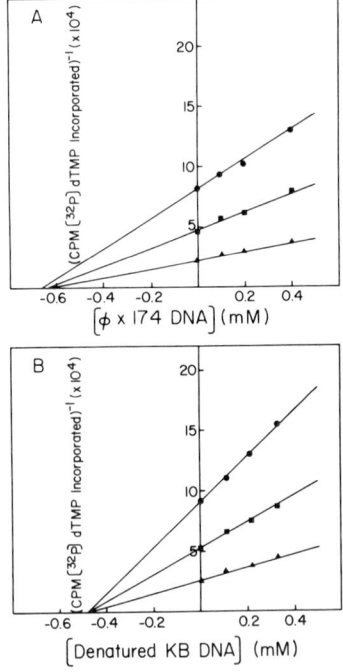

FIG. 5. Effect of single-stranded heteropolymeric ϕX174 DNA (A) and denatured KB DNA (B) on the polymerization reaction of DNA polymerase β on gapped DNA. The data are presented as Dixon plots.

both to single-stranded backbone and to the (potentially) base-pairable 3′-hydroxyl termini.

On the basis of these observations, we conclude that the major signal for primer-template recognition by DNA polymerase β is generated by a base-paired 3′ terminus. The enzyme is incapable of kinetically detectable interaction with duplex DNA, and, under optimum reaction conditions, it manifests only a weak affinity for single-stranded heteropolymeric templates that appears to be mediated through binding events that do not involve the active center of the enzyme. The lack of demonstrable interaction between polymerase β and single-stranded homodeoxypolymers cannot presently be explained, but it may suggest that the single-stranded template interaction with heteropolymers reflects, in a manner yet obscure, some capacity of the polymerase protein to respond to base composition.

III. Primer-Template Recognition by DNA Polymerase α

Kinetic analysis of the effects of the ubiquitous polyamine spermidine (21) on the reaction of polymerase α with activated DNA (9) demonstrated that spermidine produces a marked elevation both of the K_m^{app} of the polymerase for the primer-template and of the maximal reaction velocity. Moreover, the effect of the polyamine on the polymerase–nucleic acid interaction is independent both of polymerization per se or of the presence in the incubation of complementary dNTPs; i.e., the effect of spermidine appears to be quantitatively similar on both the "static affinity" (15) and the "kinetic affinity" (15) of polymerase α for gapped DNA. Two formal mechanistic explanations of these observations are (a) that spermidine decreases "nonspecific" or "nonproductive" binding of the polymerase (e.g., at sites distant from 3′-hydroxyl termini); or (b) that spermidine causes a specific destabilization of the polymerase · DNA (E · S) complex so as to facilitate both the backward (E · S → E + S) and the forward (E · S → E + P) dissociation reactions (22). To distinguish between these hypotheses, as well as to clarify some of the curious features of the polymerization reaction performed by polymerase α on gapped DNA substrates (see Section I), we undertook a detailed kinetic examination, supplemented by DNA binding studies, to illuminate the specific molecular signals involved in nucleic acid-polymerase α recognition.

DNA polymerase α appears to be completely incapable of kinetically detectable interaction with relaxed (form IV) or supercoiled (form I) duplex circular PM2 DNA, with multiply nicked PM2 DNA (Fig.

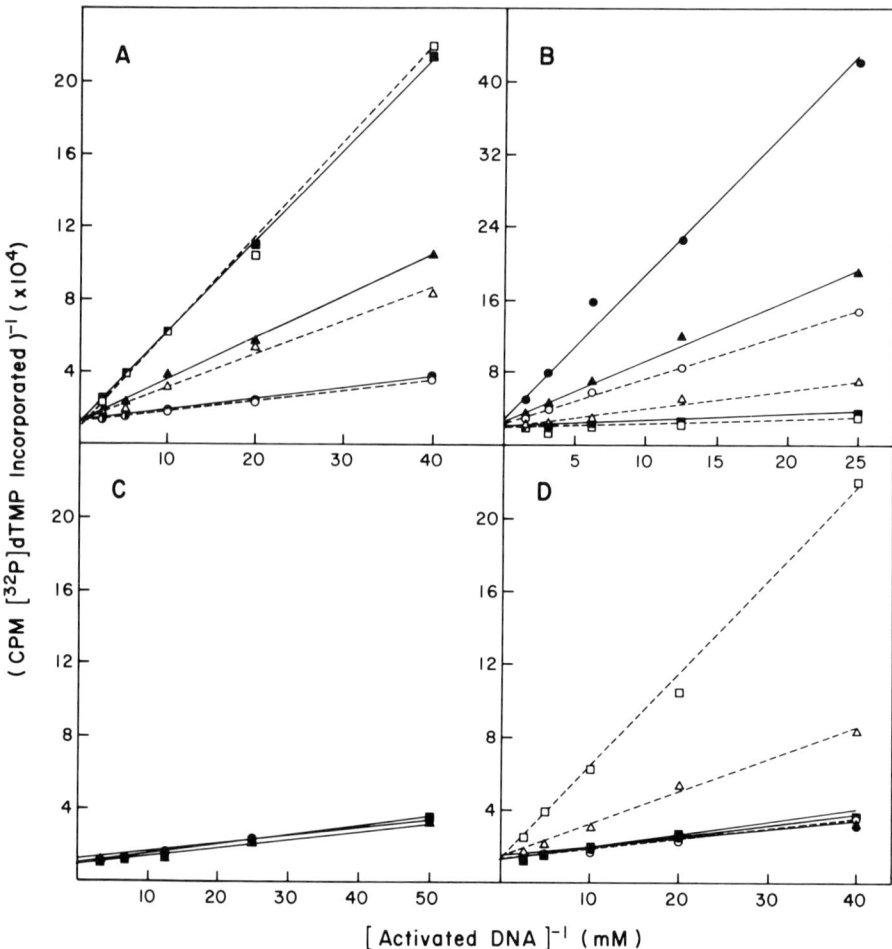

FIG. 6. Effects of duplex DNA, single-stranded DNA, and DNA termini on the polymerization reaction of DNA polymerase α on activated DNA (9). The data are presented as Lineweaver–Burk plots. The results obtained with intact circular M13 DNA are shown by the open symbols and dashed lines in panels A, B, and D. In all panels, the quantities of inhibitor DNA added to the reactions were none (control), 1×, and 2×. (A) Effect of 3'-phosphate fragments (300–400 nucleotides in average length) of M13 DNA; (B) effect of 3'-hydroxyl fragments (200 nucleotides in average length) of M13 DNA; (C) effect of flush-end duplex restriction fragments of PM2 DNA; form I and form IV PM2 DNA molecules are identically inert under these assay conditions; (D) effect of duplex PM2 DNA molecules containing an average of 9.6 nicks per molecule.

6D), or with blunt-ended linear duplex fragments of PM2 DNA (Fig. 6C). Closed-circular, single-stranded molecules of M13 DNA (Fig. 6A,B,D, dashed lines) and single-stranded fragments of M13 DNA terminated with either 3'-phosphate (Fig. 6A) or 3'-hydroxyl (Fig. 6B) residues are potent inhibitors kinetically competitive with the activated DNA substrate. These results, corroborated (9) by direct DNA binding studies, demonstrate that polymerase α possesses measurable affinity only for single-stranded, not for duplex, DNA; they suggest (9) that the enzyme does not interact with 3'-phosphate termini and indicate that the polymerase can recognize 3'-hydroxyl (primer) termini only if they are adjacent to regions of single-stranded template. The nonlinearity of the responses of the slopes of the Lineweaver–Burk plots in Fig. 6, panels A, B, and D suggested that the interaction of polymerase α with single-stranded DNA is cooperative, an interpretation substantiated by further kinetic analysis of the M13-DNA inhibi-

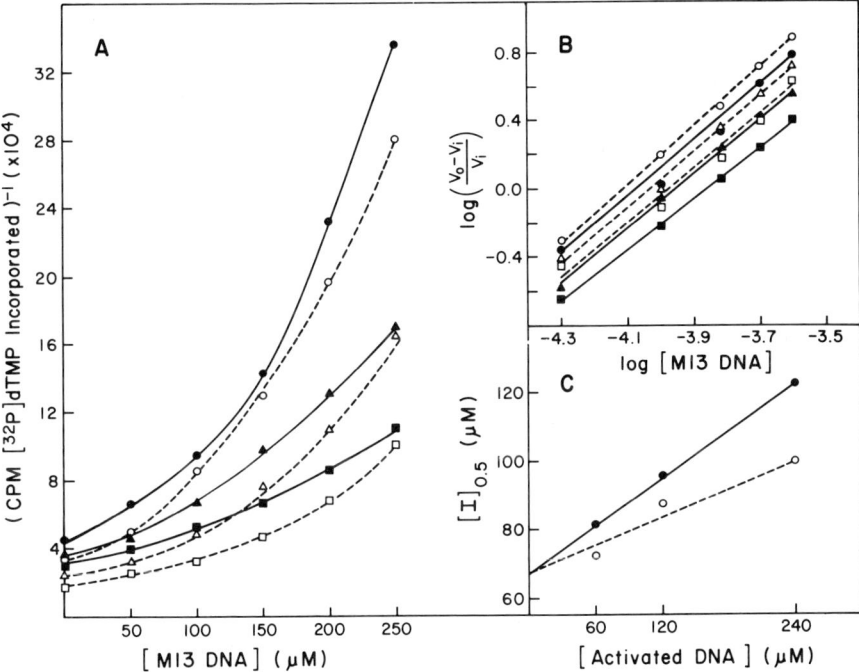

FIG. 7. Kinetic analysis of the interaction of DNA polymerase α with single-stranded circular M13 DNA in the absence (filled symbols and continuous lines) and the presence (open symbols and dashed lines) of 1 mM spermidine (9). (A) Dixon plot; (B) Hill plot; (C) plot of $[I]_{0.5}$, the concentration of inhibitor DNA required for 50% inhibition of the polymerization reaction vs concentration of activated (substrate) DNA.

tion by Dixon (Fig. 7A) and Hill (Fig. 7B) plots. The Dixon plots demonstrated marked upward concavity, and the data yield linear Hill plots with slopes of 1.5 to 1.8. Hill plots essentially identical to those in Fig. 7B, with slopes of 1.5 to 2.0, can be generated reproducibly from experiments in which either intact, closed-circular, single-stranded DNA molecules (M13 DNA, ϕX174 DNA) or 3'-phosphate- or 3'-hydroxyl-terminated linear single-stranded DNA fragments are studied as competitive inhibitors of the reaction of polymerase α with activated DNA. These results thus indicate that each catalytically active molecule of DNA polymerase α possesses at least two strongly interactive single-stranded DNA (template) binding sites.

The data in Fig. 7 also illustrate the results of a similar set of kinetic analyses performed in the presence of spermidine. In brief, the findings demonstrate that the essential features of the interaction of polymerase α with single-stranded DNA are qualitatively unaltered by the polyamine, and the interpretation of the quantitative shifts in the curves that are observed in the presence of spermidine can be most simply apprehended in Fig. 7C, from which it may be concluded (9) that the polyamine has no detectable effect whatever on the affinity of polymerase α for a single-stranded template. This analysis indicates that the effect of spermidine on the polymerization reaction cannot be explained on the basis of decreased nonproductive binding of polymerase to DNA substrate, but is consistent with the alternative hypothesis of specific destabilization of the E · S complex, most likely by intervention of the polyamine at the step of polymerase–primer terminus interaction.

During the course of these studies, we became aware of the ability of polymerase α to catalyze an incorporation reaction on single-stranded DNA fragments, and we have subsequently described the features of that synthetic reaction in detail (7). Two of the observations presented in that report are relevant to the present discussion. First, the kinetics of polymerization on a series of synthetic "hook" polymers of the structure dT_n-dA_m demonstrate that the recognition of the 3'-primer terminus by template-bound polymerase α requires that there be the potential for a minimal degree of 3'-terminal base-pairing with template ($0 \leq m \leq 5$), but that preformed, stably base-paired primer termini are not required. Second, we noted that dT_n is a particularly potent inhibitor of polymerase α, by a mechanism independent of the 3'-hydroxyl termini, and there is a profound difference in the K_i^{app} of the polymerase for dT_n vs dA_n that was not readily explained.

To pursue the latter observation, we have completed (10) an additional series of kinetic and nucleic acid binding studies that yield a quantitative description of the interaction of DNA polymerase α with a series of synthetic homopolymers, random heteropolymers, and hook polymers. Compelling evidence has been obtained in support of the interpretation that polymerase α can adopt multiple interconvertible conformations, each with a different spectrum of template binding specificities, and that each enzyme molecule possesses a minimum of two strongly interacting template binding sites.

Limitations of space permit only a brief summary of the experimental observations that have led to this conclusion. Polymerase α can be shown to exhibit a complex hierarchy of affinities both for deoxyhomopolymers (Fig. 8) [affinity for $dT_n \geq dG_n > dC_n \gg dA_n$] and for random heterotripolymers (Fig. 9) [affinity for $(dA,dG,dT)_n > (dA,dC,dT)_n > (dG,dC,dT)_n > (dA,dG,dC)_n$] that can most reasonably be explained on the basis of polymerase response to template signals

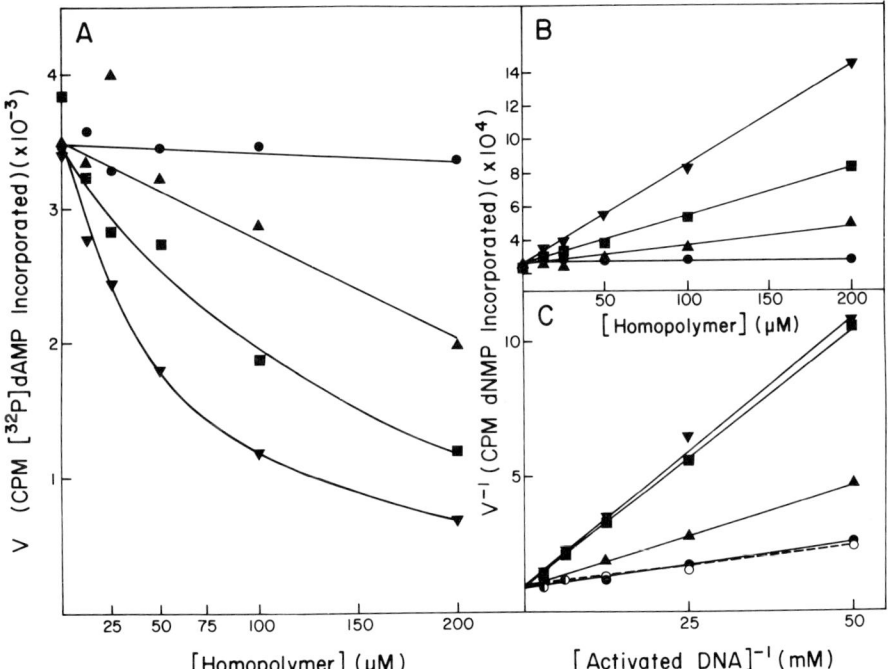

FIG. 8. Effect of unprimed synthetic homopolymers on the reaction of DNA polymerase α with activated DNA (10). (A) Poly(dA) (●); poly(dC) (▲); poly(dG) (■); poly(dT) (▼). (B) Dixon plot of the data in (A). (C) Lineweaver–Burk plot of the data in (A); ○, control reaction without homopolymer.

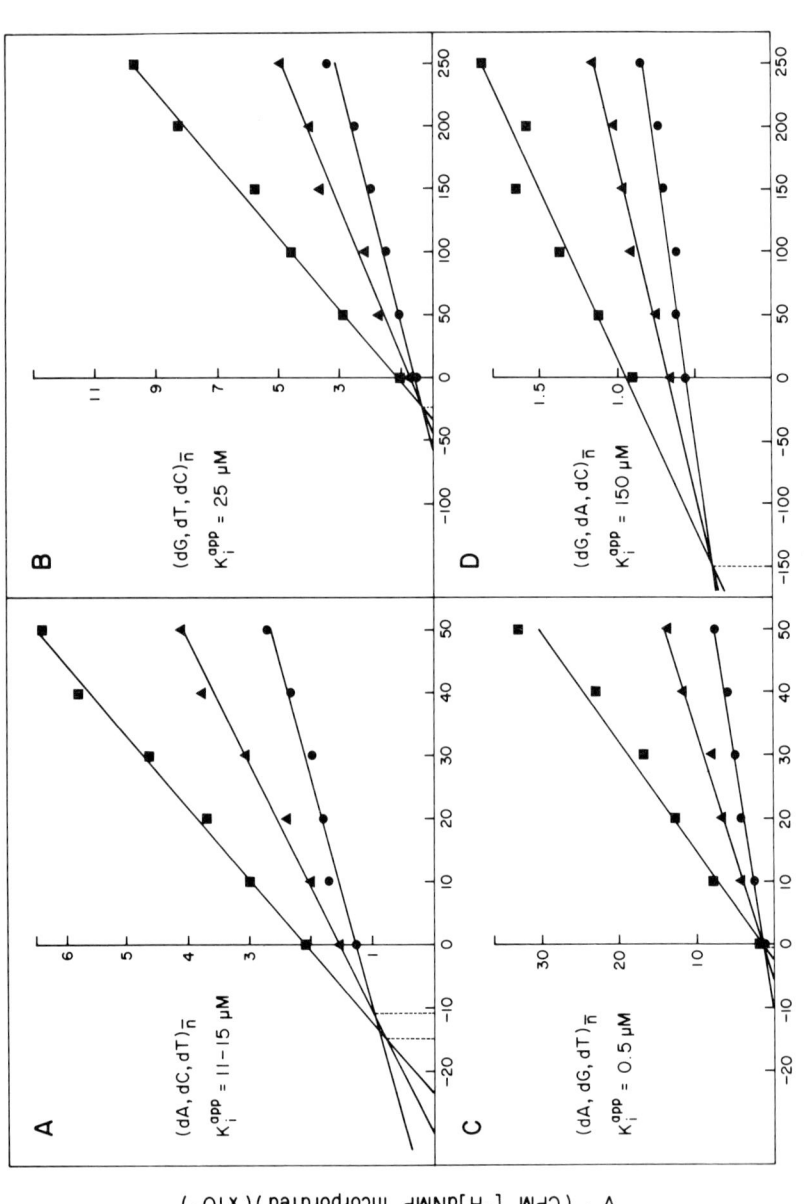

FIG. 9. Effect of synthetic heterotripolymers on the reaction of DNA polymerase α with activated DNA (10). The data are presented as Dixon plots.

that are mediated by base composition. At the extremes, values of K_i^{app} (expressed in terms of nucleotide concentration) extend over three orders of magnitude; e.g., dT_n, $K_i^{app} = 4$ μM; dA_n, $K_i^{app} \geq 750$ μM; $(dA,dG,dT)_n$, $K_i^{app} \leq 1$ μM; $(dA,dG,dC)_n$, $K_i^{app} = 150$ μM. The validity of the interpretation that these kinetic data reflect specificity of polymerase–nucleic acid binding interactions has been corroborated by semiquantitative binding assays performed by velocity gradient sedimentation (10).

It can further be demonstrated that the apparent affinity of polymerase α for a particular single-stranded polydeoxynucleotide is critically dependent upon the specific primer-template that is used to assess inhibition. Thus, while dA_n has no demonstrable effect on the replication of activated DNA ($K_i^{app} \geq 750$ μM) it competes effectively with the substrate $dA_{\overline{100}}$-$dT_{\overline{25}}$ ($K_i^{app} \approx 50$ μM); conversely, single-stranded circular φX174 DNA (Fig. 10), a moderately potent inhibitor of the replication of activated DNA ($K_i^{app} = 30$ μM), has little effect on the copying of $dT_{\overline{100}}$-$dA_{\overline{25}}$ ($K_i^{app} \geq 400$ μM) but is an extremely effective competitive inhibitor of the reaction with $dA_{\overline{100}}$-$dT_{\overline{25}}$ ($K_i^{app} \leq 1$ μM). It may be noted that, in these inhibition experiments with synthetic homo- and heteropolymers, evidence of simple cooperativity like that readily demonstrable with single-stranded DNA inhibitors and activated DNA substrate was not observed; nor was it detected in reactions in which φX174 DNA was tested as an inhibitor of hook polymer replication (Fig. 10), although kinetic complexity was clearly apparent in the presence of $dT_{\overline{100}}$-$dA_{\overline{25}}$ (Fig. 10B). We believe that this is most likely a reflection of our inability to define reaction conditions with sufficient precision to display the transition regions of the kinetic curves. However, the possibility cannot be excluded that these results may be due to interactions between polymerase and polydeoxynucleotides at an as yet unidentified class of single-strand-binding effector sites that are distinct from the template binding sites involved in catalysis.

These studies have been extended to an examination of the polymerase–primer terminus interaction, taking advantage of the considerable body of information we have developed regarding the binding and replication of hook polymers by DNA polymerase α. Mg^{2+} has a profound and positive effect on the apparent affinity of the polymerase for polydeoxypyrimidine-, but not for polydeoxypurine-, nucleotide templates. Thus, in reactions employing the homologous hook polymers as substrates, the K_i^{app} of polymerase α for dT_n and dC_n is decreased by 3- to 4-fold as the Mg^{2+} concentration is increased from 2 to

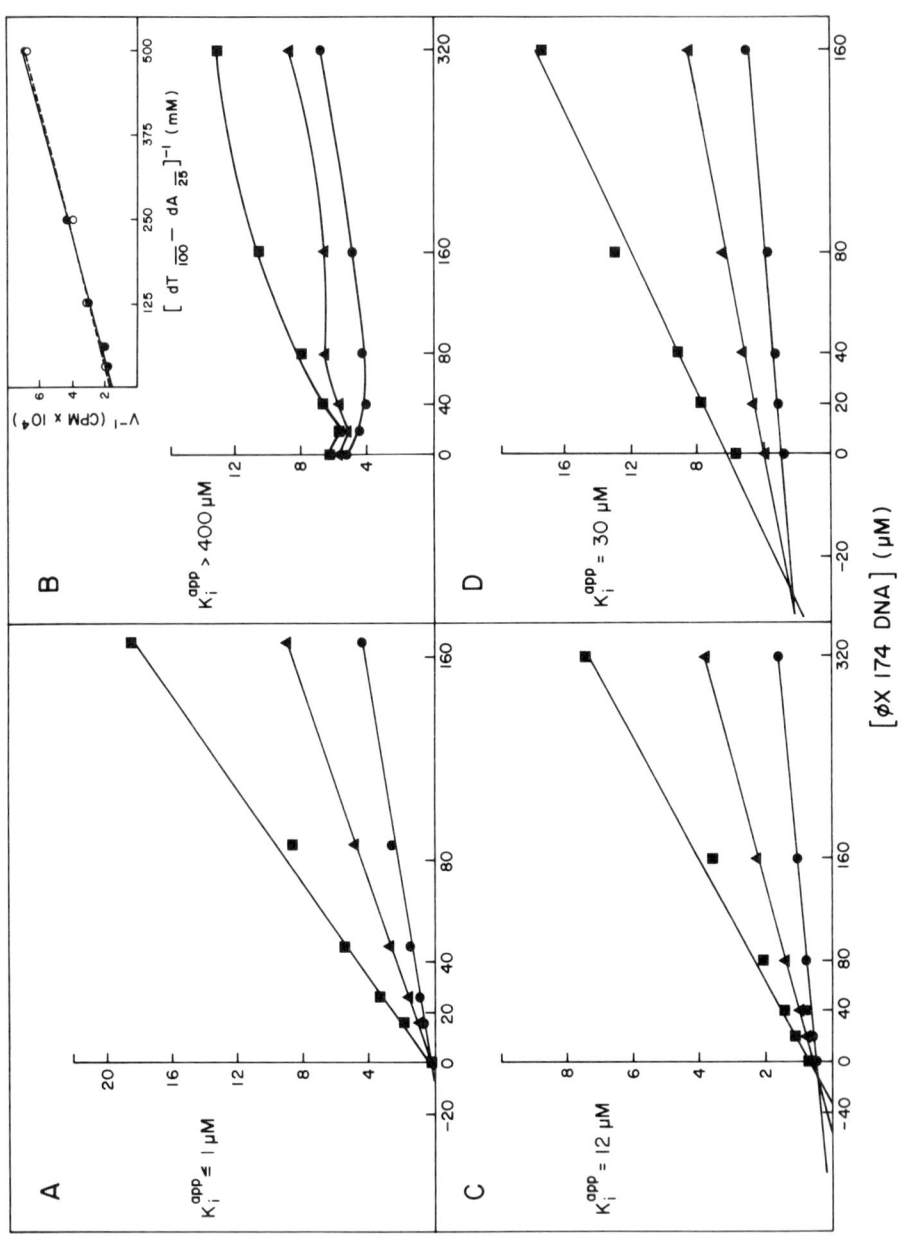

8 mM, while the K_i^{app} for dA_n is unchanged. In comparable experiments with activated DNA as substrate, K_i^{app} of polymerase α for dT_n is decreased 5-fold, and for dC_n, by 10-fold, over the same range of Mg^{2+} concentrations, while the K_i^{app} for dG_n is unchanged. In spite of this substantially increased affinity of the polymerase for polydeoxypyrimidine-nucleotide templates, the variation of Mg^{2+} concentration in this range is without detectable effect on the K_m^{app} of polymerase α for activated DNA, single-stranded synthetic heterotetrapolymer substrates, $dC_{\overline{100}}$-$dG_{\overline{25}}$, or $dT_{\overline{100}}$-$dA_{\overline{25}}$. The lack of response to Mg^{2+} variation of values of K_m^{app}, particularly for the latter two "hook" polymers, suggested that the divalent cation might also play a catalytically significant role in the polymerase–primer interaction, such that increasing Mg^{2+} concentrations resulted in a compensatory decrease in the affinity of the enzyme for the primer-terminus.

In support of this interpretation, we have shown that increasing Mg^{2+} concentrations do cause a greater than 6-fold increase in K_m^{app} for $dA_{\overline{100}}$-$dT_{\overline{25}}$. Kinetic analysis of this reaction has been particularly informative. By standard Lineweaver–Burk analysis, Mg^{2+} behaves like an inhibitor that is purely competitive with the primer-template. Given the fact that the divalent cation in this concentration range has no effect on the apparent affinity of the polymerase for poly(dA) template, this finding strongly suggests that the observed effect arises from competition by free Mg^{2+} at the primer terminus-binding site. Further examination of this phenomenon yields concave Dixon plots as well as linear Hill plots with slopes of ~ 3.8. These results, together with the purely competitive nature of the Mg^{2+} inhibition, indicate that each polymerase molecule has at least four strongly interacting Mg^{2+}-binding sites that appear to be implicated in the polymerase–primer-terminus interaction and are presumably distributed in pairs at each of the two primer-binding sites on each catalytically active enzyme molecule. We have also observed that at low concentrations of Mg^{2+} there is a stimulation of the polymerization reaction with $dA_{\overline{100}}$-$dT_{\overline{25}}$ that is progressively attenuated with increasing substrate (primer-template) concentration, a result consistent with the phenomenon of inhibitor activation described (23) with classical allosteric enzymes.

FIG. 10. Effect of primer-template on the apparent kinetic affinity of DNA polymerase α for inhibitor ϕX174 DNA (10). The data are presented as Dixon plots. The primer-templates used were: (A) $dA_{\overline{100}}$-$dT_{\overline{25}}$, (B) $dT_{\overline{100}}$-$dA_{\overline{25}}$; the inset in panel B is a Lineweaver–Burk plot of data obtained in reactions that contained ϕX174 DNA at 0 μM (●—●) or 20 μM (○---○); (C) $dC_{\overline{100}}$-$dG_{\overline{25}}$; (D) poly(dA,dG,dT,dC), a random heterotetrapolymer.

IV. Discussion and Conclusions

The studies summarized in this paper contribute substantially to the elucidation of some of the mechanisms and specific molecular signals that govern the interaction of DNA polymerases α and β with their nucleic acid substrates. These insights are of interest for at least two reasons. First, in the absence of an exploitable library of mammalian replication/repair mutants, the enzymological properties that we have documented define a mechanistic framework within which to identify and characterize putative accessory replication factors that might act either to modify primer-template structure or to modulate the inherent catalytic properties of the polymerases themselves. The mechanistic elucidation of the stimulation of DNA polymerase α by spermidine (9) provides a prototype for studies of this sort. Second, although it is presumed that the participation of both polymerases α and β in chromosome replication and repair must involve their interaction with a number of additional proteins (24–27), the inherent properties of the purified polymerase proteins would appear to be particularly appropriate to the putative roles of these enzymes in these complex and highly regulated processes.

Thus, with respect to DNA polymerase β, which is believed to play an important role in DNA repair (12, 28), the primary determinant of primer-template recognition appears to be a base-paired 3'-hydroxyl primer terminus, such as would be created by the several known repair endonucleases. The preference of this polymerase for short gaps, its ability to fill such gaps completely, and even its limited capacity to carry out strand displacement synthesis, all seem appropriate to the requirements of at least short-patch DNA-repair processes (29). In sharp contrast, the signals that govern the recognition of primer-template by DNA polymerase α are entirely different and seem, in turn, consistent with the postulated role of this enzyme in chromosome replication (9, 10). Perhaps the most intriguing and provocative finding emerging from these studies is the indication that the homogeneous polymerase α protein [which may prove ultimately to constitute the catalytically active core component (2) of more complex "holoenzyme" (30) species] exhibits many of the kinetic properties of an allosteric enzyme (23) and appears to possess at least two strongly interacting single-stranded DNA (template) binding sites, as well as a minimum of two similarly interactive primer-terminus binding sites. Both the absolute binding requirement for single strands and our more recent evidence suggesting that polymerase α is capable of responding to template base composition could provide the formal basis for a highly regulated process of polymerase–nucleic acid interaction at specific DNA sequences that may function as start and stop signals for DNA replication (10).

ACKNOWLEDGMENT

The studies described in this paper were supported by Grant CA-14835 and Training Grants GM-01922 and CA-09151 from the National Institutes of Health.

REFERENCES

1. T. S.-F. Wang, W. D. Sedwick and D. Korn, JBC **249**, 841 (1974).
2. P. A. Fisher and D. Korn, JBC **252**, 6528 (1977).
3. T. S.-F. Wang, D. C. Eichler and D. Korn, Bchem **16**, 4927 (1977).
4. D. C. Eichler, T. S.-F. Wang, D. A. Clayton and D. Korn, JBC **252**, 7888 (1977).
5. D. Korn, P. A. Fisher, J. Battey and T. S.-F. Wang, CSHSQB **43**, 613 (1977).
6. P. A. Fisher, T. S.-F. Wang and D. Korn, JBC **254**, 6128 (1979).
7. P. A. Fisher and D. Korn, JBC **254**, 11040 (1979).
8. T. S.-F. Wang and D. Korn, Bchem **19**, 1782 (1980).
9. P. A. Fisher and D. Korn, JBC **254**, 11033 (1979).
10. P. A. Fisher, J. T. Chen and D. Korn, JBC **256** (in press).
11. A. Weissbach, ARB **46**, 25 (1977).
12. T. S.-F. Wang and D. Korn (to be submitted for publication).
13. T. S.-F. Wang, W. D. Sedwick and D. Korn, JBC **250**, 7040 (1975).
14. P. A. Fisher and D. Korn, JBC **254**, 6136 (1979).
15. R. A. Bambara, D. Uyemura and T. Choi, JBC **253**, 413 (1978).
16. I. R. Lehman, Science **186**, 790 (1974).
17. C. S. Lee, R. W. Davis and N. Davidson, JMB **48**, (1970).
18. H. Kasamatsu, D. L. Robberson and J. Vinograd, PNAS **68**, 2252 (1971).
19. W. Bauer and J. Vinograd, JMB **33**, 141 (1968).
20. L. M. S. Chang and F. J. Bollum, Bchem **11**, 1264 (1972).
21. C. W. Tabor and H. Tabor, ARB **45**, 285 (1976).
22. A. Cornish-Bowden, JMB **101**, 1 (1976).
23. J. C. Gerhart and A. B. Pardee, CSHSQB **28**, 491 (1963).
24. A. Kornberg, CSHSQB **43**, 1 (1978).
25. S. H. Wickner, CSHSQB **43**, 303 (1978).
26. C. C. Richardson, L. J. Romano, R. Kolodner, J. E. LeClerc, F. Tamanoi, M. J. Engler, F. B. Dean and D. S. Richardson, CSHSQB **43**, 427 (1978).
27. C. C. Liu, R. L. Burke, U. Hibner, J. Barry and B. Alberts, CSHSQB **43**, 469 (1978).
28. J. Waser, U. Hubscher, C. C. Kuenzle and S. Spadari, EJB **97**, 361 (1979).
29. P. C. Hanawalt, P. K. Cooper, A. K. Ganesan and C. A. II. Smith, ARB **48**, 783 (1979).
30. C. McHenry and A. Kornberg, JBC **252**, 6478 (1977).

Structural and Functional Properties of Calf Thymus DNA Polymerase δ

MARIETTA Y. W. TSANG LEE[1]
CHENG-KEAT TAN
KATHLEEN M. DOWNEY[2] AND
ANTERO G. SO[3]

*Howard Hughes Medical Institute
Laboratory
Departments of Medicine and
Biochemistry and
Center for Blood Diseases
University of Miami School of
Medicine
Miami, Florida*

When the current nomenclature for mammalian DNA polymerases was proposed in 1975 (1, 1a), there were thought to be four distinct enzyme species: DNA polymerases α, β, and γ and mitochondrial DNA polymerase. The classification of these enzymes was based on such properties as size, sensitivity to sulfhydryl inhibitors, template primer preference, isoelectric point, and subcellular location. DNA polymerase γ and mitochondrial DNA polymerase are now believed to be identical (2–4).

In contrast to bacterial DNA polymerases, these mammalian enzymes are devoid of 3'-to-5' exonuclease activity (5–7), the "proofreading" activity shown to play an important role in ensuring the high fidelity of DNA replication in prokaryotes (8–11). Thus it was generally believed that the lack of proofreading activity is characteristic of eukaryotic DNA polymerases and that this represents a fundamental difference between prokaryotes and eukaryotes (7, 12–14). However, during the past 5 years DNA polymerases with associated 3'-to-5' exonuclease activity have been isolated from both lower eukaryotes (14–16) and mammalian cells (17–21).

In our laboratory, a DNA polymerase containing 3'-to-5' exonu-

---

[1] Research Associate of the Howard Hughes Medical Institute.
[2] Research Career Development Awardee (NIH K04 HL00031).
[3] Investigator of the Howard Hughes Medical Institute.

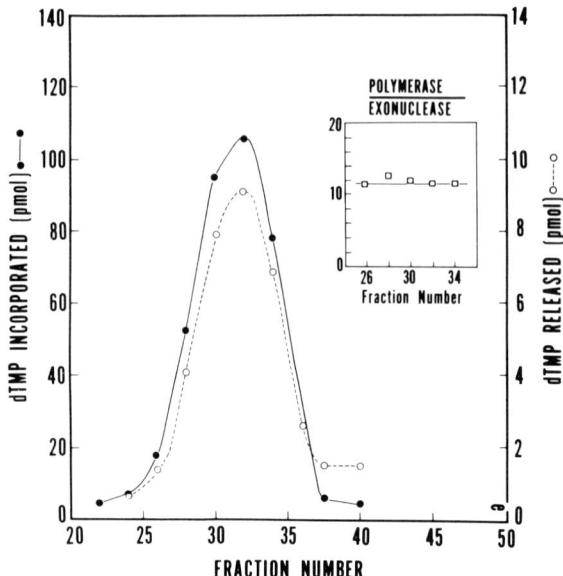

FIG. 1. Chromatography of DNA polymerase δ on Sephacryl S-200. DNA polymerase activity (●——●) was assayed with poly(dA-dT) as template/primer, and 3'-to-5' exonuclease activity (○---○) was assayed with [^{3}H]dT$_{\overline{50}}$ as substrate, as previously described (21). The ratio of polymerase to exonuclease activity (□——□) is shown in the inset.

clease activity has been isolated from rabbit bone marrow (17) and, more recently, from calf thymus (19, 21). Extending the nomenclature established for mammalian DNA polymerases, this enzyme was designated DNA polymerase δ. In this report, we describe some of our recent studies with DNA polymerase δ from calf thymus. The major topics discussed include (a) evidence that 3'-to-5' exonuclease activity is an intrinsic property of DNA polymerase δ; (b) structural properties of the enzyme; (c) evidence that the 3'-to-5' exonuclease activity has a proofreading function; and (d) inhibitor studies.

I. 3'-to-5' Exonuclease Activity: An Intrinsic Property of DNA Polymerase δ

We have reported an eight-step purification of DNA polymerase δ from fetal calf thymus (19, 21). This procedure resulted in an 8000-fold purification of the enzyme, to a specific activity of 28,000 units per milligram of protein. DNA polymerase and 3'-to-5' exonuclease activities copurified throughout the purification procedure, which in-

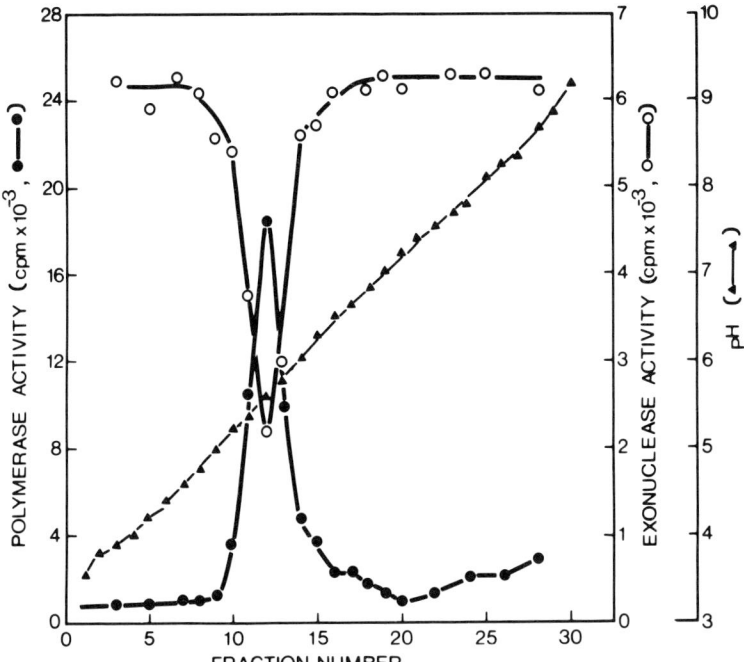

FIG. 2. Isoelectric focusing of DNA polymerase δ. DNA polymerase activity was assayed with poly(dA-dT) as template/primer, and 3'-to-5' exonuclease activity was assayed with 3'-terminally labeled $(dA-dT)_n$-$[^3H]dT_{(0.5-1.0)}$ as substrate, as previously described (21).

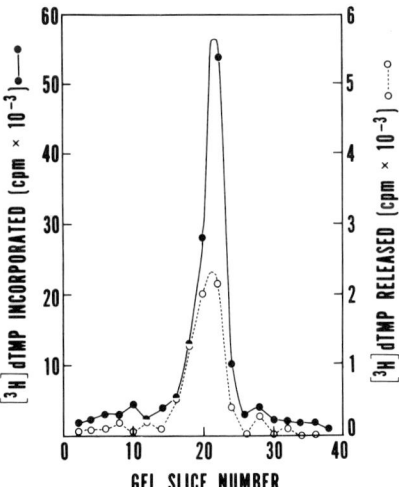

FIG. 3. Polyacrylamide gel electrophoresis of DNA polymerase δ under nondenaturing conditions. DNA polymerase δ (4 μg) was loaded on a 5% polyacrylamide gel and electrophoresed in a discontinuous system. DNA polymerase activity (●——●) with poly(dA-dT) as template/primer and 3'-to-5' exonuclease activity (○---○) with 3'-terminally labeled $(dA-dT)_n$-$[^3H]dT_{(0.5-1.0)}$ as substrate were determined in extracts of gel slices as previously described (21).

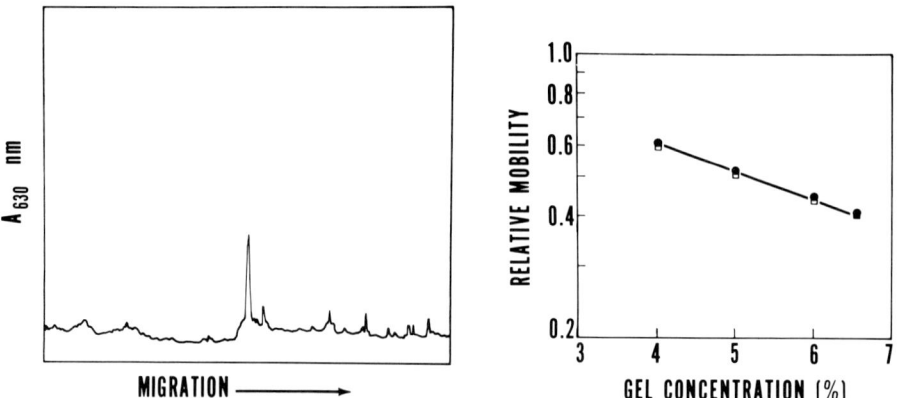

FIG. 4 (left). Densitometer scan of polyacrylamide gel. A gel identical to that described in Fig. 3 was stained with Coomassie blue and scanned with a laser densitometer.

FIG. 5 (right). Relative mobility of DNA polymerase δ as a function of gel concentration. The relative mobilities (R_m) of polymerase and 3'-to-5' exonuclease activities of DNA polymerase δ were determined at the concentrations of polyacrylamide indicated. Polymerase activity was assayed with poly(dA-dT) (●——●), and exonuclease activity was assayed with $(dA-dT)_n\text{-}[^3H]dT_{\overline{0.5-1.0}}$ (□——□) as previously described (21).

cluded ion-exchange, affinity, and gel-filtration chromatography. The final step in the purification procedure was chromatography on Sephacryl S-200, shown in Fig. 1. The elution profile of the 3'-to-5' exonuclease activity coincided with the elution profile of the DNA polymerase activity, and, more important, the ratio of polymerase activity to exonuclease activity (12:1) was constant across the peak (inset).

When DNA polymerase δ (step 7) was subjected to isoelectric focusing, both the polymerase and the 3'-to-5' exonuclease activities were sharply focused, and both had an identical pI of 5.5 (Fig. 2). Thus DNA polymerase δ, like DNA polymerases α and γ, is an acidic protein.

Analyses of DNA polymerase δ (step 8) on discontinuous 5% nondenaturing polyacrylamide gels, are shown in Fig. 3 (enzymic activities) and Fig. 4 (densitometer scan of stained gel). Both polymerase and 3'-to-5' exonuclease activities have identical electrophoretic mobilities, and both comigrate with a major protein band of R_m 0.525. To investigate the possibility of fortuitous comigration of DNA polymerase activity with a contaminating 3'-to-5' exonuclease activity, polyacrylamide gel electrophoresis under nondenaturing conditions was run at three additional gel concentrations. The exact coincidence of

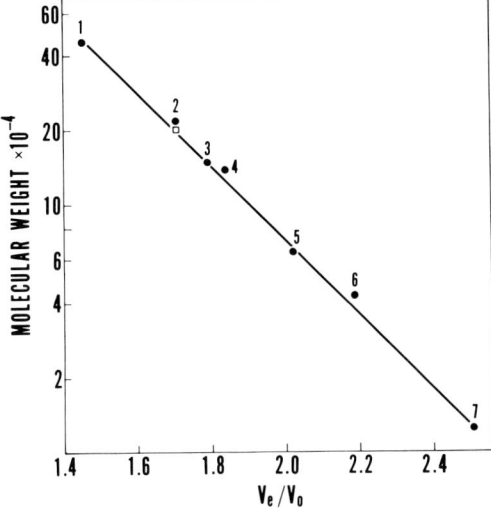

FIG. 6. Determination of molecular weight of DNA polymerase δ by gel filtration on Sephacryl S-300. Approximately 2000 units of DNA polymease δ were applied to a 2.5 × 89 cm column of Sephacryl S-300 equilibrated with 25 mM Hepes buffer (pH 7.0), 0.25 M KCl, 20% (v/v) glycerol, 0.5 mM EDTA, 0.1 mM EGTA, and 5 mM dithiothreitol. The standards are: 1, ferritin (450,000); 2, catalase (240,000); 3, yeast alcohol dehydrogenase (150,000); 4, glyceraldehyde-3-phosphate dehydrogenase (140,000); 5, bovine serum albumin (68,000); 6, ovalbumin (43,000); 7, cytochrome c (12,500). The void volume (V_0) was determined with Blue Dextran 2000. DNA polymerase δ activity (□) was assayed as previously described (21).

the two activities is shown in Fig. 5, where the log R_m of either polymerase or exonuclease activity is plotted as a function of gel concentration. The linear relationship obtained establishes that both polymerase and 3'-to-5' exonuclease activities reside on the same protein molecule or on proteins of identical shape and charge.

The purification of two species of high-molecular-weight DNA polymerase from mouse myeloma has been reported (20). One of these enzymes, designated α_1, was associated with a 3'-to-5' exonuclease activity. Whether DNA polymerase α_1 is identical to DNA polymerase δ or a distinct DNA polymerase must await further studies.

II. Structural Properties of DNA Polymerase δ

The molecular weight of DNA polymerase δ has been estimated, from gel filtration on Sephacryl S-300 in the presence of 0.25 M KCl, to be 200,000 (Fig. 6). This is somewhat higher than the 155,000 suggested by the sedimentation coefficient (7.5 S), determined by centrifugation of DNA polymerase δ in a 20 to 40% glycerol gradient containing 0.3 M KCl, with bovine serum albumin (4.35 S) and aldolase (7.35 S) as markers (data not shown).

The Stokes' radius of DNA polymerase δ was estimated to be 48.3 Å by polyacrylamide gel electrophoresis using standards of known Stokes' radii (Fig. 7). This value is in agreement with the 47.5 Å determined by gel filtration on Sephacryl S-300 with cytochrome c, ovalbumin, bovine serum albumin, yeast alcohol dehydrogenase, catalase, and ferritin as standards (data not shown). Assuming a partial specific volume of 0.71–0.75, the molecular weight of DNA polymerase δ is estimated to be 140,000–160,000. The discrepancy between molecular weight estimates based on Stokes' radius and sedimentation coefficient (140,000–160,000) and gel filtration (200,000) suggests that DNA polymerase δ may be an asymmetric molecule. A frictional ratio of 1.33 to 1.40 is also consistent with this suggestion.

The subunit structure of DNA polymerase δ was determined by gel electrophoresis in the presence of sodium dodecyl sulfate. Two protein bands were detected (Fig. 8), and the molecular weights of these subunits were estimated to be 49,000 and 60,000, using protein standards of known molecular weights (Fig. 9).

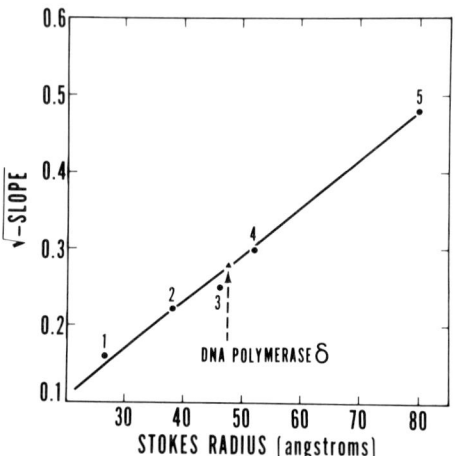

FIG. 7. Determination of Stokes' radius of DNA polymerase δ by nondenaturing polyacrylamide gel electrophoresis. Protein standards were electrophoresed on 4, 5, 6, and 6.5% polyacrylamide gels, and the log R_m was plotted as a function of gel concentration, as in Fig. 5. The negative slopes of the straight lines were thus determined (33). The Stokes' radii of the standards are: 1, ovalbumin (26 Å); 2, bovine serum albumin (37.5 Å); 3, aldolase (46 Å); 4, catalase (52 Å); 5, ferritin (79 Å).

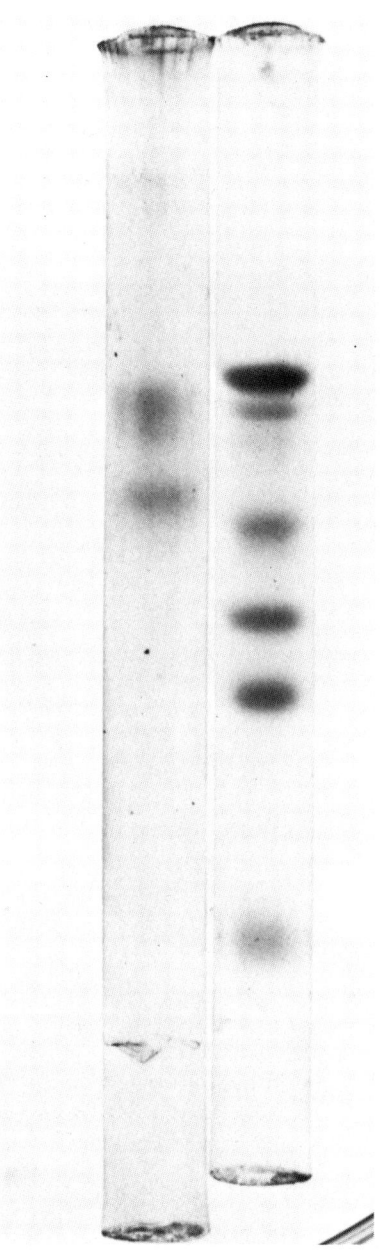

FIG. 8. Sodium dodecyl sulfate/polyacrylamide gel electrophoresis of DNA polymerase δ. An extract of the peak fraction of a preparative, nondenaturing polyacrylamide gel was simultaneously concentrated and dialyzed against 10 mM sodium phosphate (pH 7.0), 1% sodium dodecyl sulfate, 1% 2-mercaptoethanol, and 25% glycerol. The sample (200 μl) was incubated at 100°C for 15 minutes, 20 μl of 0.05% w/v bromophenol

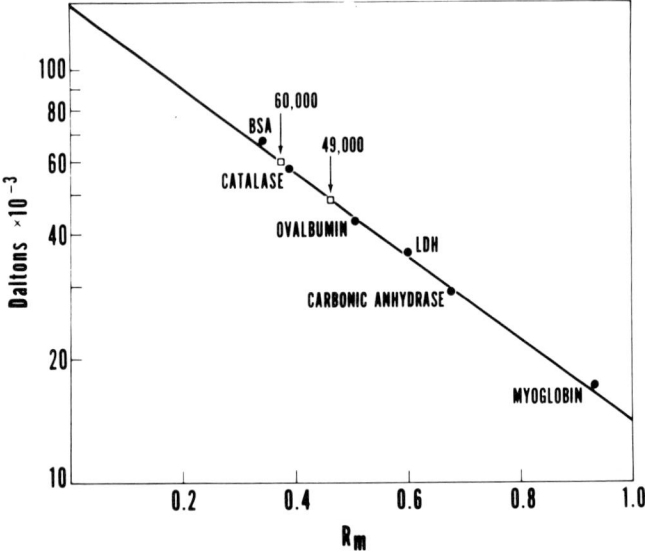

FIG. 9. Determination of subunit molecular weight of DNA polymerase δ. The R_m values of the reference proteins in Fig. 8 (●) were plotted against their known subunit molecular weights. The two subunits of DNA polymerase δ (□) were estimated to be 60,000 and 49,000.

We have also used a microgel technique to analyze the subunit structure of DNA polymerase δ. By this technique, nanogram quantities of protein can be visualized, and it was thus possible to determine the subunit structure of DNA polymerase δ in individual slices of a nondenaturing polyacrylamide gel. Figure 10 shows the protein band pattern in dodecyl-sulfate microgels of protein extracted from the individual slices (slices Nos. 20, 21, and 22) of a nondenaturing 5% polyacrylamide gel containing polymerase and exonuclease activities. Also shown are the protein standards, bovine serum albumin and *Escherichia coli* RNA polymerase. Only two subunits were observed across the peak of polymerase activity. A microdensitometer scan of microgel No. 21 is shown in Fig. 11. The stoichiometry of the subunits appears to be approximately 1:1.

---

blue were added, and the whole sample was loaded onto a 7% sodium dodecyl sulfate gel. Electrophoresis was carried out essentially according to Weber (34). The molecular weight standards are displayed in a parallel gel on the right (top to bottom): bovine serum albumin (68,000), catalase (58,000), ovalbumin (43,000), lactate dehydrogenase (36,000), carbonic anydrase (29,000), and myoglobin (17,200).

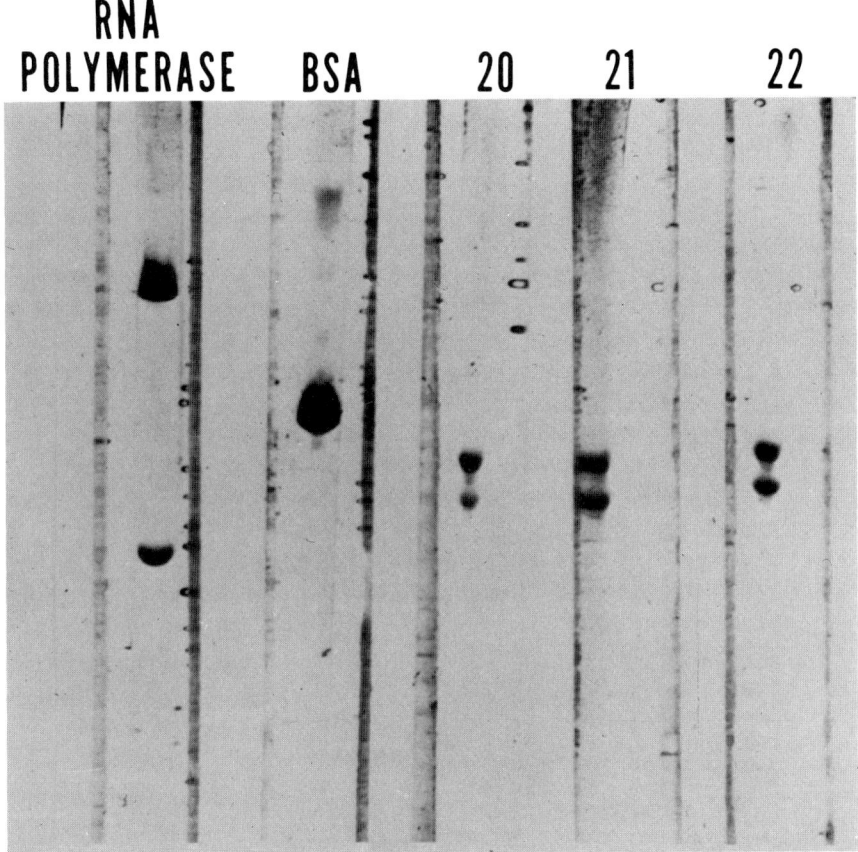

FIG. 10. Subunit structure of DNA polymerase δ by sodium dodecyl sulfate/microgel electrophoresis. Extracts from slices Nos. 20, 21, and 22 of a 5% standard nondenaturing gel, corresponding to the polymerase activity peak, were individually precipitated, dissolved in 10 μl of 0.2% sodium dodecyl sulfate, 0.2% 2-mercaptoethanol, and heated at 80°C for 15 minutes. One microliter of a solution containing 0.02% bromophenol blue, 0.7 M Tris sulfate, pH 8.4, 20% sucrose was added; 3-μl aliquots of each preparation were separately loaded onto 1 to 24% microgels and electrophoresed. The molecular weight standards are RNA polymerase (165,000, 155,000, 39,000) and bovine serum albumin (68,000).

The subunit structure of mouse myeloma DNA polymerase α_1 is similar to that of DNA polymerase δ. As analyzed by polyacrylamide gel electrophoresis in dodecyl sulfate, DNA polymerase α_1 appears to consist of two subunits of molecular weights 48,000 and 52,000 (20).

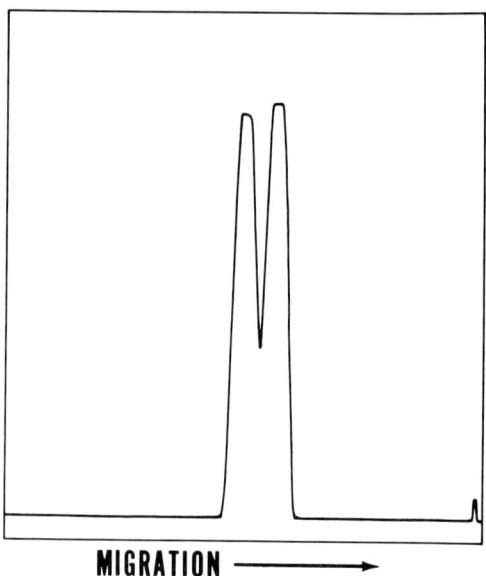

FIG. 11. Densitometer scan of microgel No. 21. The scan of microgel No. 21 was obtained from a photographic negative with a Joyce–Loebl densitometer (35).

III. Proofreading Function of the 3'-to-5' Exonuclease Activity of DNA Polymerase δ

It is characteristic of the proofreading exonuclease of bacterial DNA polymerases that a mismatched primer terminus is hydrolyzed more rapidly than a properly base-paired primer terminus. Furthermore, single-stranded polynucleotides are degraded more rapidly than double-stranded polynucleotides, reflecting the specificity of the enzyme for a non-base-paired primer terminus (11). As can be seen in Fig. 12, the rate of hydrolysis of single-stranded $(dT)_{\overline{50}}$ by the 3'-to-5' exonuclease activity of DNA polymerase δ is much more rapid than that of double-stranded poly(dA) · $(dT)_{\overline{50}}$. After 60 minutes of incubation, more than 80% of single-stranded poly(dT) was degraded, whereas less than 10% of poly(dA) · poly(dT) was hydrolyzed.

The rate of hydrolysis of complementary and noncomplementary primer termini is shown in Fig. 13. The 3'-to-5' exonuclease activity of DNA polymerase δ hydrolyses the noncomplementary dGMP primer terminus of $dA_n · (dT_{\overline{50}}$-$[^3H]dG_{\overline{1.3}})$ approximately three times faster than the properly base-paired dTMP primer terminus of $dA_n · (dT_{\overline{50}}$-$[^3H]dT_{\overline{2.3}})$. Similar results were obtained for the hydrolysis of mis-

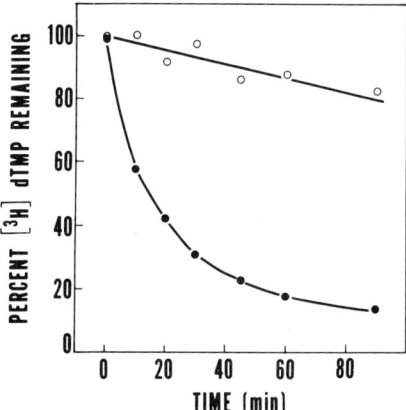

FIG. 12. Hydrolysis of single-stranded and double-stranded DNA by the 3'-to-5' exonuclease activity of DNA polymerase δ. Exonuclease activity was measured as previously described (21) with either [^{3}H]dT$_{\overline{50}}$ (●———●) or [^{3}H]dT$_{\overline{50}}$ · dA$_n$ (○———○) as substrate. The amount of DNA polymerase δ in the reaction mixture (50 μl) was 0.1 unit.

matched [^{3}H]dCMP in dA$_n$ · (dT$_{\overline{50}}$-[^{3}H]dC$_{\overline{0.8}}$). Thus the specificity of the 3'-to-5' exonuclease of DNA polymerase δ is consistent with a proofreading function of this activity.

The association of 3'-to-5' exonuclease activity with high-molecular-weight DNA polymerase from lower eukaryotes is now well established (12, 14–16). Furthermore, the 3'-to-5' exonuclease activity associated with DNA polymerases from *Ustilago maydis* and *Saccharomyces cerevisiae* has been shown to have a proofreading

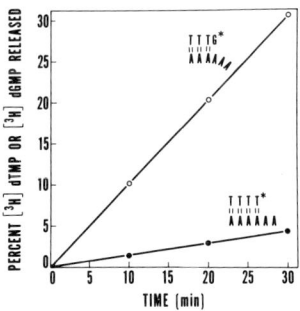

FIG. 13. Hydrolysis of complementary and noncomplementary primer termini by the 3'-to-5' exonuclease activity of DNA polymerase δ. Exonuclease assays were as previously described (21) with either dA$_n$ · (dT$_{\overline{50}}$-[^{3}H]dT$_{\overline{2.3}}$) (●———●), or dA$_n$ · (dT$_{\overline{50}}$-[^{3}H]dG$_{\overline{1.3}}$) as substrate. The amount of DNA polymerase δ in the reaction mixture (50 μl) was 0.1 unit, and the temperature of incubation was 28°C.

function (12, 14, 15). Thus, with our present demonstrations that 3'-to-5' exonuclease activity is an intrinsic property of mammalian DNA polymerase δ, and that this nuclease has a proofreading function, it would appear that one of the mechanisms by which the high fidelity of DNA replication is maintained in both higher and lower eukaryotes may be analogous to that of prokaryotes.

IV. Inhibitor Studies

Circumstantial evidence has implicated DNA polymerase α as the mammalian DNA replicase. This is based primarily on the excellent correlation between the level of DNA polymerase α activity and the cellular replication rate (13, 22–25). Several inhibitors of DNA replication have been used to further delineate the roles of DNA polymerase α, β, and γ. DNA replication is very sensitive to inhibition by aphidicolin (26–28), N-ethylmaleimide (28), and arabinofuranosyl nucleotides (27, 28), whereas it is quite resistant to dideoxythymidine nucleotides (29–32). Of the mammalian DNA polymerases studied, only DNA polymerase α had an identical pattern of sensitivity to these inhibitors (28). DNA synthesis with DNA polymerase α was strongly inhibited by aphidicolin, N-ethylmaleimide (MalNEt) and 1-β-D-arabinofuranosylcytosine triphosphate (araCTP) but relatively resistant to dideoxythymidine triphosphate (d_2TTP). In contrast, DNA polymerases β and γ were very sensitive to inhibition by d_2TTP but quite resistant to aphidicolin and araCTP. DNA polymerase β was resistant to MalNEt, while the sensitivity of DNA polymerase γ was between that of α and β.

We have done comparative studies on the relative sensitivity of DNA polymerases α and δ from calf thymus to these inhibitors. In Table I are reported the concentrations of MalNEt and aphidicolin required to produce a 50% inhibition of the rate of DNA synthesis with either DNA polymerase α or DNA polymerase δ. Also reported are the molar ratios of araATP/dATP or d_2TTP/dTTP that produce 50% inhibition of the rate of DNA synthesis. Both DNA polymerase α and DNA polymerase δ are equally sensitive to inhibition by aphidicolin and araATP, and equally resistant to d_2TTP. However, DNA polymerase δ is more sensitive to inhibition by MalNEt than is α by an order of magnitude. Thus, DNA polymerase α and δ are similar to each other and to DNA replication in their sensitivities to a variety of inhibitors. Whether these enzymes are structurally or immunologically related is not known at present.

TABLE I
INHIBITION OF DNA SYNTHESIS[a]

Inhibitor	DNA Polymerase	
	α	δ
N-Ethylmaleimide (mM)	2.0	0.2
Aphidicolin[b] (μg/ml)	2.5	3.5
araATP/dATP	0.7	0.25
d_2TTP/dTTP	125	75

[a] For each inhibitor of DNA synthesis either the concentration of inhibitor (N-ethylmaleimide and aphidicolin) or the ratio of analog (araATP and d_2TTP) to nucleotide that results in 50% inhibition of the rate of DNA synthesis is reported. DNA polymerase α was assayed with poly(dA-dT) as template/primer (21).

[b] Aphidicolin is a tetracyclic diterpenoid produced by *Cephalosporium aphidicola*.

V. Summary and Conclusions

The present studies clearly demonstrate that DNA polymerase δ has an associated 3'-to-5' exonuclease activity. This finding establishes that DNA polymerase δ is a unique enzyme, distinct from other known mammalian DNA polymerases (α, β, γ). Our studies further suggest that the 3'-to-5' exonuclease activity associated with DNA polymerase δ, similar to those of lower eukaryotes and prokaryotes, has a proofreading function. The presence of a proofreading exonuclease in both higher and lower eukaryotic DNA polymerases suggests that the mechanism by which the accuracy of DNA replication is maintained in eukaryotes may be similar to that of prokaryotes. Finally, although inhibitor studies suggest that either DNA polymerase δ or α, or both, may be involved in DNA replication, a biological role for either enzyme in DNA replication remains to be demonstrated.

ACKNOWLEDGMENTS

We thank Dr. Frans Huijing for introducing us to the technique of microgel electrophoresis, Dr. Julio Pita and Mr. Luciano Blanco for their help in the use of microdensitometer, and Dr. A. H. Todd of Imperial Chemical Industries, Macclesfield, England for a supply of aphidicolin. This research was supported by a grant from the National Institutes of Health (NIH AM 26206) and in part by funds given in memory of Joseph Peter Siegel and Stephen Andrew de Young.

References

1. A. Weissbach, D. Baltimore, F. J. Bollum, R. Gallo and D. Korn, *Science* **190**, 401 (1975).
1a. D. Gillespie, W. C. Saxinger and R. C. Gallo, This Series **15**, 88 (1975).
2. U. Bertazzoni, A. Scovassi and G. M. Brun, *EJB* **81**, 237 (1977).
3. A. Bolden, G. P. Noy and A. Weissbach, *JBC* **252**, 3351 (1977).
4. U. Hubscher, C. C. Kuenzle and S. Spadari, *EJB* **81**, 249 (1977).
5. L. M. S. Chang and F. J. Bollum, *JBC* **248**, 3398 (1973).
6. T. S.-F. Wang, W. D. Sedwick and D. Korn, *JBC* **249**, 841 (1974).
7. F. J. Bollum, This Series **15**, 109 (1975).
8. N. Muzyczka, R. L. Poland and M. L. Bessman, *JBC* **247**, 7116 (1972).
9. D. Brutlag and A. Kornberg, *JBC* **247**, 241 (1972).
10. F. D. Gillin and N. G. Nossal, *JBC* **251**, 5219 (1976).
11. A. Kornberg, "DNA Synthesis." Freeman, San Francisco, California, 1980.
12. G. T. Yarranton and G. R. Banks, *in* "DNA Synthesis, Present and Future" (I. Molineux and M. Kohiyama, eds.), p. 479. Plenum, New York, 1977.
13. A. Falaschi and S. Spadari, *in* "DNA Synthesis, Present and Future," (I. Molineux and M. Kohiyama, eds.), p. 487. Plenum, New York, 1977.
14. L. M. S. Chang, *JBC* **252**, 1873 (1977).
15. G. T. Yarranton and G. R. Banks, *EJB* **77**, 521 (1977).
16. E. Winterberger, *EJB* **84**, 167 (1978).
17. J. J. Byrnes, K. M. Downey, V. L. Black and A. G. So, *Bchem* **15**, 2817 (1976).
18. J. J. Byrnes and V. L. Black, *Bchem* **17**, 4226 (1978).
19. M. Y. W. T. Lee, C.-K. Tan, K. M. Downey and A. G. So, *FP* **38**, 779 (1979).
20. Y.-C. Chen, E. W. Bohn, S. R. Planck and S. H. Wilson, *JBC* **254**, 11678 (1979).
21. M. Y. W. T. Lee, C.-K. Tan, A. G. So and K. M. Downey, *Bchem* **19**, 2096 (1980).
22. L. M. S. Chang, M. K. Brown and F. J. Bollum, *JMB* **74**, 1 (1975).
23. S. Spadari and A. Weissbach, *JMB* **86**, 11 (1974).
24. A. Weissbach, *ARB* **46**, 25 (1977).
25. M. G. Sarngadharan, M. Robert-Guroff and R. C. Gallo, *BBA* **516**, 419 (1978).
26. M. Ohashi, T. Taguchi and S. Ikegami, *BBRC* **82**, 1084 (1978).
27. E. Wist and H. Prydz, *NARes* **6**, 1583 (1979).
28. H. Krohan, P. Schaffer and M. L. DePamphilis, *Bchem* **18**, 4431 (1979).
29. H. J. Edenberg, S. Anderson and M. L. DePamphilis, *JBC* **253**, 3273 (1978).
30. M. A. Wagar, M. J. Evans and J. A. Huberman, *NARes* **5**, 1933 (1978).
31. E. Wist, *BBA* **562**, 62 (1979).
32. S. Seki and T. Oda, *BBA* **606**, 246 (1980).
33. D. Rodbard, and A. Chrambach, *in* "Electrophoresis and Isoelectric Focusing in Polyacrylamide Gel" (R. C. Allen, and H. R. Maurer, eds.), Chapter 22. de Gruyter, Berlin and New York, 1974.
34. K. Weber, *in* "Methods in Enzymology" (C. H. W. Hirs and S. N. Timasheff, eds.), Vol. 25, p. 3. Academic Press, New York, 1972.
35. J. C. Pita, F. J. Muller, S. M. Morales and E. J. Alarcon, *JBC* **254**, 10313 (1979).

II. Mechanisms of Transcription
Chairman: ELLIOT VOLKIN
Summarizer: ROBERT K. FUJIMURA

Summary   99
R. K. FUJIMURA

Regulatory Circuits of Bacteriophage Lambda   103
S. L. ADHYA, S. GARGES, AND D. F. WARD

Summary

R. K. FUJIMURA

Biology Division
Oak Ridge National Laboratory
Oak Ridge, Tennessee

The papers presented in this session dealt with the control of transcription by bacterial RNA polymerases.

Chamberlin described the development of an *in vitro* assay system employing T7 and T3 DNA templates that can be used to detect and analyze factors affecting the interaction of bacterial RNA polymerases with promoter or terminator sites (1), or altering the rate or efficiency of one of the other steps of the transcription cycle (2). Studies with the promoter and terminator test systems revealed that RNA polymerases from a wide range of evolutionarily diverse bacterial species, including *Escherichia coli*, *Bacillus subtilis*, *Mycobacterium smegmatis*, *Rhodospirillum rubrum*, all utilize the same collection of T7 promoter sites and the T7 early terminator.

However, there were substantial differences in the rate with which different T7 promoters were used by different bacterial RNA polymerases, as indicated by such quantitative parameters as rate of binding and transition temperature.

Chamberlin defined two parameters with which to characterize polymerase–promoter interactions: promoter identity, which describes the property shared by all promoters utilized by a particular RNA polymerase; and promoter strength or avidity, which describes the relative rate of utilization of a promoter in competition with other promoters sharing a common identity. It appears that the structural elements of the bacterial RNA polymerase protein that govern promoter identity are strongly conserved in prokaryotic evolution (1). However, changes in RNA polymerase that alter relative promoter strength are quite commonly detected between different bacterial species, and at least one rif^R $rpoB$ mutation in *E. coli* also alters the relative promoter strength of the purified RNA polymerase.

A novel form of RNA polymerase isolated from vegetative cells of *B. subtilis* utilizes, *in vitro*, a promoter site on T7 DNA completely different from other bacterial RNA polymerases (3), and does not use the normal T7 promoters at a detectable rate. This RNA polymerase therefore has an altered specificity for promoter identity as compared to the known *B. subtilis* RNA polymerase holoenzyme. The enzyme contains the normal *B. subtilis* β, β' and α subunits as well as a new peptide of 28,000 daltons, but lacks the *B. subtilis* σ subunit; this also appears to be the case in the systems studied by Pero (4) and Losick (5).

Studies of the biochemical properties of the RNA chain termination reaction at the T7 and T3 early terminator sites show that the termination reaction is relatively insensitive to changes in reaction conditions (6). Termination at the weak T3 terminator is enhanced at low NTP concentrations by elevated NaCl concentrations and by alterations in the RNA polymerase protein. Both *B. subtilis* polymerase and RNA polymerases from two different rif^R *rpoB* mutants of *E. coli* utilize the weak T3 terminator well (6). In contrast, no conditions have been found that block utilization of the strong T7 terminator, including the presence of sufficient RNase in the reaction to degrade continually the RNA product. Chamberlin suggested a model involving a pausing of RNA polymerase at termination sites where translocation of the enzyme depends on displacement of the nascent RNA strand from the hybrid intermediate and re-formation of the DNA duplex. Factors that block strand displacement and template renaturation, for example, the presence of a (G + C)-rich hybrid sequence or the potential for base-pairing to form a stem-loop structure in the nontranscribed strand, would halt the enzyme and make it available for the RNA chain release reaction to occur either spontaneously or catalyzed by a release factor such as rho.

Pero reported on the temporal control of gene expression by sigma-like regulatory proteins that interact with bacterial RNA polymerase. She described *B. subtilis* phage SPO1, which has become the prototype for studies of gene control by regulatory proteins that alter promoter selection by RNA polymerase. This large virulent phage codes for sigma-like proteins that bind sequentially to the bacterial core polymerase, thereby directing a temporally defined program of phage gene expression. Promoters for phage "early" genes are recognized by *B. subtilis* RNA polymerase containing the host sigma factor; these promoters are also actively transcribed *in vitro* by *E. coli* RNA polymerase. Not surprisingly, the nucleotide sequence of an SPO1 early gene promoter closely resembles that of *E. coli* promoters, differing by only one base from the most common sequences in both the Pribnow box and −35 region.

Phage "middle" gene promoters are activated by a sigma-like protein encoded by regulatory gene 28. Middle gene promoters differ from promoters bound by sigma-containing polymerases in both the

SUMMARY

−35 region and the Pribnow box (4, and unpublished results). Based on the nucleotide sequences of five middle gene promoters, the most common sequences in the −35 region and the −10 region are AG-GAGA and TTTNTTT, respectively (italicized bases are present in all five promoters; others in four of the five sequences).

Two middle genes, regulatory genes 33 and 34, code for sigma-like proteins that act synergistically to direct transcription of phage "late" genes. Late gene promoters have been located on the phage genome and their nucleotide sequence is being determined.

Pero further suggested that control of gene expression by sigma-like regulatory proteins may be a general mechanism applicable to bacterial as well as phage systems. She described a bacterial-coded sigma-like protein of 37,000 daltons that has been isolated (5, and unpublished results) and shown to bind to core RNA polymerase to cause the specific recognition of a promoter under sporulation control. This brought to four the number of different sigma-like proteins that have been shown to bind to B. subtilis core RNA polymerase and change its specificity of promoter recognition.

Johnson reviewed the work from Ptashne's laboratory on mechanisms of gene regulation by the λ repressor and cro protein. This is reported in more detail elsewhere (7) and discussed by Adhya (see below). Bacteriophage λ encodes two repressor proteins, one encoded by the *cI* gene and called λ repressor, the other encoded by the *cro* gene and called cro. λ repressor is required for lysogeny, and cro is required for lytic phage growth. λ repressor turns off early genes including *cro* to maintain a lysogen, and in addition turns on transcription of *cI*. At the higher concentrations, λ repressor also turns off transcription of *cI*. For lytic growth, cro performs two negative functions: it prevents transcription of *cI*, and it turns down the early genes including *cro* after an initial burst of transcription. According to Ptashne's laboratory, these controls are mediated by repressor and cro proteins differentially occupying three contiguous sites (o_{R1}, o_{R2}, and o_{R3}) in the right operator of the phage chromosome. There are two promoters that overlap o_R. One of these (p_R) directs rightward transcription of a set of early genes, including *cro*. The other promoter (p_{RM}) directs leftward transcription of *cI*. In a lysogen, λ repressors bind preferentially and cooperatively to sites o_{R1} and o_{R2} and thereby repress p_R and activate p_{RM}. Only at higher concentrations does the repressor fill o_{R3} and repress p_{RM}. The cro protein recognizes the same three sites in o_R but binds noncooperatively and with reverse affinity, that is, $o_{R3} > o_{R1} = o_{R2}$. Thus, the cro protein turns off p_{RM} first and then turns off p_R. This model of regulation is supported by a variety of experiments, both biochemical and genetic, performed by the Ptashne group.

Adhya reviewed the mechanisms of expression of λ gene in phage development, in both its lytic and lysogenic phases. He showed that λ gene expressions are also controlled by regulating transcription termination at signals present in the viral DNA for the lytic pathway of viral development. As presented earlier for SPO1 phage, λ expresses its genome in a sequential manner for lytic growth. The switch from expression of immediate-early to delayed-early to late genes requires the λ N-gene product, which acts by suppressing the transcription termination signals located between various gene clusters. Adhya summarizes his studies in more detail in these proceedings.

These papers show that control of transcription in prokaryotes can occur both at initiation and termination sites. The known mechanisms are highly diverse. Those discussed include new sigma-like subunits of RNA polymerase that affect its affinity for various promoter sites, or turn on and turn off of transcription by change in concentration of repressors, and factors that affect terminator sites.

REFERENCES

1. J. L. Wiggs, J. W. Bush and M. J. Chamberlin, Cell **16**, 97 (1979).
2. M. J. Chamberlin, W. C. Nierman, J. L. Wiggs and N. Neff, JBC **254**, 10061 (1979).
3. J. Jaehning, J. L. Wiggs and M. J. Chamberlin, PNAS **76**, 5470 (1979).
4. C. Talkington and J. Pero, PNAS **76**, 5465 (1979).
5. W. G. Haldenwang and R. Losick, Nature **282**, 256 (1979).
6. N. Neff and M. J. Chamberlin, Bchem **19**, 3005 (1980).
7. M. Ptashne, A. Jeffrey, A. D. Johnson, R. Maurer, B. J. Meyer, C. O. Pabo, T. M. Roberts and R. T. Sauer, Cell **19**, 1 (1980).

Regulatory Circuits of Bacteriophage Lambda

S. L. ADHYA
S. GARGES AND
D. F. WARD

Laboratory of Molecular Biology
National Cancer Institute
Bethesda, Maryland

Phage λ is a temperate bacterial virus that thrives on a dual, but mutually exclusive, life style. After infection of its host *Escherichia coli*, it either multiplies and produces progeny viruses by lysing the host or it integrates its own DNA into the host chromosome with concomitant repression of the viral genes by a virus-specific repressor. In the latter case, the inserted viral chromosome is called a "prophage" and is replicated passively as part of the host chromosome. The cell is said to become lysogenic and, because of the presence of λ repressor, is immune to superinfection by a second λ particle.

This chapter summarizes the regulatory systems of λ to point out that both of its developmental pathways follow finely tuned temporal control mechanisms far more sophisticated than those conceived in the 1950s and early 1960s. We emphasize the salient features of the complex processes, such as: how a multivalent operator makes transcription initiation from two divergent promoters mutually exclusive; how a transcription system is modified to ignore punctuation signals; how the expression of unlinked operons is coordinated; how a viral chromosome integrates into its host chromosome; and how a biological "decision" may be made. Many of these provide new concepts and directions in studying regulatory biology of other systems, both prokaryotic and eukaryotic.

I. Early Development

Expression of λ genes during its development, like that of many bacteriophages, is manifested in two distinct temporal phases. The early phase is characterized by the expression of the phage replication and recombination functions, and the late phase is characterized by the appearance of functions required for packaging and maturation.

Immediately after infection, only a small portion of the genome is expressed (1, 2). Following this, the remaining early genes are turned on. A major interest in the early development of λ is the mechanism by which the expression of the delayed-early genes is turned on. In many other phages, this is achieved by activating new promoters, either by the production of a phage-specific RNA polymerase or by a phage-coded modification of the host RNA polymerase resulting in an altered specificity (see 3). With λ, however, there are only two promoters controlling early expression (4). The delayed-early genes, located downstream with respect to the immediate-early genes, are turned on by the process of anti-termination of transcription (5, 6). This is illustrated in Fig. 1. For example, transcription initiated at the early promoter, p_L, enables only the N gene to be expressed, since termination of transcription at the t_{L1} terminator precludes extension of the transcript into the delayed-early genes. However, the product of the N gene (pN) is able to modify the transcription machinery so that t_{L1} is no longer effective as a terminator. Transcription is now able to proceed beyond t_{L1}, and the delayed-early genes are expressed.

It is still not understood exactly how pN is able to prevent termination of transcription. Termination at many terminator sites, including t_{L1} and t_{R1} (Fig. 1), requires the action of the host transcription-termination factor Rho (5). pN is not, however, an anti-Rho factor since it is also able to prevent termination at sites active even in the absence of Rho (7). Therefore, it appears that pN modifies the specificity of the transcription complex, and prevents the recognition of terminators.

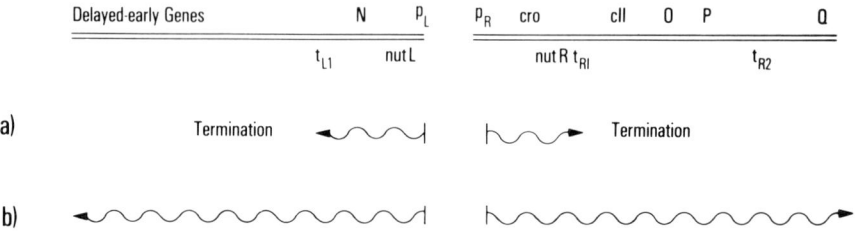

FIG. 1. Early transcription in phage λ. λ has two operons that are expressed immediately after infection. The p_L operon codes for the N gene and several of the delayed-early genes. The p_R operon includes the regulatory genes *cro*, *cII*, and *Q*, together with the DNA replication genes *O* and *P*. (a) In the absence of pN, transcription initiates at p_L and p_R but stops at the terminator sites t_{L1} and t_{R1}, respectively. Thus, only the N and *cro* genes are expressed. (b) Once pN has accumulated, transcription-complexes become modified at the *nutL* and *nutR* sites, respectively, so that the terminator sites are no longer recognized. Thus, transcription can be extended, and the delayed-early genes are now expressed.

Evidence supporting an interaction between pN and RNA polymerase comes from the isolation of several mutants that affect pN action. For example, mutations called *ron* map within the *rpoB* gene, which specifies the RNA polymerase beta subunit (8). These mutants block the growth of certain λ variants carrying alterations in the N gene, suggesting that pN interacts with the beta subunit of RNA polymerase. Several other mutations in *rpoB* have also been found to affect λ N function (9, 10).

That the interaction of pN with RNA polymerase is not a sufficient explanation of the anti-termination role of pN is apparent from the finding that pN is able to prevent transcription termination only within phage operons, not within bacterial operons (11–13). This specificity is not a reflection of differences between phage and bacterial terminator sites, since fusions of p_L, or p_R, to bacterial genes show that pN is able to suppress termination signals in the bacterial DNA. It is now accepted that pN-mediated modification of the transcription complex can occur only at specific sites in the phage operons, and these are distinct from the sites of pN action, i.e., the terminators (14). The sites of pN recognition (*nut*) have been identified and defined for the leftward p_L operon by the isolation of mutants defective for pN-mediated anti-termination (15). Transcription of the N gene from p_L is normal in these mutants, but, since pN can no longer act, transcription promptly stops at the t_{L1} terminator site. DNA sequencing shows that these mutations are located after the start point of transcription, but before the start of the structural gene N (16). pN can also anti-terminate transcription if a *nut* site is fused to a bacterial promoter (17).

Although it is not known exactly what events occur at the *nut* site, anti-termination is not the consequence of a simple interaction between pN and RNA polymerase at *nut*. It is also not known whether pN recognition occurs at a DNA sequence or on an RNA transcript.

The finding of several host mutations [*nusA* (18), *nusB* (19, 20), *nusE* (D. I. Friedman, personal communication), for example], that preclude pN function, but do not map in genes coding for subunits of RNA polymerase, reveals the complexity of the process and shows that several other proteins are necessary for pN action. Genetic and biochemical evidence suggests that one of the target sites for pN is the *nusA* gene product (D. I. Friedman, J. Greenblatt, personal communications). Indeed, binding of pN to RNA polymerase *in vitro* has not been observed, although it could be argued that such an interaction could occur only at the *nut* site. It does appear, however, that the *nusA* gene product binds to RNA polymerase core enzyme *in vitro*, suggesting that the proposed action of pN on RNA polymerase may involve the *nusA* gene product. Since NusA is identical to L-factor, which is required for optimal expression of *lacZ* in a coupled *in vitro* transcription-translation system (*21*), it may be that NusA is an integral component of gene expression in *E. coli*.

Recently, evidence has begun to accumulate indicating an involvement of ribosomal proteins in pN function. The *nusE* mutation, which prevents pN action, is located in the gene encoding ribosomal protein S10 (D. I. Friedman, personal communication). It is not known whether the ribosomes are directly involved in pN-mediated anti-termination, or whether ribosomal protein(s) directly participate in the transcription process, analogous to participation of ribosomal protein S1 in phage $Q\beta$ replicase (*22*).

It is apparent that, although we understand the consequences of pN action, a great deal more work is required before we will understand the mechanism by which pN prevents normal transcription termination.

The p_R operon is controlled similarly to that of the p_L operon (Fig. 1). In the presence of the N function, transcription initiated at p_R fails to terminate at the t_{R1} and t_{R2} termination sites, thereby permitting expression of the delayed-early genes *cII*, *O*, *P*, and *Q* (Fig. 1). As in the case with transcription from p_L, the ability of pN to prevent termination of transcription at these terminator sites is dependent on a pN recognition site, *nutR* (*16, 17, 23*). In contrast to *nutL*, which is located prior to any of the genes in the p_L operon, the *nutR* site is located after the *cro* gene. Thus the transcription complex can be modified by pN long after RNA chain initiation. It is interesting to speculate whether this location of *nutR* has any special significance for regulation of λ. Control of Repressor synthesis, required for lysogenic development, is achieved by transcription directed in opposition to that from p_R (see Section III). Modification of transcription from p_R might interfere with such opposing transcription.

All the genes in the p_R operon have important roles in λ development. The products of the *cro* and *cII* genes play a critical role in the decision between the lysogenic and lytic developmental pathways (see Section V). pQ is the positive activator of late gene expression and is also discussed in Section II. The products of genes *O* and *P* are required for replication of λ DNA (*24*). Early in infection, replication proceeds bidirectionally (*25*) from a single origin located within the *O* gene (*26*). This origin is activated by transcription through the site (*27*). Thus the prevention of termination by pN at t_{R1} serves the dual purpose of producing the replication gene functions and activating the replication origin. Replication of the genome results in an elevated rate of expression of the regulatory proteins, Cro and cII. The mechanism of replication and the host proteins involved have recently been reviewed in detail (28).

II. Lytic Development

Irrespective of the final outcome of infection, the events occurring during the early stages of λ infection (see Section I) are common to both the lysogenic and lytic pathways. The factors determining which pathway is followed are discussed in Section V. Commitment to one pathway is not reversible; thus the decision point represents a "switch" between the two courses of λ development.

After the decision to enter lytic development, two sequences of events must occur: (a) expression of the genes required for the alternate lysogenic pathway must be prevented; and (b) the gene products required for the maturation of progeny virions must be produced. The blockage of the lysogenic functions is not required just to prevent synthesis of unnecessary gene products, but also to prevent any interference with expression of the lytic functions. The gene *cII* and *cIII* proteins inhibit synthesis of the late gene products (29). Turn-off of the lysogenic functions is achieved by Cro, which partially represses transcription from the early λ promoters, p_L and p_R (30). Consequently, the synthesis of the products of genes *cII* and *cIII* is inhibited, and since the *cII* product is an unstable protein, it decays rapidly (C. Epp, M. Jones, I. Herskowitz and M. Pearson, personal communications). cII and cIII functions positively control lysogeny (see Section III); hence by removing them, the lysogenic response is blocked. The reduced rate of p_R transcription provides sufficient levels of the O, P, and Q gene products for lytic growth.

The reduction in transcription from p_R by Cro is also important for optimal replication of the genome. Early in infection, replication proceeds bidirectionally via the θ structure, but switches at later times to the more efficient rolling-circle mode (28). The reason for the greater efficiency of the rolling-circle mode, as compared to θ replication, lies in the mechanism of maturation of λ DNA. λ can package its DNA only from linear concatemers, not from circular molecules. Packaging of monomeric circles of λ DNA, produced in the early mode of replication, can be achieved only by recombination, resulting in the production of multimeric concatemers. As opposed to θ replication, high-level transcription through the origin of replication may interfere with the rolling-circle mode of replication. This may be one of the reasons why cro^- phages replicate abnormally (31). Consequently, the Cro-mediated repression of transcription from p_R is essential for efficient phage DNA maturation.

The gene products required for the maturation of progeny virions are all synthesized late in infection. Their expression is directed by the late promoter $p_{R'}$ (32) (see Fig. 2). It should be noted that shortly after infection, the linear λ DNA molecule circularizes via the cohesive ends (see Section III). Thus genes R and A, which are distant on mature λ DNA, are in close proximity during expression of the λ genome. Transcription from the $p_{R'}$ promoter is constitutive, but is normally terminated at a Rho-independent site yielding the so-called 6 S transcript (33). Expression of the late genes is positively controlled by the Q gene product which is, in turn, synthesized from the p_R transcript. By analogy to the pN function, it has been suggested that pQ-mediated activation of late gene expression is achieved by anti-termination of the 6 S transcript (33, 34).

Several lines of evidence support this anti-termination thesis. The discrete 6 S transcript disappears in the presence of pQ function, indicating either a blockage in its synthesis or an extension in its size (A. Oppenheim, D. L. Court, and M. E. Gottesman, personal communication). All the late genes are transcribed as a single polycistronic mRNA molecule up to at least 20,000 nucleotides in length (35). In such a large operon, nonsense mutations would be expected to have a strong polar effect on expression of all distal genes. This is not observed (36). A direct demonstration of the ability of pQ to act as an anti-terminator comes from studies on a lysogen in which the $p_{R'}$ pro-

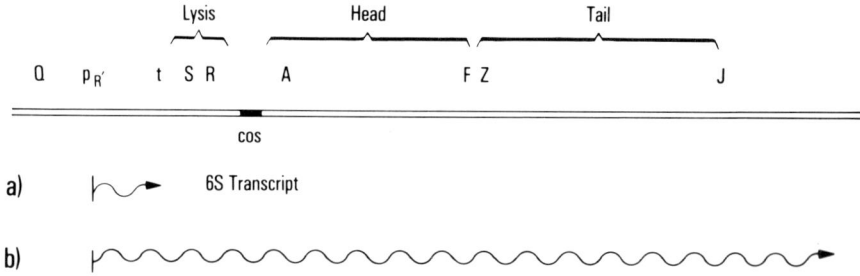

FIG. 2. Late transcription during lytic development. The late genes of λ include the lysis functions S and R, together with the genes concerned with production of the head and tail components of the progeny virions. All these genes are contained in a single operon transcribed from the late promoter $p_{R'}$. (a) In the absence of pQ, transcription initiated at $p_{R'}$ terminates at t to give the 6 S transcript. (b) In its presence, termination does not occur and transcription extends into the late genes.

moter of the prophage is oriented in the same direction as the *gal* operon (D. Forbes, personal communication). Escape synthesis of *gal* from $p_{R'}$ was found to be dependent on pQ function. Polar mutations in the $p_{R'}$–*gal* fusion were suppressed, indicating that pQ must have been acting as an anti-termination factor.

III. Lysogenic Response

If λ does not follow a lytic pathway after infection, the cell is converted into a lysogen. Establishment of lysogeny requires: (*a*) expression of the phage regulatory and structural genes to provide ultimately high levels of two λ proteins, "Repressor" (37) and "Integrase" (38); (*b*) integration of the phage DNA into the host chromosome by Integrase; and (*c*) permanent repression of phage lytic promoters, p_L and p_R, by Repressor. *cI* and *int*, the structural genes for Repressor and Integrase, respectively, belong to two different operons (for review, see 39). To follow the lysogenic pathway, Repressor and Integrase should be made more or less coordinately at high levels without onset of lytic functions. For the decision-making processes, see Section V.

The *cI* gene can be transcribed from two promoters (Fig. 3). To establish repression after infection, it is transcribed from the distal promoter, p_{RE} (37), whereas the proximal promoter, p_{RM}, is responsible for *cI* transcription in an established lysogen (40). p_{RM} is discussed in Section IV. Similarly, *int* gene can be transcribed from p_L (41) as well as from its cognate promoter, p_I (42) (Fig. 3). λ makes Integrase for prophage insertion mostly from p_I (43, 44). p_{RE} and p_I, the two promoters that control lysogeny, are positively activated by the action of delayed-early gene products cII and cIII (37, 42–45). This makes coordination of Repressor and Integrase synthesis possible. Besides, the

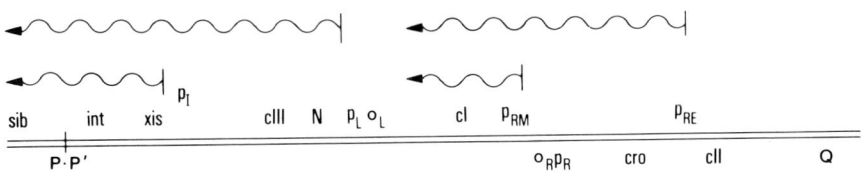

FIG. 3. Early transcription in λ required to establish lysogeny. After infection, cII protein made from the p_R operon at high multiplicity of infection turns on the p_{RE} and p_I promoters. p_{RE} directs λ Repressor synthesis from the *cI* gene, and p_I controls Integrase synthesis from the *int* gene. Repressor then represses the lytic promoters p_L and p_R and induces its own synthesis from p_{RM}, the promoter that continues to make Repressor in a lysogen. Integrase catalyzes insertion of the lambda chromosome into the bacterial chromosome, as shown in Fig. 4.

rates at which proteins cII and cIII direct the synthesis of Repressor and Integrase are much higher than their synthesis from p_{RM} and p_L, respectively. A high level of Repressor would facilitate repression of more than one molecule of phage DNA; λ may have undergone limited rounds of replication before repression ensues. A high level of p_{RE} transcription per se is an additional help to lysogenic establishment. Increased p_{RE} transcription delays the expression of phage late genes (46). The delay may be due to reduced rate of p_R transcription. Transcriptions from p_{RE} and p_R oppose and overlap each other (Fig. 3).

The structure of the p_{RE} and p_I promoters are known by mutational changes and by sequence analyses (47–49). Since cII protein alone can promote correct initiation of transcription by RNA polymerase from p_{RE} and p_I, it is now believed that cII activates the promoters

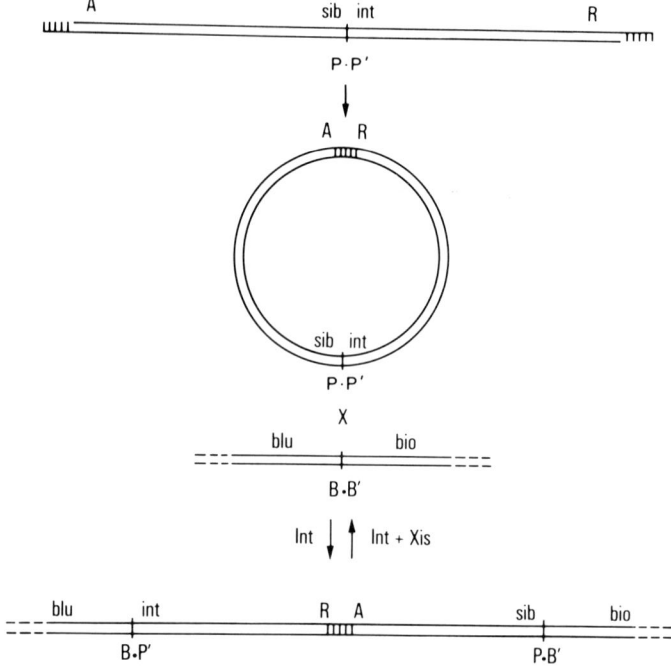

FIG. 4. λ integration and excision by the Campbell model (51). The vegetative and physical map of λ is linear as shown in the top line. The linear DNA circularizes by joining the single-stranded ends, which are complementary to each other. A reciprocal recombination occurs between the phage attachment site P · P' (located between *sib* and *int*) and host attachment site, B · B' (located between host *blu* and *bio* genes). The integration reaction is catalyzed by Integrase. The prophage map, shown by the bottom line in the figure, becomes a cyclic permutation of the vegetative map. The reverse reaction, prophage excision, requires Integrase and Xis protein.

directly (H. Shimatake and M. Rosenberg, personal communication). cIII provides an additional control at the level of cII activity. Control of cII activity by cIII and other host factors is discussed further in Section V.

Prophage Insertion. One aspect of λ lysogeny that is now understood in greater detail is the site-specific integration of the phage DNA into the host chromosome. After infection, λ DNA circularizes by first pairing and then ligating the two 12-nucleotide single-stranded free ends (Fig. 4), which are complementary to each other (50). Then linear prophage insertion results from reciprocal recombination between specific sites on the phage DNA (designated P · P') and bacterial DNA (B · B'), as predicted by Campbell (51). As a result, the prophage map becomes a cyclic permutation of the vegetative map. The integration reaction is catalyzed by Integrase, which is a DNA "nicking and closing" enzyme (51a), and several host proteins —the products of *gyrA* (*nalA*), *gyrB* (*cou*), *himA*, and *hip* genes (52, 53, 53a). The products of the *gyrA* and *gyrB* genes constitute "DNA gyrase," which converts a covalent DNA circle into a negatively supertwisted form (54, 54a, 55). The functions of the *himA* and *hip* genes are unknown. The P · P' and B · B' segments include a common (core) DNA sequence 15 nucleotide pairs long (56). The crossover event involves staggered cuts at unique positions within the core sequence (K. Mizuuchi, personal communication). There are four Integrase binding sites located in the P · P' sequence (spanning 234 nucleotide pairs) and only one located in the B · B' partner (57). These sites play a critical role in recombination.

IV. Prophage Maintenance

Once a stable lysogen is established, the λ prophage is replicated passively as part of the host chromosome, and all phage functions required for phage lytic growth (including those for active λ DNA replication) are repressed by Repressor. The low level of λ transcription that does take place in the prophage state comes from the p_{RM} operon, which includes the Repressor gene *cI* (58) (Fig. 5). The presence of Repressor in a lysogen not only stabilizes the prophage, but also makes the cell specifically immune to subsequent superinfection with λ. The mechanism of λ Repressor action is remarkable (for review, see 59). In a prophage state, the *cI* operon is transcribed entirely from the proximal promoter p_{RM}, but not from p_{RE} (40). Since p_{RM} does not need cII for activation, one may assume that constitutive synthesis of Repressor from p_{RM} and simple reversible binding of Repressor to the

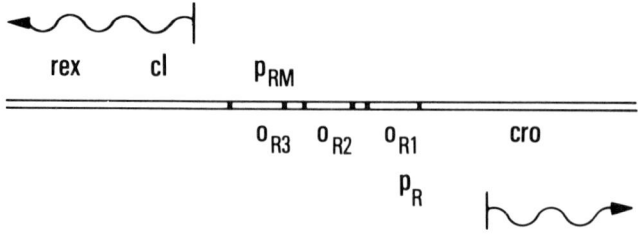

FIG. 5. Control of the divergent promoters, p_{RM} and p_R. The operator site, o_R, is divided into three domains: o_{R1}, o_{R2}, and o_{R3}. To maintain the lysogenic state, Repressor, the cI gene product, binds cooperatively to o_{R1} and o_{R2} to turn off p_R operon transcription and to facilitate transcription originating from p_{RM}. Cro protein binds preferentially to o_{R3}, preventing p_{RM} transcription, thus blocking Repressor synthesis and allowing the lytic cycle to proceed. At higher concentrations, Cro binds to o_{R1} and o_{R2}, turning off p_R.

phage operators, o_L and o_R, are essential and sufficient for maintenance of lysogeny. In practice, this is not the case. To suit the efficiency of both repression and derepression, Repressor serves two functions (59): (a) it turns off p_L and p_R promoters by binding to the cognate o_L and o_R operators; and (b) its own synthesis from p_{RM} is turned on at low concentrations and is turned off at high concentrations of Repressor.

In view of this dual control on p_R and p_{RM} (Fig. 5), Repressor action at o_R has been studied in detail. Repressor protein (MW = 27,000) acts as a dimer. Active Repressor binds to o_R at three similar, but not identical sites, called o_{R1}, o_{R2}, and o_{R3} (60, 61). The intrinsic affinity of Repressor for o_{R1} is 10 times that for o_{R2} and o_{R3} (60). Repressor bound to o_{R1} helps another Repressor molecule to bind to o_{R2}. This cooperative binding, at physiological concentrations of Repressor, results in o_{R1} and o_{R2} being filled and o_{R3} remaining vacant. At very high concentrations of Repressor, o_{R3} is also filled. Binding at o_{R2} and o_{R3} is also cooperative if, and only if, o_{R1} is vacant. These conclusions are derived from *in vitro* binding studies using wild-type and mutant operators. The cooperativity is the result of interactions between Repressor dimers, and not because of conformational change in one operator unit (e.g., o_{R2}) by Repressor binding to another (e.g., o_{R1}).

Occupation of o_{R1} or o_{R2}, or both, represses p_R by inhibiting RNA polymerase binding (62) (Fig. 5). Repressor present at o_{R2} stimulates transcription from p_{RM}, presumably by an allosteric contact with RNA polymerase at p_{RM} (63). However, occupation of o_{R3} by Repressor prevents RNA polymerase binding to p_{RM} and thus inhibits its own synthesis (64).

What is the significance of this three-unit operator system? It seems that a hypothetical two-unit system—a tight-binding o_{Rt} and a weak-binding o_{Rw}—is sufficient to maintain the lysogenic system: Repressor at o_{Rt} would turn off p_R and turn on p_{RM} (by a contact with RNA polymerase). But Ptashne et al. (59) argue that a three-unit operator structure protects the system from deleterious mutations. They suggest that λ must grow both as a prophage and as a phage to save the temperate system. Mutations that would affect the dual life-style of λ would destroy the lysogenic system. For example, in the hypothetical two-unit operator system, a single mutation in o_{Rt} would eliminate both p_R repression and p_{RM} stimulation and, in combination with an o_L mutation, would give virulence.[1] In the three-unit system, a single mutation in o_{R1} or o_{R2}, when combined with an o_L mutation, does not generate virulence. In an o_{R1} mutant, Repressor binds to o_{R2} and o_{R3} cooperatively, turning off both p_R and p_{RM}. An o_{R2} mutation only slightly affects o_{R1} binding and, therefore, p_R repression. It also does not show stimulation of RNA polymerase activity at p_{RM}.

V. Decision

One of the more intriguing aspects of λ involves the "decision" made by the bacteriophage upon infection whether to grow lytically or to make the host cell lysogenic. The protein required for the lytic cycle acts to preclude the lysogenic course, and vice versa, as discussed later. Since both of the regulatory proteins for the lysogenic and lytic cycles come from the same operon, how is it that one event occurs over the other? The relative amounts of the regulatory proteins cII and Cro are important in the decision, but what determines the amounts and the activities of these proteins is not yet clear. *Escherichia coli* genes that influence the decision process have been identified; these gene products may act to regulate the amounts of cII and Cro.

For λ to lysogenize, the p_R and p_L operon gene products, cII and cIII, are necessary for the coordinate activation of two promoters, p_I for Integrase synthesis and p_{RE} for Repressor synthesis (42–45). This coordinacy is important since activation of p_{RE}, without that of p_I, would result in repression without prophage integration (66). Without p_{RE} activation, the phage would always follow a lytic course (37, 67, 68). When both promoters are simultaneously activated, integration

---

[1] λ phages, which carry Repressor insensitive mutations in o_{R1}, o_{R2}, and o_{L1} sites, can grow in a λ lysogen and are called λ virulent mutants (65).

occurs and Repressor shuts off p_R and p_L, preventing any lytic functions from being expressed.

For λ to grow lytically, Cro protein must be made (31). The Cro protein has at least two functions: (a) to prevent Repressor being made from p_{RM}, the promoter for repressor maintenance, by preferentially binding to o_{R3} (69) (Fig. 5); and (b) to act as a weak repressor of early promoters (70), so that replication can proceed efficiently (see Section II) and perhaps so that transcription from $p_{R'}$ can initiate properly. Thus, if the effective concentration of cII is higher than that of Cro, lysogeny will be preferred, and if Cro levels are high relative to cII, the lytic cycle will prevail.

When the multiplicity of infection is low, the lytic cycle is preferred over lysogeny; however, if it is high, lysogeny is more frequent (46). This phenomenon has been ascribed to active levels of cII and cIII. cII protein is unstable; it is believed that cIII may act to stabilize it. When there is a higher multiplicity of infection, there is a higher dosage of the *cII* and *cIII* genes, resulting in more of their products, and less chance for the unstable cII to be degraded. However, since *cII* and *cro* are on the same operon, there is no reason a priori to believe that Cro levels would not also be increased in a higher multiplicity of infection, and the balance tipped in favor of lytic growth. There must be other factors involved in regulating the cII:Cro ratio.

Escherichia coli genes that also affect λ lysogeny have been identified by mutations in the *himA, hip, cya,* and *hfl* genes, among others (52, 71, 72). Of these, *himA* and *hip* have been shown to have an enzymic role in prophage insertion (see Section III). In addition, a strain that has a deletion mutation in *himA* makes less Repressor and Integrase (H. I. Miller, personal communication), and so is multiply defective for lysogeny. Since Repressor and Integrase synthesis for lysogeny requires activation by cII and cIII, the *himA* gene protein may affect the activity or synthesis of cII or cIII. Mutations in *cya*, which encodes adenylate cyclase, the protein responsible for cyclic AMP synthesis, and in *crp*, the gene for the cyclic-AMP-receptor protein (CRP), cause a lower frequency of lysogeny (71). Cyclic AMP and CRP have roles in catabolite repression; what their involvement is in λ lysogeny is unknown. In contrast to these genes, mutation in the *hfl* gene increases lysogeny, even in the presence of a λ *cIII* mutation (72). It has been proposed that the *hfl* protein inhibits or inactivates cII, and cIII normally antagonizes this inhibition (73). A mutation in *cII* that allows the phage to form lysogens even in the absence of cIII protein has been isolated (74). This cIII-independent altered cII protein may be inherently more stable than wild-type cII, even in the presence of *hfl* activity. Another role for cIII may be that of an effector of cII such that it renders the cII more efficient in activating p_{RE} and p_I. By this model, the cIII-independent cII is always in a favorable state for activation.

An *E. coli* protein located in the membrane is implicated in regulation of lysogeny by acting through Cro (75). A mutation that makes the cell defective for this protein results in wild-type λ giving a Cro⁻ phenotype. Whether the protein directly or indirectly interacts with Cro, or is responsible for its synthesis, is not known.

VI. Induction

The prophage state of λ is maintained only through continued presence of active Repressor (see Section IV). If the Repressor is inactivated, transcription starts from both p_R and p_L, and through a site-specific recombination event catalyzed by the products of the phage genes *int* and *xis* (Fig. 3), excision of the phage from the chromosome can occur (Fig. 4) (for review, see 39). The phage then replicates and produces progeny. This process of changing from the prophage state to the lytic mode, through derepression, is known as prophage induction. In this section, we discuss two aspects of phage induction, Repressor inactivation and control of excision.

Ultraviolet light or chemicals that cause damage to DNA bring on a variety of cellular responses, among them an increase in the cell's RecA protein level and the induction of a lambda prophage (for review, see 76). RecA protein has two capacities: it can function in DNA recombination and repair by pairing homologous strands of DNA (77, 78); it can also function as a protease that can cleave, and thus inactivate, λ Repressor (79). In undamaged cells, there is a low level of RecA protein. This protein probably has no activity for proteolysis, since λ Repressor is not cleaved during lysogeny. When damage to DNA occurs, the RecA protein apparently becomes active as a protease since λ Repressor is cleaved at this time. It has been proposed that the single-stranded regions of gapped DNA resulting from deleterious treatment act as a signal to activate the recA protein so that it can function as a protease (79). Presumably, as long as the "signal" for RecA protease activation is present, λ Repressor will continue to be cleaved, and repression of λ will be lifted.

An alternative model suggested for prophage induction does not primarily involve cleavage of Repressor (80). Since gapped *E. coli* DNA competes with λ operator DNA for Repressor binding, it "soaks up" the Repressor, making it unavailable for maintaining lysogeny.

Repressor bound to the gaps could be more susceptible to RecA cleavage.

Once the λ Repressor has been inactivated or is otherwise made unavailable, several mechanisms ensure that induction will be irreversible. Two events further prevent Repressor synthesis: transcription from p_{RM}, the autoregulated promoter for Repressor maintenance, ceases; and transcription starts from p_R and p_L. If Repressor could again be made available for p_{RM}, theoretically this promoter could be reactivated to make more Repressor, but this is not the case. The Cro protein made from p_R binds to λ DNA at operator site o_{R3} (Fig. 5) and prevents Repressor being made from p_{RM} (69). Since intact Repressor is necessary for p_{RM} activation (63), any further Repressor would have to be made from the establishment promoter, p_{RE}. For its activation, p_{RE} requires high levels of cII and cIII proteins, made from p_R and p_L operons, respectively. Since cII and cIII cannot be made in large amounts from the single copy prophage (46), p_{RE} will not be activated. Thus, Repressor is made neither from p_{RM} nor from p_{RE}, and this prevents reestablishment of repression.

The actual excision of the prophage from the chromosome is catalyzed by Integrase and Xis protein. Although both of these proteins are necessary *in vivo*, Integrase alone at a certain salt concentration can catalyze excision *in vitro* (K. Abremski, personal communication). The Xis protein might act to alter Integrase so that an enzyme state more favorable for excision than for integration is achieved; Xis is also known to make the excision reaction less sensitive to heat (81), so Xis might act as a stabilizer of Integrase.

Integrase synthesis has an additional control. When λ is growing lytically, Integrase is not needed, and the amount made from p_L is very low (43, 44). This has been attributed to the *cis*-acting *sib* site located to the left of the phage attachment site, P · P' (82; C. Montanez and G. Guarneros, personal communication). The mechanism of the *sib* inhibition on Integrase synthesis is unknown. When λ is in the prophage state, *sib* and *int* are no longer linked (Fig. 4), thus there is no *sib* interference and, consequently, sufficient Integrase for excision can be made from p_L. This is necessary because p_I is not active under conditions of induction since cII and cIII levels are low.

References

1. H. Echols, in "The Bacteriophage λ" (A. D. Hershey, ed.), p. 247. Cold Spring Harbor Laboratory, Cold Spring Harbor, New York, 1971.
2. I. Herskowitz, *Annu. Rev. Genet.* **7**, 289 (1973).
3. R. Losick and M. Chamberlin, eds., "RNA Polymerase." Cold Spring Harbor Laboratory, Cold Spring Harbor, New York, 1976.

4. S. F. Heinemann and W. G. Spiegelman, *CSHSQB* **35**, 315 (1970).
5. J. Roberts, *Nature* **224**, 1168 (1969).
6. H. A. Lozeron, J. A. Dahlberg and W. Szybalski, *Virology* **71**, 262 (1976).
7. M. E. Gottesman, S. Adhya and A. Das, *JMB* **140**, 57 (1980).
8. A. Ghysen and M. Pironio, *JMB* **65**, 259 (1972).
9. M. F. Baumann and D. I. Friedman, *Virology* **73**, 128 (1976).
10. N. Sternberg, *Virology* **73**, 139 (1976).
11. T. Segawa and F. Imamoto, *JMB* **87**, 741 (1974).
12. N. C. Franklin, *JMB* **89**, 33 (1974).
13. S. Adhya, M. Gottesman and B. deCrombrugghe, *PNAS* **71**, 2534 (1974).
14. D. I. Friedman, G. S. Wilgus, and R. J. Mural, *JMB* **81**, 505 (1973).
15. J. S. Salstrom and W. Szybalski, *JMB* **124**, 195 (1978).
16. M. Rosenberg, D. Court, H. Shimatake, C. Brady, and D. L. Wulff, *Nature* **272**, 414 (1978).
17. B. deCrombrugghe, M. Mudryj, R. DiLauro and M. E. Gottesman, *Cell* **18**, 1145 (1979).
18. D. I. Friedman, C. T. Jolly and R. Mural, *Virology* **51**, 216 (1973).
19. D. I. Friedman, M. Baumann and L. S. Baron, *Virology* **73**, 119 (1976).
20. F. Keppel, C. P. Georgopoulos and H. Eisen, *Biochimie* **56**, 1503 (1974).
21. J. Greenblatt, J. Li, S. Adhya, D. I. Friedman, L. S. Baron, B. Redfield, H. Kung and H. Weissbach, *PNAS* **77**, 1991 (1980).
22. R. I. Kamen, *in* "RNA Phages" (N. Zinder, ed.), p. 203. Cold Spring Harbor Laboratory, Cold Spring Harbor, New York, 1975.
23. E. Flamm, Ph.D. Thesis, University of Michigan, Ann Arbor, 1980.
24. A. Joyner, L. N. Isaacs, H. Echols and W. S. Sly, *JMB* **19**, 174 (1966).
25. M. Schnos and R. Inman, *JMB* **51**, 61 (1970).
26. M. E. Furth, F. R. Blattner, C. McLeester and W. F. Dove, *Science* **198**, 1046 (1977).
27. W. F. Dove, H. Inokuchi and W. F. Stevens, *in* "The Bacteriophage λ" (A. D. Hershey, ed.), p. 747. Cold Spring Harbor Laboratory, Cold Spring Harbor, New York, 1971.
28. S. Mitra, *Annu. Rev. Genet.* **14**, 347 (1980).
29. R. McMacken, N. Mantei, B. Butler, A. Joyner and H. Echols, *JMB* **49**, 639 (1970).
30. L. F. Reichardt, *JMB* **93**, 267 (1975).
31. A. Folkmanis, W. Maltzman, P. Mellon, A. Skalka, and H. Echols, *Virology* **81**, 352 (1977).
32. I. Herskowitz and E. R. Signer, *JMB* **47**, 545 (1970).
33. J. W. Roberts, *PNAS* **72**, 3300 (1975).
34. F. R. Blattner and J. E. Dahlberg, *Nature NB* **237**, 227 (1972).
35. P. Gariglio and M. H. Green, *Virology* **53**, 392 (1972).
36. J. S. Parkinson, *Genetics* **59**, 311 (1968).
37. L. Reichardt and A. D. Kaiser, *PNAS* **68**, 2185 (1971).
38. A. Campbell, *PNAS* **73**, 887 (1976).
39. R. A. Weisberg, S. Gottesman, and M. E. Gottesman, *in* "Comprehensive Virology" (H. Fraenkel-Conrat and R. R. Wagner, eds.), Vol. 8, p. 197. Plenum, New York, 1977.
40. K. Yen and G. Gussin, *Virology* **56**, 300 (1973).
41. E. Signer, *Virology* **46**, 624 (1970).
42. N. Katzir, A. Oppenheim, M. Belfort and A. B. Oppenheim, *Virology* **74**, 324 (1976).
43. D. Court, S. Adhya, H. Nash and L. Enquist, *in* "DNA Insertion Elements, Plasmids and Episomes" (A. Bukhari, J. Shapiro, and S. Adhya, eds.), p. 389. Cold Spring Harbor Laboratory, Cold Spring Harbor, New York, 1977.

44. M. Kotewics, S. Chung, Y. Takeda and H. Echols, PNAS **74**, 1511 (1977).
45. M. O. Jones, R. Fischer, I. Herskowitz and H. Echols, PNAS **76**, 150 (1979).
46. D. Court, L. Green and H. Echols, Virology **63**, 484 (1975).
47. J. Abraham, D. Mascarenhas, R. Fischer, M. Benedik, A. Campbell and H. Echols, PNAS **77**, 2477 (1980).
48. R. H. Hoess, C. Foeller, K. Bidwell and A. Landy, PNAS **77**, 2482 (1980).
49. D. L. Wulff, M. Beher, S. Izumi, J. Beck, M. Mahoney, H. Shimatake, C. Brady, D. Court and M. Rosenberg, JMB **138**, 209 (1980).
50. M. Yarmolinsky, in "The Bacteriophage λ" (A. D. Hershey, ed.), p. 97. Cold Spring Harbor Laboratory, Cold Spring Harbor, New York, 1971.
51. A. Campbell, Adv. Genet. **11**, 101 (1962).
51a. Y. Kikuchi and H. Nash, PNAS **76**, 3760 (1979).
52. H. I. Miller, A. Kikuchi, H. A. Nash, R. A. Weisberg and D. I. Friedman, CSHSQB **43**, 1121 (1979).
53. Y. Kikuchi and H. Nash, CSHSQB **43**, 1099 (1979).
53a. K. Mizuuchi, M. Gellert and H. Nash, JMB **121**, 375 (1978).
54. M. Gellert, M. H. O'Dea, T. Itoh and J.-I. Tomizawa, PNAS **73**, 4473 (1976).
54a. A. Sugino, C. L. Peebles, K. N. Kreuzer and N. R. Cozzarelli, PNAS **74**, 4767 (1977).
55. M. Gellert, K. Mizuuchi, M. H. O'Dea, T. Itoh and J.-I. Tomizawa, PNAS **74**, 4772 (1977).
56. A. Landy and W. Ross, Science **197**, 1147 (1977).
57. P.-L. Hsu, W. Ross and A. Landy, Nature **285**, 85 (1980).
58. L. N. Isaacs, H. Echols and W. S. Sly, JMB **13**, 963 (1965).
59. M. Ptashne, A. Jeffrey, A. D. Johnson, R. Maurer, B. J. Meyer, C. O. Pabo, T. M. Roberts and R. T. Sauer, Cell **19**, 1 (1980).
60. A. Johnson, B. J. Meyer and M. Ptashne, PNAS **76**, 5061 (1979).
61. Y. Takeda, JMB **127**, 177 (1979).
62. B. J. Meyer, R. Maurer and M. Ptashne, JMB **139**, 163 (1980).
63. B. J. Meyer and M. Ptashne, JMB **139**, 195 (1980).
64. R. Maurer, B. J. Meyer and M. Ptashne, JMB **139**, 147 (1980).
65. M. Ptashne, K. Backman, M. Z. Humayun, A. Jeffrey, R. Maurer, B. Meyer and R. T. Sauer, Science **194**, 156 (1976).
66. T. Ogawa and J. Tomizawa, JMB **23**, 225 (1967).
67. A. D. Kaiser, Virology **3**, 42 (1957).
68. H. Echols and L. Green, PNAS **68**, 2190 (1971).
69. A. Johnson, B. J. Meyer and M. Ptashne, PNAS **75**, 2190 (1971).
70. H. Eisen, P. Brachet, P. deSilva and F. Jacob, PNAS **66**, 855 (1970).
71. T. Grodzicker, R. R. Arditti and H. Eisen, PNAS **69**, 366 (1972).
72. M. Belfort and D. L. Wulff, Virology **55**, 183 (1973).
73. M. Belfort and D. L. Wulff, PNAS **71**, 779 (1974).
74. M. O. Jones and I. Herskowitz, Virology **88**, 199 (1978).
75. A. Oppenheim, A. Honigman and A. B. Oppenheim, Virology **61**, 1 (1974).
76. E. W. Witkin, Bacteriol. Rev. **40**, 869 (1976).
77. A. J. Clark, Annu. Rev. Genet. **7**, 67 (1973).
78. T. Shibata, C. DasGupta, R. Cunningham and C. Radding, PNAS **76**, 1638 (1979).
79. N. L. Craig and J. W. Roberts, Nature **283**, 26 (1980).
80. R. Sussman, J. Resnick, K. Calame and J. Baluch, PNAS **75**, 5817 (1978).
81. G. Guarneros and H. Echols, Virology **52**, 30 (1973).
82. G. Guarneros and J. M. Galindo, Virology **95**, 119 (1979).

III. Chromatin Transcription and Replication

Chairman and Summarizer: RONALD L. SEALE

Summary RONALD L. SEALE	121
Site of Histone Assembly RONALD L. SEALE	123
Chromatin Replication in *Tetrahymena pyriformis* A. T. ANNUNZIATO AND C. L. F. WOODCOCK	135
Role of Chromatin Structure, Histone Acetylation, and the Primary Sequence of DNA in the Expression of SV40 and Polyoma in Normal or Teratocarcinoma Cells G. MOYNE, M. KATINKA, S. SARAGOSTI, A. CHESTIER, AND M. YANIV	151

Summary

RONALD L. SEALE

*Scripps Clinic and Research Foundation
La Jolla, California*

Recent advances in understanding the nucleosomal structure of chromatin has led to a unified theory for the repeating DNA-protein subunit structure of the eukaryotic chromosome. Extrapolation of this knowledge to explain how chromatin functions has not followed. Certain simplified ideas brought forth early in this period (e.g., that all nucleosomes are identical either in composition or in shape, that nucleosomes are randomly situated on DNA without regard to nucleotide sequence, that core histones are not translocated or displaced during transcription and replication once they are assembled on DNA, and so forth) are presently being reevaluated. Frequently, observations indicate that these structures are not passive and randomly situated, DNA-packaging elements, but that they may be directly involved with chromatin metabolism from the level of composition to higher order packing.

The four papers in this session addressed various metabolic states of nucleosomes, and a summary statement of the evidence presented is that nucleosomes are actively and nonrandomly engaged in replication and transcription functions. Three of these reports concerned the behavior of nucleosomes during replication. The observations of Seidman *et al.* confirm and extend previous evidence indicating the nonrandom dispersal of nucleosomes at the replication fork and indicate passage of the parental nucleosomes to the leading side of the replication fork. Thus, either polarity is implicated in nucleosome structure or a gating mechanism is operant in conjunction with DNA replication, such that nucleosomes follow the direction of DNA at opposite sides of the fork, in the trans configuration.

The trans configuration of nucleosome partitioning on the replication bubble may have a further function, as these authors present evidence that the recipient DNA strand is also the coding strand, presumably receiving epigenetic information in the parental nucleosomes.

Closely allied to this study is the report of Yaniv, who showed restriction-enzyme digestion combined with electron microscopic evidence for nucleosome phasing on SV40. In corroboration of several previous reports, a 250–300 base-pair region containing the SV40 origin was shown to be accessible to nucleolytic attack and was visibly devoid of typical nucleosomal structures. Correspondence between promoter sites and replication origins may be further borne out by the evidence that embryocarcinoma deletions have created a new promoter site.

Further evidence was presented toward identification of the sites for nucleosome assembly. Seale and Annunziato and Woodcock showed that the newly synthesized histones are assembled into nucleosomes on nonreplicating DNA. This necessitates that nucleosomes be redistributed on chromatin, and possible mechanisms were presented by these workers.

This brings us to the point of reconciliation of two lines of evidence, a most important problem for future investigation. If nucleosomes are inherited by the coding strand of DNA for the purpose of stable inheritance of epigenetic information carried by these particles, how does one fit in the data for core histone redistribution on DNA, a process that would obliterate epigenetic sequence-specific information (if such exists)? Perhaps both are true in degrees, and control is achieved by more temporal mechanisms acting in concert with a subclass of nucleosomes, i.e., in the sense of prokaryotic operons, than is presently suspected. Nucleosome phasing in coding sequences, the SV40 origin of replication, the *Xenopus* 5 S genes, and on newly replicated DNA are short-range effects. The evidence implies either nucleosome sliding or site-specific deposition; long-range nucleosome redistribution has not been investigated to date.

Site of Histone Assembly

RONALD L. SEALE

Scripps Clinic and Research
Foundation
La Jolla, California

Since the late 1960s, it has been known that DNA synthesis and histone synthesis are temporally coupled during the S phase of the cell cycle (1–7). The doubling of the mass of DNA is matched by a doubling of histone protein. Since an obvious temporal deficiency in histone protein occurs at the replication fork, it is a resonable assumption that newly synthesized histone protein is assembled into nucleosomes directly upon the vacancies on daughter-strand DNA.

This seemingly straightforward problem has attracted the attention of numerous investigators (reviewed in 8). Although there is evidence that new histone protein is assembled on newly replicated DNA (9–13), other studies indicate that there is no preferential deposition on new DNA (14–18).

I. Evidence for Nucleosome Assembly on Newly Replicated DNA

The first study on the site of histone deposition employed the incorporation of BrdU into new DNA and [^{14}C]leucine into chromatin proteins of rat liver (9). Nuclei, and then chromatin, were isolated and irradiated to fragment the new BrdU-DNA. The chromatin was adjusted to pH 12.2 to denature the DNA and then sedimented in alkaline gradients in order to follow the sedimentation behavior of the new proteins versus the old or new single-stranded DNA. On the basis of cosedimentation of labels, it was concluded that new protein followed preferentially the new DNA strand of daughter chromatids. The method employed in this study is suspect, for, at pH values above 10, chromatin proteins are quantitatively dissociated from DNA (Seale, unpublished). However, this result was reinforced by a more recent report (10) that employed cross-linking prior to sedimentation analysis.

Studies on nucleosome assembly sites on SV40 "minichromatin" (11) showed that pulse-labeled histone proteins are enriched by 40–60% in the region of sucrose gradients containing replicating intermediates, marked by a thymidine pulse-label. This result, for several reasons, does not necessarily support the interpretation that histone is assembled on new DNA. First, other metabolically active SV40 chromatin species, namely, transcriptionally active minichromatin, are found in the same region of the gradient and comprise 95% of the chromatin of that region. Second, deposition of labeled histone protein on newly replicated DNA versus unreplicated DNA was not determined; the replicative intermediates ("theta structures") are composed of both types of chromatin.

A feature of viral chromatin that has received little emphasis is that metabolically active species are physically distinct from metabolically inactive minichromatin. Some minichromatin molecules appear to be metabolically inert, while others are either continuously transcribed or replicated. Further, viral chromatin is greatly amplified during replication; therefore, virtually *all* viral histones are newly synthesized and may be continuously deposited on the same continuously replicating molecules, as opposed to the strictly limited doubling of cellular chromatin. One must be cautious, then, in extrapolation of results derived from studies on DNA viruses, because such properties peculiar to the virus can indeed be misleading for extrapolation to like cellular functions.

A third report favoring new histone deposition on new DNA also used cosedimentation of pulse-labeled histones and DNA as the sole criterion for physical association (12). When *Drosophila* cells were pulse-labeled with either amino acids or thymidine, both pulse labels were released by micrococcal nuclease to slowly sedimenting regions of sucrose gradients. The very interesting observation was that the pulse-labeled histones did not sediment together; the H3–H4 pair and the H2A–H2B pair sedimented at different positions in the gradient. The authors assumed that this observation represented different times of deposition of the histone pairs during assembly. An alternative interpretation is that the new chromatin-associated histones were not yet assembled into nucleosomes. In this vein, one notes that only 10–50% of the total labeled protein was investigated; the remaining 50–90% of the label pelleted and was not further analyzed. A significant fraction ($\leq 50\%$) of newly synthesized histone protein that isolates with chromatin is not assembled into nucleosomes, and these proteins may be released from chromatin under conditions that leave mature nucleosomes intact (Seale, unpublished).

The most compelling evidence for selective assembly of new histones on new DNA used a density-labeling protocol to separate new and old chromatins. Newly synthesized total chromosomal protein is associated with both new and nonreplicating chromatin species; measurements of the ratio of ^{3}H (new) histone to ^{14}C (uniformly labeled) histone proteins in gels showed that the new DNA was enriched for new histone protein, relative to the nonreplicating chromatin (13). Enrichment for the histone pairs was not uniform, as also found in the previously mentioned investigations (12, 20; Seale, unpublished); H2A and H2B were twice as prevalent on new DNA as on nonreplicating DNA, and the H3–H4 pair was enriched about fivefold. The H3–H4 pair and the H2A–H2B pair were not stoichiometric as predicted by another study (19). While the first study (13) indicated an enrichment of newly synthesized histones on newly synthesized DNA, it also indicated a substantial ($\geq 50\%$) deposition of new proteins on the nonreplicating DNA.

II. Evidence against Preferential Nucleosome Assembly on New DNA

The results of a second group of investigations indicate that the new histones are not selectively deposited on or near the replication fork (14–18; Seale, unpublished). Several studies (14–16; Seale, unpublished) were based on buoyant density separation of new vs old chromatin. Newly synthesized chromatin DNA was marked by the incorporation of BrdU (15; unpublished) IdU (14, 16), and newly synthesized proteins were labeled with tritiated amino acids. Upon resolution of chromatin (14, 15), or salt-washed mononucleosomes (16; unpublished), the primary deposition of radiolabeled protein was with the native density, nonreplicating chromatin DNA.

The earlier of these papers employed whole chromatin for analysis (14, 15), obviating identification of the radioactive proteins. These results have been confirmed using isolated core particles in which the labeled protein was specifically histone (16–18; Seale, unpublished). These investigations have been criticized in that the formaldehyde fixation step may induce protein redistribution. One notes that each of the investigations included controls for this eventuality, and no protein exchange was detected.

Owing to the extensive peak overlap between density-labeled and native density chromatins in isopycnic gradients, it has not been possible to exclude that some new histone is assembled on replicating DNA. It would thus be useful to see different methods used in the investigation of this question.

While certain studies on either side of the assembly site question have serious shortcomings, there is a residue of conflicting evidence on both sides that cannot be reconciled. With the advent of new methods for fractionation and characterization of nucleosomes, it is a propitious time to reinvestigate this aspect of the assembly problem.

III. Results

The present investigation into the site of deposition of newly synthesized histone protein involved two methodologies. One was a reevaluation of earlier work employing the density labeling of new DNA in order to resolve newly replicated, hybrid-density chromatin from native-density, bulk chromatin in buoyant density gradients. Second, the intrinsic nuclease sensitivity of newly synthesized chromatin was followed as a method to distinguish new from old chromatin DNA. In both cases, histones were pulse-labeled in order to relate their properties to those of either newly replicated or bulk chromatin.

For buoyant density studies, randomly growing HeLa cells were incubated with [^{14}C]BrdU plus [^{3}H]lysine for 90 minutes. During this labeling period, both isotopes were incorporated in a linear fashion without lag and without exhaustion of either isotope. Additional controls demonstrated that the BrdU-substituted DNA was in nucleosomes. In 1.5 hours, about 15% of the total chromatin is synthesized in HeLa cells. Thus, the prediction is that the newly synthesized histones and density-labeled DNA should be associated if they are exclusively assembled together into nucleosomes by the cell.

After the labeling period, nuclei were isolated and digested briefly with micrococcal nuclease. The entire nuclear digest was adjusted to 0.5 M NaCl at 4°C and sedimented in sucrose gradients containing 0.5 M NaCl (Fig. 1A). The monosome peak (bracket) was collected, dialyzed, and cross-linked with formaldehyde. This peak contains core particles freed from nonhistones and Hl; only the four nucleosomal histones plus 140–160 base-pair DNA were present. Upon banding in CsCl, the peak of protein radioactivity was not coincident with the hybrid density (newly replicated) chromatin, but predominated at the position of native density chromatin.

In these gradients, hybrid density DNA can be completely resolved from native density DNA. However, because of associated chromatin proteins, the density difference between the DNAs is obscured when whole chromatin is banded (15). If chromatin stripped of nonhistone proteins and Hl is examined, as in Fig. 1 (16), sufficient resolution between the two species is obtained for designation of the predominant position of protein radioactivity with one or the other chromatin species.

SITE OF HISTONE ASSEMBLY 127

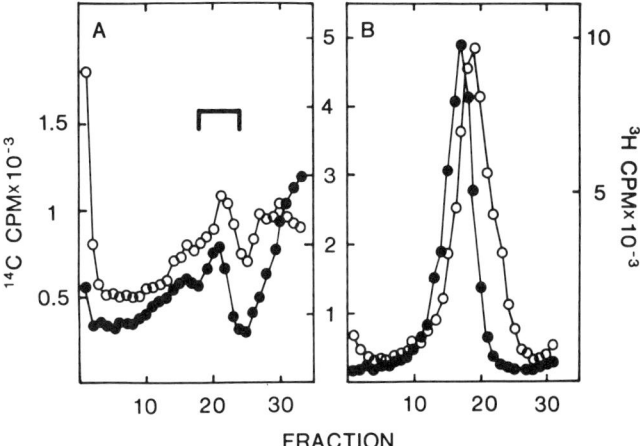

FIG. 1. Micrococcal nuclease-digested nuclei were adjusted to 0.5 M NaCl and centrifuged in sucrose gradients containing 0.5 M NaCl at 4°C (panel A). The monomer peak (bracket) was dialyzed against 10^{-5} M triethanolamine and cross-linked with 1% neutralized formaldehyde at 0°C. The fixed monomers were then centrifuged to equilibrium in CsCl guanidine gradients at 4°C (15). Centrifugation is right to left. ○——○, [^{3}H]lysine; ●——●, [^{14}C]BrdU.

The chromatin density-labeling procedure was also utilized in an experiment designed to investigate the source of proteins mobilized for retroactive packaging of DNA synthesized during a block of protein synthesis. Under these conditions, one-half of the newly synthesized DNA is nucleosome-free and one-half is complexed with parental histones presumably inherited at the fork (21-24). Both cycloheximide and [^{14}C]BrdU were added to cells for 15 minutes. At this time, cycloheximide and [^{14}C]BrdU were removed and [^{3}H]arginine plus hydroxyurea (to prevent further DNA synthesis) were added for an additional 15 minutes. Control experiments showed that the DNA synthesized in cycloheximide was more nuclease sensitive than control chromatin and that at the end of the second incubation the labeled DNA had a nuclease sensitivity equal to that of mature chromatin. In this experiment, one predicts that the extensive regions of DNA (depleted in nucleosomes during DNA replication in cycloheximide), assembled following drug removal, should be associated with the labeled histone if new histone is the source of protein for such assembly.

As in Fig. 1A, the stripped core monosome peak was collected and examined by equilibrium centrifugation (Fig. 2). The [^{14}C]BrdU hybrid density nucleosome peak was clearly resolved from the [^{3}H]nu-

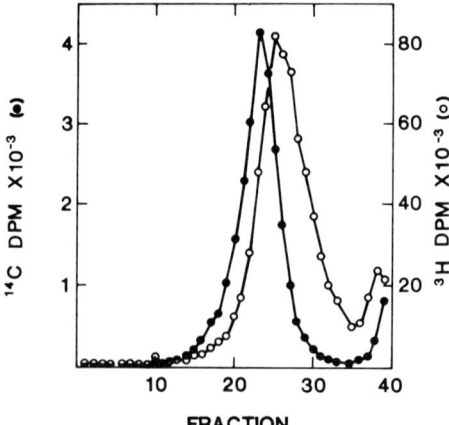

FIG. 2. Assembly of cycloheximide-induced, nucleosome-depleted chromatin. [^{14}C]BrdU-substituted, nucleosome-depleted chromatin was assembled into nucleosomes in the presence of [^{3}H]arginine (see text). Stripped monomers were prepared and centrifuged to equilibrium as in Fig. 1. Centrifugation is right to left. ○——○, [^{3}H]arginine; ●——●, [^{14}C]BrdU.

cleosome peak, again indicating the lack of exclusive association of newly synthesized histone with the DNA synthesized during the protein synthesis block.

A second approach to the study of the site of nucleosome assembly used to advantage the intrinsic properties of newly synthesized chromatin DNA. New chromatin DNA is more nuclease sensitive (21, 25), and the nucleosomal multimers are slightly smaller than their counterparts in whole chromatin (26, 27; Annunziato and Seale, unpublished). These properties can be observed in electrophoretically separated nucleosomes (Fig. 3A) as well as in stripped nucleosomes (Fig. 3B), composed of only the four nucleosomal histones and DNA. In both preparations, note the more rapid disappearance of high-molecular-weight, pulse-labeled chromatin and the shorter multimeric subunits. Thus, the experimental objective was to pulse-label histones and then determine whether they followed the profile of total chromatin or of new chromatin.

The profiles of uniformly labeled nucleosomal DNA and of pulse-labeled chromatin proteins are shown in Fig. 4. Unexpectedly, the total pulse-labeled chromatin proteins followed neither type of DNA. A single major ^{3}H-labeled peak was evident on the leading edge of the chromatin dimer peak, and, notably, radioactivity fell precipitously in the monomer region. The peak of pulse-labeled proteins was not ob-

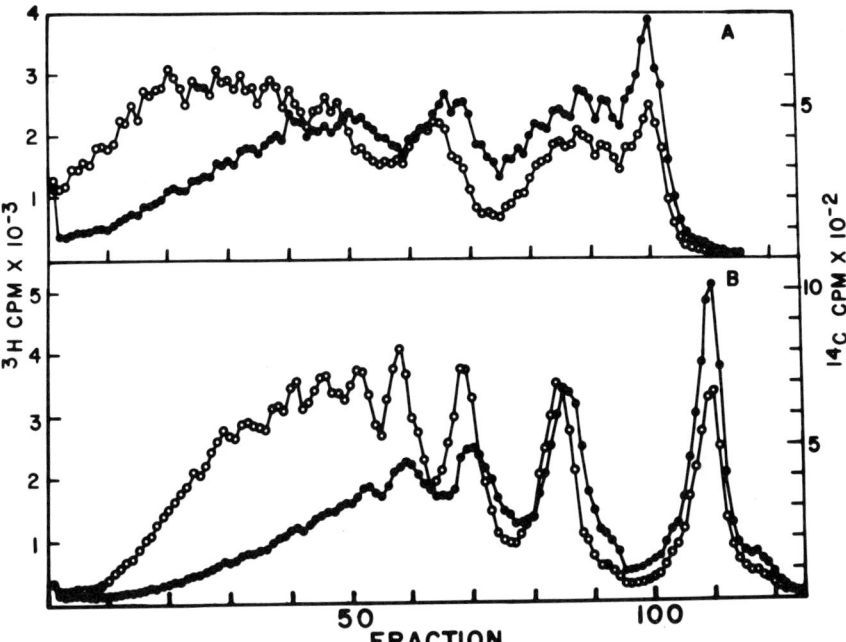

FIG. 3. Electrophoretic properties of newly synthesized chromatin. Cells prelabeled for one generation with [^{14}C]thymidine were incubated with [^{3}H]thymidine for 2.5 minutes. Micrococcal nuclease-solubilized chromatin was separated electrophoretically (28) as whole chromatin (panel A) or as 0.45 M NaCl-washed chromatin (panel B). O——O, [^{14}C]thymidine; ●——●, [^{3}H]thymidine.

served when uniformly labeled proteins in chromatin were so analyzed; in this case the radioactivity profile corresponded to the DNA profile in whole and in stripped chromatin.

The nature of the proteins comprising the radioactivity spectrum of Fig. 4 was ascertained by performing a second dimension of electrophoresis in dodecyl sulfate followed by autoradiography (Fig. 5). The histone proteins were not the proteins responsible for the large peak of radioactivity; rather, a series of nonhistone proteins and apparent "high-mobility-group" (HMG) proteins migrate in this region. The reason for the concentration of pulse-labeled protein species in a limited region of the gel has not yet been determined.

In order to correlate the electrophoretic properties of newly synthesized nucleosomal histones with either chromatin species, the nuclease-solubilized chromatin was stripped in 0.45 M NaCl and separated by electrophoresis as in Fig. 3B (Fig. 6). Both the kinetics of

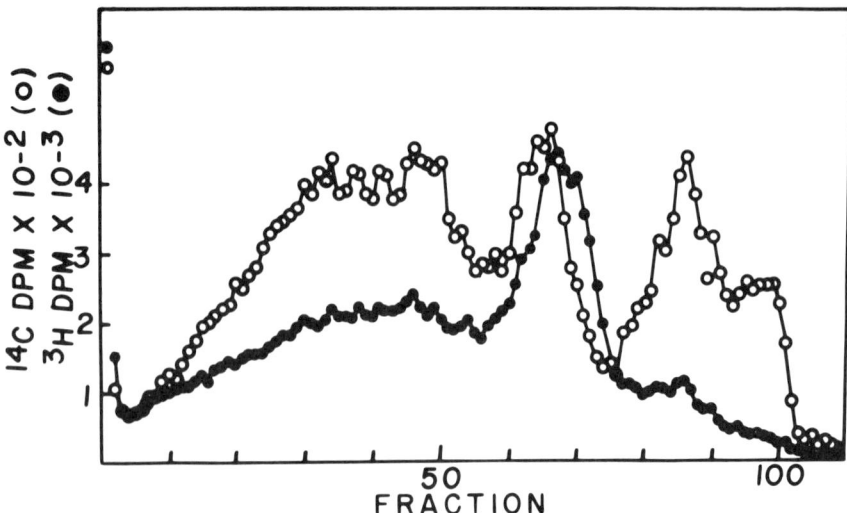

FIG. 4. Electrophoretic characteristics of total pulse-labeled chromatin protein. Cells incubated with [^{14}C]thymidine for 18 hours were pulse-labeled with [^{3}H]arginine plus [^{3}H]lysine for 2.5 minutes. Micrococcal nuclease-sheared chromatin was examined as in Fig. 3A. ○——○, [^{14}C]thymidine; ●——●, [^{3}H]arginine and [^{3}H]lysine.

digestion and the subunit size of the newly synthesized nucleosomal histones corresponded closely to those parameters of bulk chromatin. There was a slight phasing difference evident in the dimer and trimer, as expected for assembly on new DNA. We conclude that the new histones are assembled on both old and new DNA and that the [^{3}H]histone profile of Fig. 6 represents the sum of these two overlapping radioactivity spectra. A more analytical treatment of this observation will be presented elsewhere.

IV. Discussion

The deposition (assembly) of newly synthesized histone protein into nucleosomes on newly replicated DNA versus nonreplicating DNA was investigated by methods employing the density-labeling of new DNA and by the nuclease sensitivity of new chromatin DNA relative to that of total chromatin. In all cases examined, no evidence was found favoring predominant deposition of new histones on new DNA. Rather, new histones appear to be assembled on both newly replicated DNA as well as on DNA that was not replicated during the experimental period.

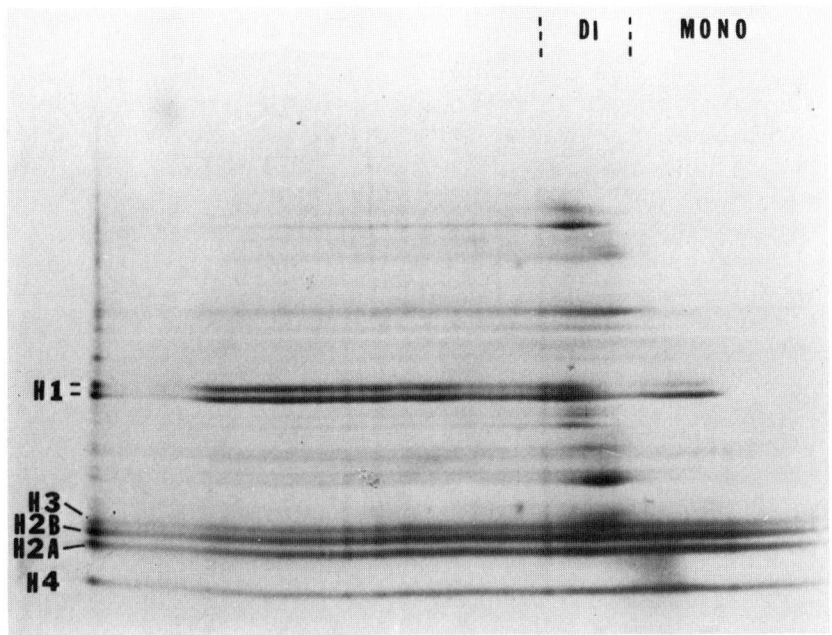

FIG. 5. Two-dimensional separation of new chromatin proteins. Cells were incubated for 2.5 minutes with [³H]arginine plus [³H]lysine. Nuclease-solubilized chromatin was resolved in agarose–acrylamide gels (28), and the protein species were separated in dodecyl sulfate on polyacrylamide gels in the second dimension.

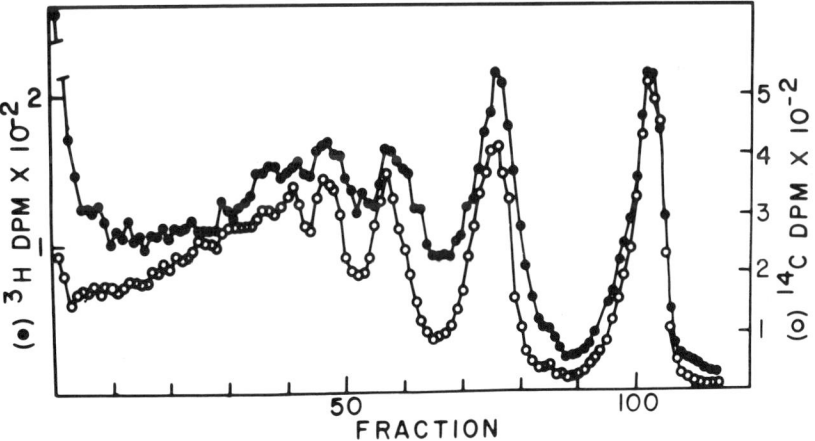

FIG. 6. Electrophoresis of stripped chromatin. Cells grown for 18 hours in [¹⁴C]thymidine were incubated for 2 minutes with [³H]arginine plus [³H]lysine. Nuclease-solubilized chromatin was stripped in 0.45 M NaCl and resolved in agarose–acrylamide gels. O——O, ¹⁴C; ●——●, ³H.

The lack of correspondence between the properties of newly synthesized histone proteins and newly replicated DNA conflicts with other evidence (12). One feature of newly synthesized histones, the lack of assembly of a significant fraction of the new histones (Seale, unpublished), may explain this observation. Only 10–50% of the newly synthesized protein was examined, and these proteins were found in the low-molecular-weight region of sucrose gradients, as would be expected for proteins dissociated from chromatin. The H3–H4 pair was not coincident with the H2A–H2B pair, as predicted for assembly of nucleosomes exclusively from newly synthesized protein (19). Last, the stoichiometry of H3 and H4 was not constant across the gradient fractions, as it must be if these proteins represent nascent nucleosomes. The lack of stoichiometry of newly deposited H3 and H4 is also evident in other results (20) showing initial deposition of H3 and H4 on SV40 replication intermediates. The ratio of pulse-labeled H3 to H4 was approximately twice that expected for equimolar representation, after correcting for methionine content.

Since newly synthesized histone protein does not accumulate exclusively on newly replicated DNA, it follows that previously existing histones are redistributed to the replication fork. Alternatively, newly made protein is assembled on fork DNA, but is not found there shortly after assembly. In either case, nucleosomes must move on chromatin, and must do so rather extensively.

The mechanism for nucleosome redistribution is obscure at this time. Two mechanisms seem equally probable. In one, the octameric core dissociated from chromatin and relocates elsewhere on DNA. For this to occur, one must also postulate an escort factor (29) to hold the core proteins intact, and to account for their lack of mixing during cell growth (19). A second possibility is that nucleosome core proteins do not entirely leave the chromatin matrix, but migrate along it to the next available opening, as postulated by Stein (30).

If nucleosomes redistribute on chromatin, there must be a compelling reason for them to do so. One suspicious function is transcription. It is by no means clear that transcribable nucleosomes remain bound to DNA while the polymerase makes two turns about every nucleosome, trailing an enormous transcript behind it. It seems entirely possible that nucleosomes are dislodged in the process of transcription and subsequently reassemble on DNA after the RNA polymerase has passed. If reassembly occurred at any available site on DNA, rather than at the same site, then nucleosome nearest-neighbor mixing would occur.

The preceding paragraphs are, of course, highly speculative, but may explain several observations in the current literature. Nucleosomes on transcribable chromatin have a distinct conformation that renders them DNase I susceptible (31). The HMG-14 and -17 proteins have been implicated in this compositional and conformational change (32, 33); one notes that only two presently known proteins participate in the active conformation of thousands of genes. This common conformational state may represent the ability of these particles to be dislodged during transcription. This may also account for morphological differences in actively transcribed and nontranscribed regions (34). Transcriptional dislocation of nucleosomes would create a pool of octameric cores, which would compete with newly synthesized histones for any given association site. Thus, the inability to label newly synthesized DNA selectively with new proteins could result from such an isotope dilution effect.

The question of nucleosome movement, or redistribution, seems to be one of the more urgent problems in chromatin metabolism. Theories for stable and heritable epigenetically defined chromatin configurations depend on fixed nucleosome–DNA associations. Nucleosome movement requires that these ideas be redefined to allow more temporal modes of control that allow or disallow transcription. Evidence is mounting for nonrandom siting of nucleosomes on several specific sequences that have been tested, including coding sequences (35), the SV40 origin (36, 37), and the 5 S genes (38). Short-range nucleosome movement has also been observed on newly replicated DNA (A. T. Annunziato and R. Seale, unpublished; 26, 27) and lends credence to the idea that long-range redistribution may also occur.

Acknowledgments

This work was supported by grants from the National Institutes of Health and was performed in part at the National Jewish Hospital, Denver, Colorado. I am grateful to Drs. C. A. Thomas, J. Gottesfeld, and A. T. Annunziato for stimulating discussions.

References

1. E. Robbins and T. W. Borun, PNAS **57**, 409 (1967).
2. T. W. Borun, M. D. Scharff and E. Robbins, PNAS **58**, 1977 (1967).
3. D. Gallwitz and G. C. Mueller, JBC **244**, 5947 (1969).
4. L. R. Gurley, R. A. Walters and R. A. Tobey, ABB **148**, 633 (1972).
5. M. Breindl and D. Gallwitz, EJB **32**, 381 (1973).
6. R. P. Perry and D. E. Kelley, JMB **79**, 681 (1973).
7. T. W. Borun, F. Gabrielli, K. Ajiro, A. Zweidler and C. Baglioni, Cell **4**, 59 (1975).
8. R. L. Seale, in "The Cell Nucleus" (H. Busch, ed.), Vol. 4, p. 155. Academic Press, New York, 1978.
9. R. Tsanev and G. Russev, EJB **43**, 257 (1974).

10. G. Russev and R. Tsanev, *EJB* **93**, 123 (1979).
11. C. Cremisi, A. Chestier and M. Yaniv, *Cell* **12**, 947 (1977).
12. A. Worcel, S. Han and M. L. Wong, *Cell* **15**, 969 (1978).
13. T. Senshu, M. Fukada and M. Ohashi, *J. Biochem.* **84**, 985 (1978).
14. V. Jackson, D. K. Granner and R. Chalkley, *PNAS* **73**, 2266 (1976).
15. R. L. Seale, *PNAS* **73**, 2270 (1976).
16. R. Hancock, *PNAS* **75**, 2130 (1978).
17. A. T. Annunziato and C. L. F. Woodcock (this volume).
18. A. T. Annunziato, Ph.D. Thesis, Univ. of Massachusetts (1979).
19. I. M. Leffak, R. Grainger and H. Weintraub, *Cell* **12**, 837 (1977).
20. C. Cremisi and M. Yaniv, *BBRC* **92**, 117 (1980).
21. R. L. Seale, *Cell* **9**, 423 (1976).
22. R. L. Seale and R. T. Simpson, *JMB* **94**, 479 (1975).
23. H. Weintraub, *Cell* **9**, 419 (1976).
24. M. M. Seidman, A. J. Levine and H. Weintraub, *Cell* **18**, 439 (1979).
25. R. L. Seale, *Nature* **255**, 247 (1975).
26. R. L. Seale, *PNAS* **75**, 2717 (1978).
27. R. F. Murphy, R. B. Wallace and J. Bonner, *PNAS* **75**, 5903 (1978).
28. J. O. Todd and W. T. Garrard, *JBC* **252**, 4729 (1977).
29. R. A. Laskey, B. M. Honda, A. D. Mills and J. T. Finch, *Nature* **275**, 416 (1978).
30. A. Stein, *JMB* **130**, 103 (1979).
31. H. Weintraub and M. Groudine, *Science* **93**, 848 (1976).
32. S. Weisbrod and H. Weintraub, *PNAS* **76**, 631 (1979).
33. S. Weisbrod, M. Groudine and H. Weintraub, *Cell* **19**, 289 (1980).
34. U. Scheer, *Cell* **13**, 535 (1978).
35. C. Wu, P. M. Bingham, K. J. Livak, R. Holmgren and S. C. R. Elgin, *Cell* **16**, 797 (1979).
36. A. J. Varshavsky, O. Sundin and M. Bohn, *Cell* **16**, 453 (1979).
37. B. A. J. Ponder and L. V. Crawford, *Cell* **11**, 35 (1977).
38. J. M. Gottesfeld and L. S. Bloomer *Cell* **21**, 751 (1980).

Chromatin Replication in *Tetrahymena pyriformis*

A. T. ANNUNZIATO[1] AND
C. L. F. WOODCOCK

*Department of Zoology
University of Massachusetts
Amherst, Massachusetts*

The replication of eukaryotic chromatin must rank as one of the most complex cellular processes. DNA replication alone, as discussed elsewhere in this volume, requires a surprisingly elaborate, multicomponent mechanism, and it is upon this already complicated system that the segregation and reassembly of the chromosomal proteins must be superimposed. Chromosomal protein assembly also provides an opportunity for the transcriptional status of a particular segment of the genome to be propagated or altered (1). A third level of complexity arises from the necessity to reestablish any specific higher-order structures of chromatin (2).

Although considerable progress has been made in recent years in understanding the events that take place during the segregation and assembly of chromosomal proteins, a number of findings remain controversial, and the results are limited to the nucleosomal histone proteins. Very little is known about the behavior of nonhistone chromosomal proteins during replication, although their contribution to the structural and functional modification of chromatin may well be crucial. It is now well established from nuclease digestion studies that nascent DNA is organized into nucleosome-like structures (presumably by complexing with histones) within 30 seconds of its formation (3, 4). In agreement with this, electron micrographs of replicating chromatin show nucleosome-like structures within 50 nm of the fork itself (5), although it should be noted that this work utilized the rapidly replicating *Drosophila* embryo system, where the usual negligible free-histone pools, and tight couplings between DNA and histone synthesis, do not obtain. There is also compelling evidence, again from nuclease digestion studies and from electron microscopy, that the maternal nucleosomes (or, rather, maternal histone octamers) segregate to one daughter strand, presumably leaving the other temporar-

---

[1] Present address: Department of Cellular Biology, Scripps Clinic and Research Foundation, 10666 N. Torrey Pines Rd., La Jolla, California 92037.

ily devoid of histones (6-8). Nucleosome assembly would therefore be confined to one daughter strand, while DNA replication occurs on both daughter strands. This distinction needs to be borne in mind in considering experiments using pulses of radioactive thymidine to label nascent chromatin; in this case, it can be expected that only half of the label will become associated with newly assembled nucleosomes.

Since DNA and histone synthesis are closely coupled in most eukaryotic systems (9), it might be anticipated that the newly synthesized histones would assemble into nucleosomes on the temporarily histone-free daughter strands. However, a consensus on this point has yet to be reached, with evidence presented both for the specific association of newly synthesized histones with nascent DNA (10-13), and for a random mode of histone deposition (14-16).

We have examined the histone deposition question using extremely mild experimental conditions that should not in themselves influence the system. The results suggest that, at least in the protozoan *Tetrahymena*, newly synthesized histones are not deposited on newly replicated DNA, but are deposited at random on the chromatin as a whole. This finding poses the question of how histones are transferred within the nucleus to form "new" nucleosomes, and some possible mechanisms for achieving this are considered.

With regard to the sites of replication within the nucleus, two distinct possibilities can be envisaged. First, the replication forks with their attendant proteins could be essentially free in the nucleoplasm and physically move along the chromatin strand from origin to termination. The opposite situation would be fixed points of replication within the nucleus, with the chromatin fibers passing over or through these fixed points during replication (17, 18). Such a model would be somewhat analogous to the system in bacteria, except that most evidence excludes a membrane attachment site for eukaryotic replication. Our investigation of chromatin replication in *Tetrahymena* provided some unexpected evidence in favor of an altered conformation of chromatin in the immediate area of the replication fork, which would be consistent with the involvement of fixed structural entities at the fork region.

I. Materials and Methods

Conditions for growth and double labeling of *Tetrahymena pyriformis* cells with thymidine and lysine in defined media were as described (19). Nuclei were isolated by the method of Gorovsky *et al.*

(20) with slight modifications (19) which included the presence of 0.5 mM phenylmethylsulfonyl fluoride (α-toluenesulfonyl fluoride, αTosF). Nuclei were digested with micrococcal nuclease (1–1.5 units/50 μg of DNA) for 18 hours at 0°C in 0.3 M sucrose and 25 μM $CaCl_2$ in 5 mM sodium phosphate buffer (pH 6.75). The digestions were terminated by adding EDTA to 2 mM. After pelleting undigested material, the supernatants were layered onto "low-salt" 5 to 30% sucrose gradients containing 10 mM Pipes, and 1 mM EGTA, pH 7.0 or "high-salt" 5 to 20% gradients containing 0.5 M NaCl and 2 mM EDTA, pH 7.0. All gradients contained a 60% sucrose shelf at the bottom. After centrifugation for 15 hours at 28,500 rpm (SW41 rotor), the gradients were pumped from below through an absorbance monitor, and fractions were collected and assayed for acid-insoluble radioactivity.

To determine which proteins were labeled during a lysine pulse, cells were grown without a ^{14}C label, and, after digestion and fractionation, each gradient was divided into "premonomer," monomer to multimer (fraction 25), and shelf regions; the pooled fractions were dialyzed, concentrated, and electrophoresed on 18% polyacrylamide gels containing dodecyl sulfate. After Coomassie-Blue staining, the gels were processed for fluorography.

II. Deposition of Newly Synthesized Histones

In a number of systems, newly replicated chromatin is differentially susceptible to digestion by staphylococcal nuclease (3, 4, 13, 21), and this method of discriminating between old and new chromatin in Tetrahymena pyriformis (amicronucleate strain GL) was investigated. Isolated Tetrahymena macronuclei are rapidly attacked by the nuclease, but under very mild digestion conditions with micromolar amounts of Ca^{2+}, and carrying out the reaction at 0°C, it is possible to obtain the desired ratio of mononucleosomes to polynucleosomes.

Figure 1 shows the optical density profile of the solubilized products from a typical nuclease digestion experiment. Similar absorbance profiles were obtained for all the digestions presented below, but have been omitted from the figures for the sake of clarity. Such digests from cells labeled for several generations with [^{14}C]thymidine and then pulsed for 1 minute with [^3H]thymidine show radioactivity profiles similar to those in Fig. 2A. The ^{14}C counts derived from "bulk" chromatin follow the typical pattern, but the ^3H counts from newly replicated material show small but highly reproducible differences. These differences can be seen more easily in the $^3H/^{14}C$ ratios, where

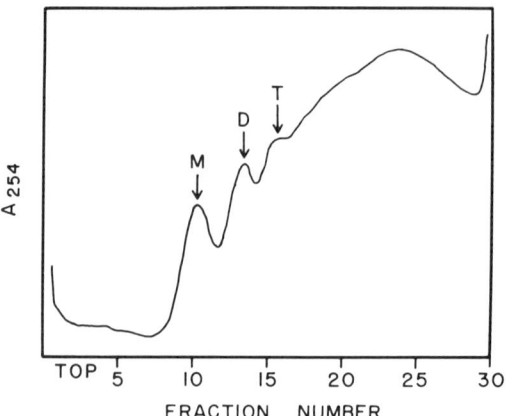

FIG. 1. Sucrose gradient profile of *Tetrahymena* chromatin after digestion with micrococcal nuclease at 0°C. Thirty A_{260} units of macronuclei were incubated for 18 hours at 0°C with 45 units of micrococcal nuclease, as detailed in Section I. The digest supernatant was layered onto a linear 8 to 30% sucrose gradient and centrifuged in a Beckman SW41 rotor for 15 hours at 4°C. The gradient was fractionated from the top and monitored at 254 nm. M, D, and T denote the position of monomeric, dimeric, and trimeric nucleosomes, respectively.

the heightened susceptibility of newly synthesized chromatin is seen as an increased ratio in the mononucleosome and premononucleosome regions. There is also a substantial enrichment for newly synthesized chromatin in the heavy sucrose shelf at the bottom of the gradient. (The rationale for including a sucrose shelf, and the interpretation of these data, are discussed below.) A 24-hour double label (open triangles, Fig. 2A), produces the expected uniform $^3H/^{14}C$ ratio across the gradient. If the 1-minute thymidine pulse is followed by a 10-minute chase with cold thymidine, the increased susceptibility to digestion of the new chromatin is reduced (Fig. 3A), but it is clear that maturation is not yet complete, as there is still a significant enrichment for new chromatin in the mononucleosome region.

This differential digestibility of newly replicated chromatin provides a suitable way to investigate the site of deposition of newly synthesized histones. If new histones associate exclusively with new DNA, then, after a mild nuclease digestion, the new histones and new DNA should migrate together, and there should be an increased susceptibility of new histone-containing chromatin. To test this, *Tetrahymena* cells were double-labeled with [^{14}C]- and [^3H]lysine, and nuclei were isolated and digested as before. Figure 4 shows that, under these conditions, there is no differential susceptibility of chromatin containing newly synthesized histones: the $^3H/^{14}C$ ratios are essentially con-

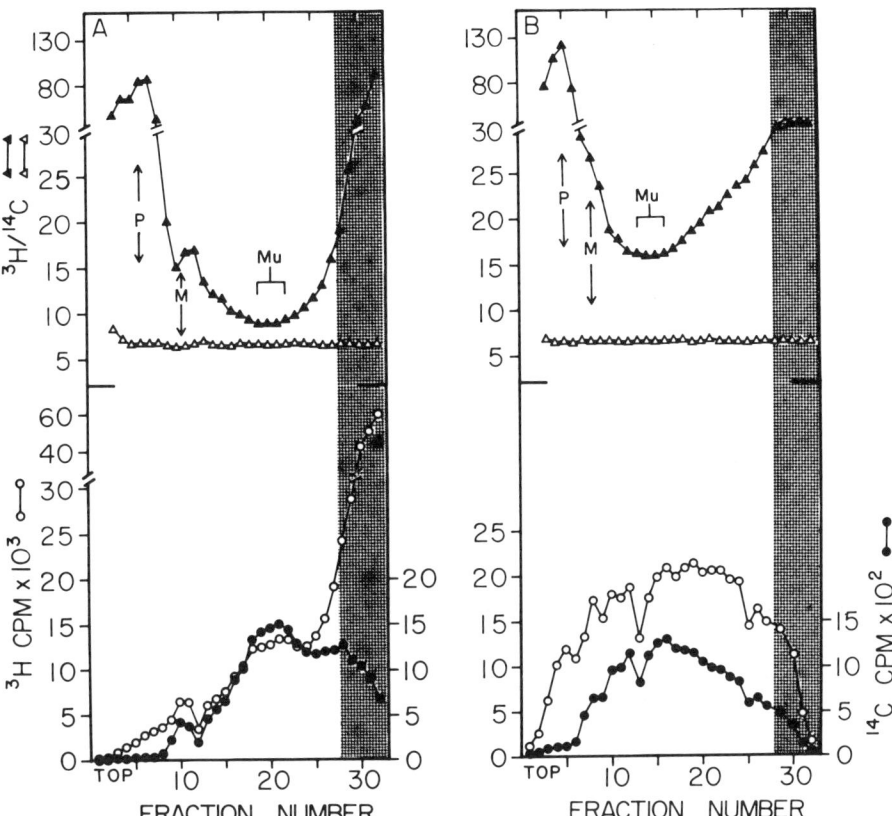

FIG. 2. Analysis of newly replicated *Tetrahymena* chromatin on sucrose gradients of low and high ionic strength. *Tetrahymena* cells were grown for 24 hours in the presence of [^{14}C]thymidine and subsequently pulse-labeled for 1 minute with [^{3}H]thymidine. Isolated nuclei were digested with micrococcal nuclease, and the solubilized chromatin was layered onto linear sucrose gradients as described in Section I. (A) Sucrose gradient containing 10 mM Pipes, 1 mM EGTA (pH 7.0). (B) Sucrose gradient containing 0.5 M NaCl, 2 mM EDTA (pH 7.0). The shaded area denotes the "shelf" region. ○——○, ^{3}H-labeled (new) chromatin; ●——●, ^{14}C-labeled (bulk) chromatin; ▲——▲, ^{3}H/^{14}C ratio of pulsed chromatin; △——△, ^{3}H/^{14}C ratio of a control gradient (not shown) containing chromatin labeled for 24 hours with both [^{3}H]thymidine and [^{14}C]thymidine. M marks the position of the monomer peak as determined from the absorbance scan (not shown). P denotes the premonomer region; Mu denotes multimer. The ^{3}H/^{14}C ratios in fraction numbers one and two have been omitted, since the very low ^{14}C counts in these fractions render the ratios unreliable.

stant, except for a slight elevation in the shelf region, which is probably from a nonhistone component (see below). In order to verify that the histones were the predominant labeled proteins under these conditions, samples from the premononucleosome, mononucleosome to

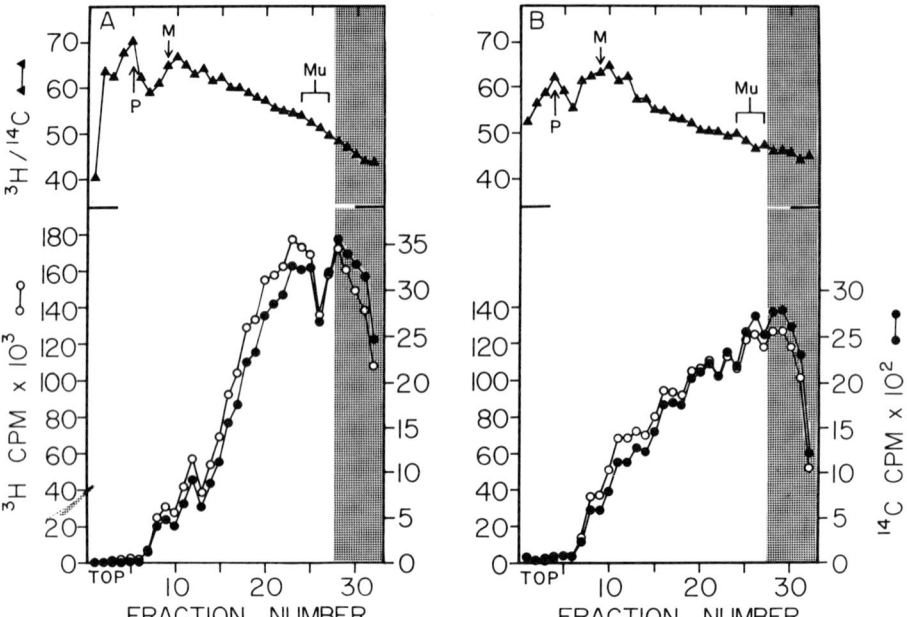

FIG. 3. Analysis of pulse-chase *Tetrahymena* chromatin on sucrose gradients of low and high ionic strength. *Tetrahymena* cells were grown for 24 hours in the presence of [^{14}C]thymidine, pulsed for 1 minute with [^{3}H]thymidine, and subsequently "chased" for 10 minutes with 1 mM nonradioactive thymidine. Isolated nuclei were digested with micrococcal nuclease, and the solubilized chromatin was layered onto linear sucrose gradients. (A) Sucrose gradient containing 10 mM Pipes, 1 mM EGTA (pH 7.0). (B) Sucrose gradient containing 0.5 M NaCl, 2 mM EDTA, (pH 7.0). The shaded area denotes the "shelf" region. ○——○, ^{3}H-labeled (pulse-chased) chromatin; ●——●, ^{14}C-labeled (bulk) chromatin; ▲——▲, ^{3}H/^{14}C ratio. M marks the position of the monomer peak as determined from the absorbance scan (not shown). P denotes the premonomer region; Mu denotes multimer.

multinucleosome, and shelf regions of a gradient prepared from [^{3}H]lysine-pulsed nuclei were electrophoresed in dodecyl sulfate, and the labeled products were identified by fluorography. As seen in Figs. 5 and 6, the core histones (which migrate as three bands in *Tetrahymena*), were the principal labeled proteins except in the shelf region (Fig. 6c), which contains several newly synthesized nonhistone proteins, thereby accounting for the elevated ^{3}H/^{14}C ratios in this part of the gradient (Fig. 4).

This result suggests that newly synthesized histones do not associate with new DNA, but instead become deposited at random within the nucleus. The possibility was considered that newly synthesized

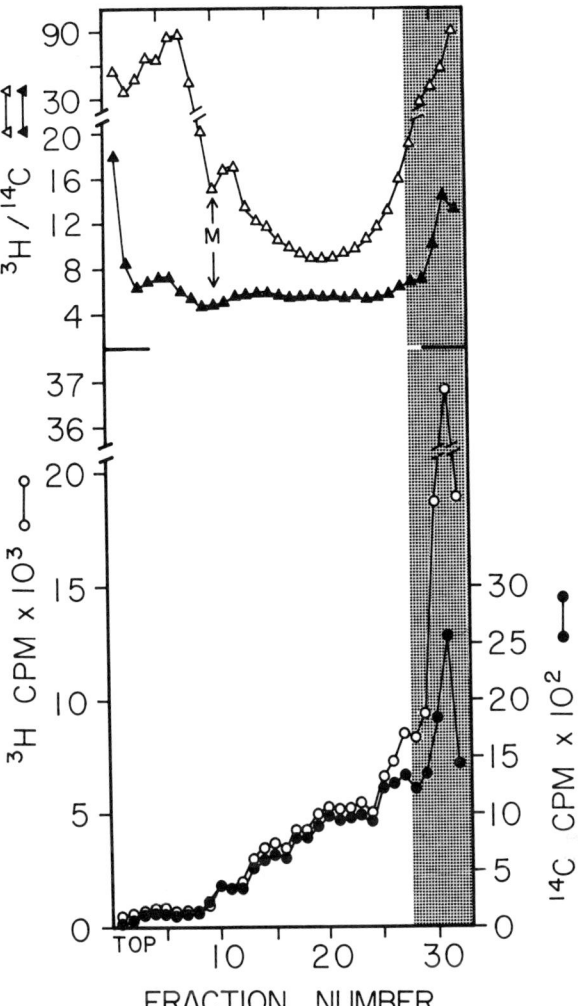

FIG. 4. Sucrose-gradient analysis of lysine-labeled chromatin. *Tetrahymena* cells were grown for 24 hours in the presence of [^{14}C]lysine and subsequently pulse-labeled for 3 minutes with [^{3}H]lysine. Isolated nuclei were digested with micrococcal nuclease, and the solubilized chromatin was layered onto linear sucrose gradients containing 10 mM Pipes, 1 mM EGTA, pH 7.0 (5 to 30% sucrose, 60% sucrose shelf), The shaded area denotes the shelf region; ○——○, ^{3}H-labeled chromatin (new histone); ●——●, ^{14}C-labeled chromatin (old histone); ▲——▲, ^{3}H/^{14}C ratio of lysine-pulsed chromatin; △——△, ^{3}H/^{14}C ratio of a control gradient (from Fig. 2A) containing chromatin labeled for 1 minute with [^{3}H]thymidine. M marks the position of the monomer peak as determined from the absorbance scan of the gradient at 254 nm (not shown).

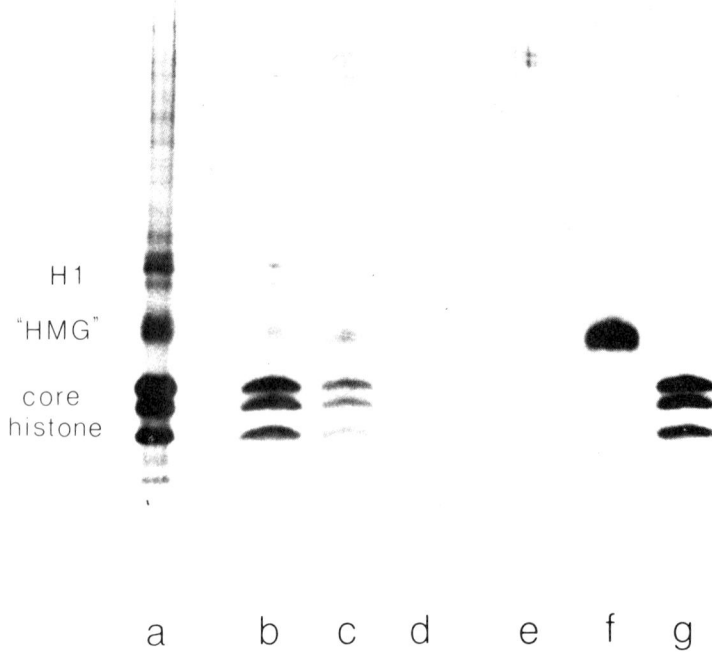

FIG. 5. Bulk proteins released during micrococcal nuclease digestion of *Tetrahymena* macronuclei. Cells were labeled for 3 minutes with [^{3}H]lysine; nuclei were isolated and digested with micrococcal nuclease, and the digest supernatant was layered onto gradients of low and high ionic strength. Each gradient was divided into three fractions as described in Sections I and II; chromatin was recovered from each fraction, and chromatin proteins were separated by acrylamide gel electrophoresis in dodecyl sulfate; the gel was stained with Coomassie Blue to show bulk proteins. a, Unlabeled control showing total nuclear proteins from undigested macronuclei; b, low-salt "middle"; c, low-salt "shelf"; d, low-salt "top"; e, high-salt "shelf"; f, high-salt "top"; g, high-salt "middle." "HMG," high-mobility group. Under these electrophoresis conditions, the four "core" histones of *Tetrahymena* migrate as three bands (M. Gorovsky, personal communication).

histones might not be deposited on new DNA immediately, but only after a lag period (perhaps due to initial deposition in the fork area, but on nonreplicated DNA). In this case, an enrichment for new histone would be expected in the mononucleosome region of a gradient in pulse-chase experiments. However, after a 10-minute chase there is still no enrichment for new histone anywhere in the gradient (Fig. 7). Thus, we are led to the conclusion that it must be "old" histones that associate with newly synthesized DNA to form "new" nucleosomes.

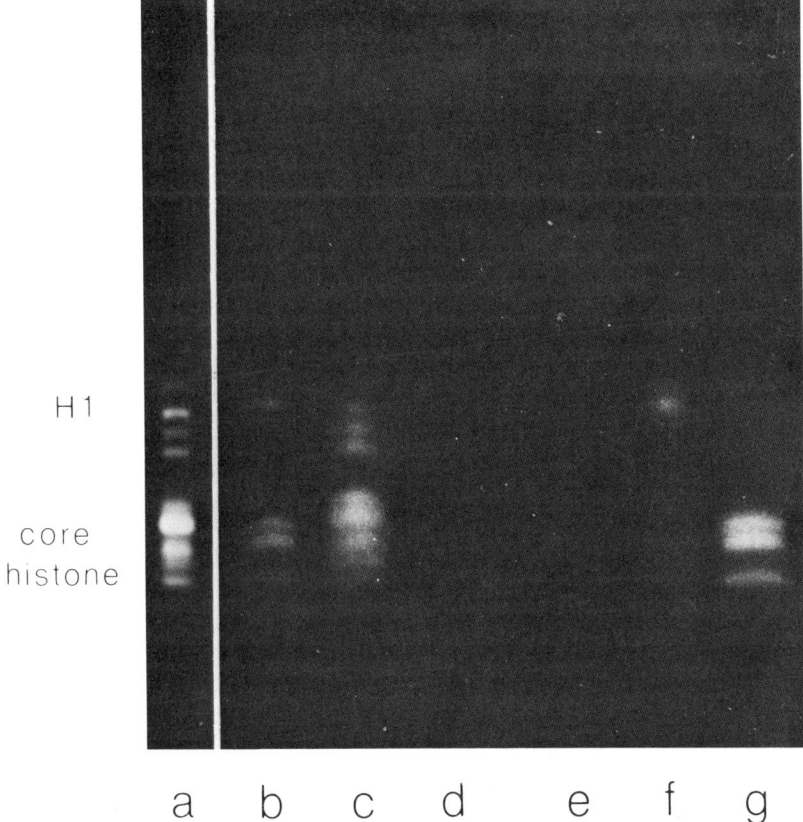

FIG. 6. Fluorograph of the stained gel in Fig. 5 (nascent proteins). Lanes b–g are the same as in the legend to Fig. 5. Lane a is the residual nuclear pellet after nuclease digestion and washing with 2% Triton X-100, 0.2 mM EDTA (as described for thymidine-labeled nuclei, in Materials and Methods).

III. Conformation of Replicating Chromatin

When the soluble staphylococcal nuclease digestion products of double-thymidine-labeled *Tetrahymena* nuclei were run on standard 5% to 20% sucrose gradients, a substantial portion of the label pelleted. Moreover, examination of the pellets showed a strong enrichment for newly replicated chromatin. We were curious as to why the newly replicated material, which was preferentially digested under our conditions (Fig. 2) was, at the same time, sedimenting as though it were in the form of large polynucleosomes. In subsequent experi-

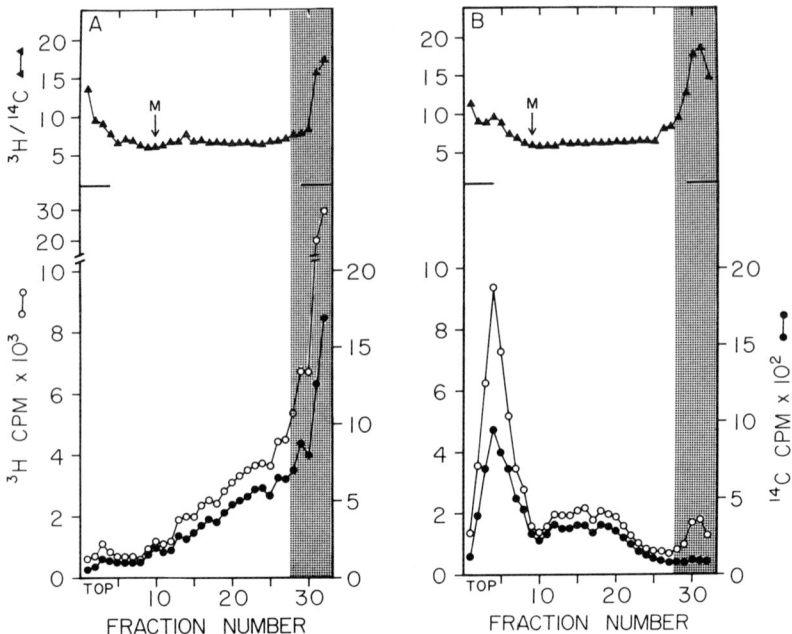

FIG. 7. Analysis of pulse-chased *Tetrahymena* chromatin on sucrose gradients of low and high ionic strength. *Tetrahymena* cells were grown for 24-hours in the presence of [^{14}C]lysine, pulsed for 5-minutes with [^{3}H]lysine, and subsequently incubated for 10 minutes in the presence of 2 mM nonradioactive lysine. Isolated nuclei were digested with micrococcal nuclease, and the solubilized chromatin was layered onto linear sucrose gradients. (A) Sucrose gradient containing 10 mM Pipes, 1 mM EGTA (pH 7.0). (B) Sucrose gradient containing 0.5 M NaCl, 2 mM EDTA (pH 7.0). The shaded area denotes the shelf region. ○——○, ^{3}H-labeled (pulse-chased) chromatin; ●——●, ^{14}C-labeled (bulk) chromatin; ▲——▲, ^{3}H/^{14}C ratio. M marks the position of the monomer peak as determined by the absorbance scan of the gradient at 254 nm (not shown).

ments, a 60% sucrose shelf, which trapped most to this rapidly sedimenting material, was placed at the bottom of the gradients. (The "shelf" region is indicated by the shaded areas in the gradient profile figures.) From numerous experiments designed to explore the nature of the "shelf" material, the following conclusions were reached.

1. After a 1-minute [^{3}H]thymidine pulse, the shelf material is the component most highly enriched in newly synthesized DNA.
2. The newly synthesized shelf material for the most part does not consist of large polynucleosomes, since, if the digestion products are run on a gradient containing 0.5 M NaCl, the shelf enrichment is much reduced, the radioactivity formerly in the

shelf redistributing throughout the gradient (Fig. 2B). This conclusion is further supported by the finding that the shelf material contains DNA lengths of all sizes, from mononucleosome up (data not shown).
3. After a 1-minute pulse with [^{3}H]thymidine followed by a 10-minute chase, there is no longer any enrichment for new DNA in the shelf (Fig. 3). This finding indicates that the shelf phenomenon is much shorter lived than the preferential digestion to mononucleosomes, which is still apparent after a 10-minute chase (Fig. 3).

We therefore conclude that newly replicated chromatin has two

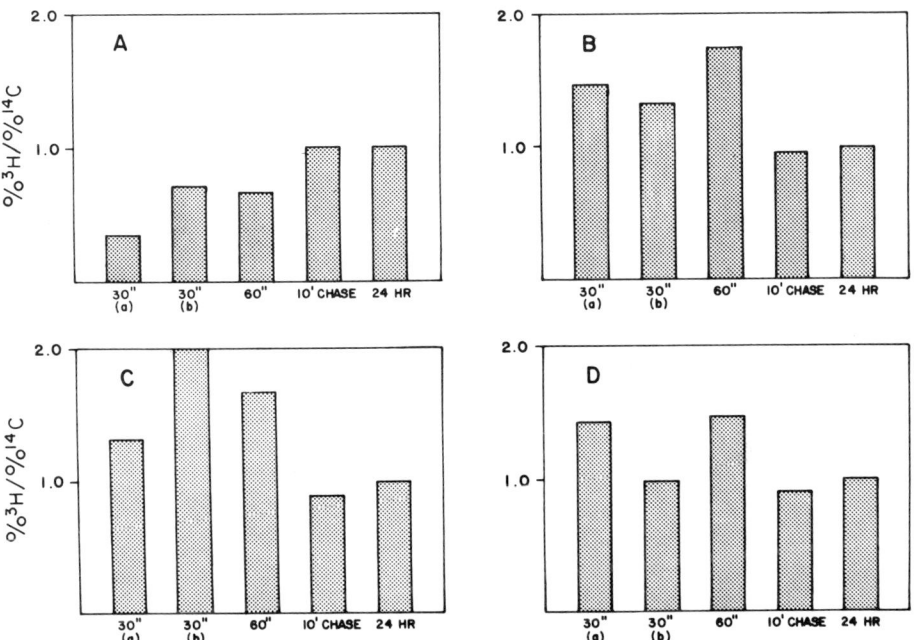

FIG. 8. Ratio of percentage of ^{3}H released divided by percentage of ^{14}C released during nuclear fractionation. Radioactivity released at each step has been expressed as a percentage of the total recovered radioactivity (including the final pellet). For all experiments, *Tetrahymena* chromatin was uniformly labeled for 24 hours with [^{14}C]thymidine. Labeling times shown designate the length of the [^{3}H]thymidine incubation. The pulse-chased sample was incubated for 10 minutes in 1 mM nonradioactive thymidine following a 1-minute pulse with [^{3}H]thymidine. The 30-second (a) sample was digested until approximately 37% of the bulk [^{14}C]chromatin was solubilized. All other samples were digested until approximately 60% of the bulk [^{14}C]chromatin was solubilized. (A) Micrococcal nuclease digest supernatant; (B) 2% Triton X-100, 0.2 mM EDTA supernatant; (C) 0.5 M NaCl, 2.0 mM EDTA supernatant; (D) final solubilized pellet.

distinct properties, a preferential digestibility to mononucleosomes and subnucleosomal fragments and a much more short-lived tendency to aggregate into fast-sedimenting material. Newly synthesized and deposited histones show neither of these properties. Further information concerning the aggregation phenomenon was provided by examining the release of labeled material from nuclei after digestion. Nuclei were labeled, isolated, and digested as before, after which they were subjected to a sequential series of extractions. After each extraction, the material released was collected and the acid-precipitable $^3H/^{14}C$ ratio was determined. After the initial digest, it was found that the released chromatin was depleted in newly synthesized material compared to the starting nuclei, but this depletion was lost after a 10-minute chase or after longer-term labeling (Fig. 8A). The newly replicated material retained in the nucleus had, however, been cut by the nuclease, since it could be preferentially released by treatment with Triton X-100, or by 0.5 M NaCl (Fig. 8B,C). We believe that the retention of newly synthesized chromatin within the digested nucleus is closely related to the rapid sedimentation of new chromatin; both result from an aggregation type of phenomenon that, in the first instance, impedes release from the nucleus, and, in the case of material that is released, causes its aberrant sedimentation. As discussed in Section IV, this behavior of new chromatin is very similar to the observed association between newly replicated DNA and the residual nuclear "matrix," suggesting that new chromatin may transiently assume an altered conformation or composition, or both, that tends to retard its release after digestion.

IV. Discussion

The evidence presented above clearly suggests a random deposition of newly synthesized histones onto chromatin. This is in agreement with several other studies that have reached the same conclusion (14–16). However, still other results have seemingly suggested the opposite conclusion, that new histones are deposited on new DNA (11–13). Some of these latter studies can be criticized on the grounds of the harsh treatments used, which could affect the results, and a complete discussion of these aspects has been represented elsewhere (19). One of the more perplexing discrepancies is that between the results presented here, and a recent study of histone deposition in *Drosophila* culture cells in which essentially the same strategy of distinguishing between old and new chromatin was used (13). The *Drosophila* cells showed a similar heightened digestibility of newly synthesized DNA, but in this case, the lysine- or methionine-labeled histones followed the DNA pattern, at least as far as could be determined from the fluorographic technique employed. We are at present unable to offer any satisfactory explanation for this discrepancy: the two obvious possibilities, that *Tetrahymena* and *Drosophila* differ in replicative mechanism, or that small differences in salt concentration, buffer type, etc., could cause a dramatic rearrangement of deposited histones, are both unattractive. We are currently examining histone deposition in other systems in an attempt to provide further insight into the question.

If we assume, for the sake of argument, that random deposition of newly synthesized histones does occur, then the question arises as to how "new" nucleosomes are assembled on the histone-deficient daughter strand. There seem to be two main possibilities.

1. Sliding of histone octamers from unreplicated regions to the nucleosome-deficient chromatin. This may not be energetically out of the question in the ionic milieu of the nucleus (22, 23), but the topological problems associated with large-scale sliding of octamers over spans of several replicons make such a mechanism unattractive.

2. "Jumping" of histone octamers from old chromatin to the fork region. While *in vitro* studies show that the jumping of octamers from one piece of DNA to another requires salt concentrations of about 0.8 M (24), more recent work has indicated that nucleosome assembly (and presumably disassembly) may be facilitated at physiological ionic strengths by proteinaceous "assembly factors" (25, 26) or by polyanions that effectively compete with DNA (27). Certain acidic nonhistone chromosomal proteins may have these same properties. Thus the transfer of octamers within chromatin may be energetically feasible at physiological ionic strengths. If such a mechanism were present, it would be expected that free DNA added to a nucleus in S phase would be able to compete with the nuclear chromatin for histones. So far as we are aware, no evidence for this or any other sort of labile histone pool within the nucleus has been obtained. However, if, as discussed below, there are specific replication sites within the nucleus, then the components necessary for octamer transfer might be similarly localized and require carefully designed experiments for their detection.

A final point that may be pertinent to the question of octamer transfer during replication is the ability for nucleosomes *in vitro* to accept extra octamers (28–30). Although these nucleosome-octamer complexes appear to require an ionic strength of about 0.5 M for stability, this value could be lower *in vivo* owing to the stabilizing influence of other nuclear components, especially the acidic nonhistone proteins

(27). Octamer migration might provide a mechanism for newly synthesized histones entering the nucleus to equilibrate rapidly with existing chromatin, so that the octamers associating with new DNA would be a random sample of new and old.

Conformation of Replicating Chromatin. The transient "aggregation" and preferential retention of newly synthesized chromatin indicates that it differs in some way from bulk chromatin. This phenomenon is not confined to *Tetrahymena,* but has also been observed in HeLa cells (Annunziato, unpublished observation). These observations could have a trivial explanation in terms of nascent chromatin being more "sticky" than bulk chromatin, and while this possibility cannot be completely excluded, reconstitution experiments performed on a related system (*18*) failed to support such an explanation, and other considerations (*19*) also make the trivial case unlikely. Another possibility is that chromatin in the immediate vicinity of the replication fork has an altered conformation, perhaps being bound in a large "replication complex" that contains the necessary enzymes as well as any structural components needed for the spatial organization of replication. In line with this suggestion, parallels can be drawn between our observations and those of others that suggest an association between nascent DNA and the nuclear "matrix." The matrix has been defined as the residual proteinaceous material left after removal of most of the membrane, DNA, and salt-soluble protein from the nucleus (*18, 31*). Although work in this area needs to be carefully assessed because of the possibility of a nonspecific precipitation of nonhistone chromosomal proteins during the preparation of nuclear matrices, it is clear that the residue exhibits the same structural features in thin sections as the original chromatin (*31*). More recently, evidence has been presented in support of specific replication centers within the nucleus, anchored on the matrix itself (*18, 32*), and such a suggestion is certainly consistent with our finding that nascent chromatin is associated with large nuclear complexes. The idea of fixed replication points for eukaryotes as well as prokaryotes is clearly becoming more attractive and is deserving of further study.

Acknowledgments

Supported in part by the U. S. Public Health Service (RR07048 to the University of Massachusetts and GM 25305 to C. L. F. W.).

References

1. R. L. Seale, *in* "The Cell Nucleus" (H. Busch, ed.), Vol. 4, p. 155. Academic Press, New York, 1978.
2. C. Wu, P. M. Bingham, K. J. Livak, R. Holmgren and S. C. R. Elgin, *Cell* **16**, 797 (1979).

3. C. E. Hildebrand and R. A. Walters, *BBRC* **73**, 157 (1976).
4. A. Levy and K. M. Jakob, *Cell* **14**, 259 (1978).
5. S. L. McKnight, M. Bustin and D. L. Miller, *CSHSQB* **42**, 741 (1977).
6. R. L. Seale, *Cell* **9**, 423 (1976).
7. H. Weintraub, *Cell* **9**, 419 (1976).
8. D. Riley and H. Weintraub, *PNAS* **76**, 328 (1979).
9. H. J. Edenberg and J. A. Huberman, *Annu. Rev. Genet.* **9**, 245 (1975).
10. C. Cremisi, A. Chestier and M. Yaniv, *Cell* **12**, 947 (1977).
11. R. Tsanev and G. Russev, *EJB* **43**, 257 (1974).
12. T. Senshu, M. Fukada and M. Ohashi, *J. Biochem.* **84**, 985 (1978).
13. A. Worcel, S. Han and M. L. Wong, *Cell* **15**, 969 (1978).
14. R. L. Seale, *PNAS* **73**, 2270 (1976).
15. R. Hancock, *PNAS* **75**, 2130 (1978).
16. V. Jackson, D. Granner and R. Chalkley, *PNAS* **73**, 2266 (1976).
17. C. W. Dingman, *J. Theoret. Biol.* **43**, 187 (1974).
18. D. M. Pardoll, B. Vogelstein and D. S. Coffey, *Cell* **19**, 527 (1980).
19. A. T. Annunziato, Ph.D Thesis, Univ. of Massachusetts (1979).
20. M. A. Gorovsky, M. C. Yao, J. B. Keevert and G. L. Pleger, *in* "Methods in Cell Biology" (D. M. Prescott, ed.), Vol. 9, p. 311. Academic Press, New York, 1975.
21. K. H. Klempnauer, E. Fanning, B. Otto, and R. Knippers, *JMB* **136**, 359 (1980).
22. M. J. Smerdon and M. W. Lieberman, *PNAS* **75**, 4238 (1978).
23. P. Beard, *Cell* **15**, 955 (1978).
24. J. E. Germond, M. Bellard, P. Oudet, and P. Chambon, *NARes* **3**, 3173 (1976).
25. R. A. Laskey, B. M. Hondo, A. D. Mills and J. T. Finch, *Nature* **275**, 416 (1978).
26. T. Nelson, T. S. Hsieh and D. Brutlag, *PNAS* **76**, 5510 (1979).
27. A. Stein, J. P. Whitlock and M. Bina, *PNAS* **76**, 5000 (1979).
28. A. Stein, *JMB* **130**, 103 (1979).
29. G. Voordouw and H. Eisenberg, *Nature* **273**, 446 (1978).
30. C. L. F. Woodcock, L.-L. Y. Frado and J. S. Wall, *PNAS* **77**, 4818 (1980).
31. R. Berezney, *in* "The Cell Nucleus" (H. Busch, ed.), Vol. 7, p. 413. Academic Press, New York, 1978.
32. P. A. Dijkwel, H. F. Mullenders and F. Wanka, *NARes* **6**, 219 (1979).

Role of Chromatin Structure, Histone Acetylation, and the Primary Sequence of DNA in the Expression of SV40 and Polyoma in Normal or Teratocarcinoma Cells

G. Moyne
M. Katinka
S. Saragosti
A. Chestier and
M. Yaniv

Department of Molecular Biology
Pasteur Institute
Paris, France

The DNA tumor viruses of the papova group, mainly SV40 and polyoma, have been extensively studied during the last decade. The complete nucleotide sequence of three of them, SV40 (*1*, *2*), polyoma (*3*), and BK (*4*, *5*) are established. The circular double-stranded DNA molecule of about 5200 base-pairs can be divided into two regions (Fig. 1). Half of the viral genome codes for the early functions (T antigens)—proteins that are expressed before the onset of DNA replication. These proteins participate in the control of viral DNA replication and of viral RNA transcription and in the establishment and maintenance of cellular transformation (*6*). The opposite strand of the second half of the DNA molecule codes for the late functions—the virion structural proteins VP1, VP2, and VP3. Upon the onset of viral DNA replication, late RNA is being synthesized without cessation of early RNA synthesis. The rate of synthesis and the concentration of late mRNA in the infected permissive cell exceed by roughly 10- to 20-fold the amount of early mRNA (*7*).

In nonpermissive cells (e.g., hamster or rat), the penetration of the viral DNA is followed by the expression of early functions; however, no DNA replication and no late proteins can be detected. In some of these cells, the viral DNA integrates in the cellular genome and continues to express early proteins, and the cell becomes transformed (*6*).

Early work on the internal proteins of the virions indicated that the

Progress in Nucleic Acid Research
and Molecular Biology, Vol. 26

ISBN 0-12-540026-8

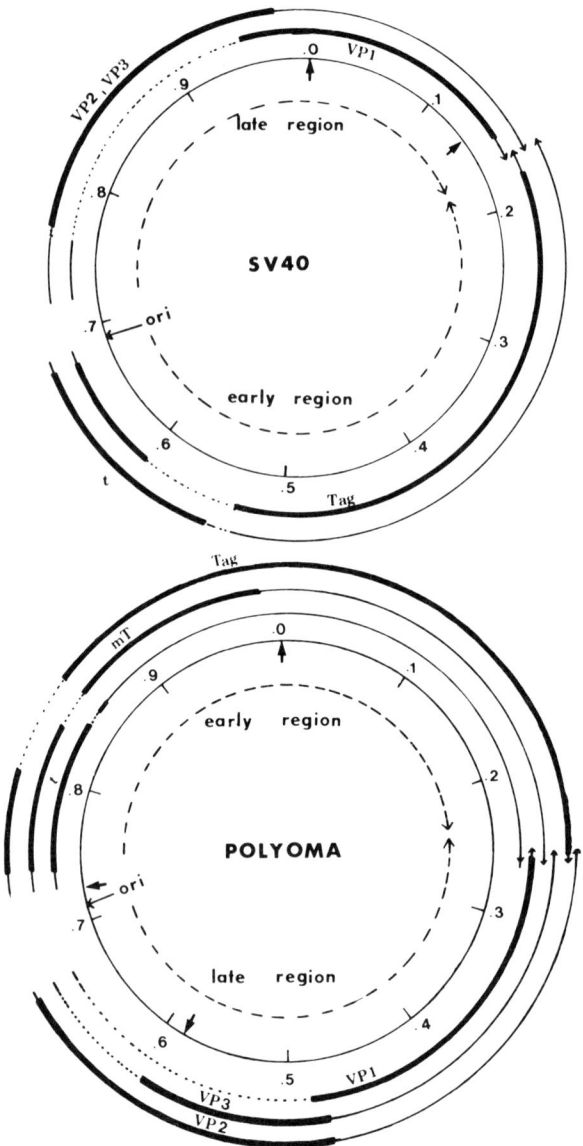

FIG. 1. The structure of SV40 and polyoma genomes. The physical map is divided into 1.0 map units with the *Eco*RI site at 0.0 map units. The arrows (↑) indicate the *Bam*HI cleavage site for SV40 (0.14) and the *Bam*HI cleavage site for polyoma (0.59) and the *Bgl*I site for polyoma (0.72). *Ori* designates the site of the origin of bidirectional DNA replication. The structure of the viral mRNA given in the figure includes coding regions (━━), parts of mRNA not coding for proteins (———), and sequences that are removed by splicing (---); t is small T antigen, mT is middle T antigen, and Tag is large T antigen.

viral DNA is associated with the four core histones H2A, H2B, H3, and H4, and that these histones are of cellular origin (8–10). Following the discovery by electron microscopy of the beaded structure of the cellular chromatin (11), analysis of the intranuclear SV40 DNA–protein complex (12) revealed a circular, bead-on-a-string structure, similar to the linear array of beads seen in cellular chromatin. Further analysis by electron microscopy and enzymic digestions showed that the DNA of all the members of the papova group (SV40, polyoma, or papilloma) have a chromatin-like structure either in the infected nuclei or inside the virion (13–15). Chromatin reconstitution experiments (16) show that the superhelicity of the circular viral DNA is a direct consequence of the formation of nucleosomes by the association of the viral DNA with the four core histones.

The most plausible scheme for the structure of the nucleosome is the wrapping of the DNA externally around the protein core formed by two molecules of each of the four histones (17). The scanning transmission electron micrograph of an SV40 minichromosome shown in Fig. 2 clearly demonstrates the presence of heavily stained DNA on the outside of the nucleosome. The length of the DNA in the nucleosome measured on such images is 163 base-pairs; roughly two turns of double-stranded DNA have to surround the central histone core to give a particle of 90 Å diameter. The fact that such an arrangement generates only 24, not 48, superhelical turns after deproteinization (18) is still unexplained.

The viral chromatin leaks out of infected nuclei under hypotonic conditions or in the presence of mild nonionic detergents like Triton X-100. At ionic strengths below 0.2 M NaCl, this viral chromatin contains histone H1 in addition to the four core histones already mentioned (19). However, recent experiments show that rapid exchange of histone H1 can occur between the chromatin of noninfected nuclei and viral minichromosomes extracted from infected nuclei (20). Such an exchange, but not a bona fide association, could explain the presence of H1 in the viral chromatin. The isolated viral minichromosomes contain replicative intermediate complexes that can be elongated *in vitro* (21). Similar preparations also contain transcription complexes that permit the elongation of RNA chains *in vitro* (22, 23).

Study of the replication of the SV40 chromatin *in vivo* or *in vitro* has permitted the elucidation of several steps in the process of chromosome replication in eukaryotic cells. The leading DNA strand is synthesized continuously (24–26), whereas the lagging strand is synthesized in short pieces of length 150–250 bases with an RNA primer present at the 5' end. Concomitantly with the replication of the DNA,

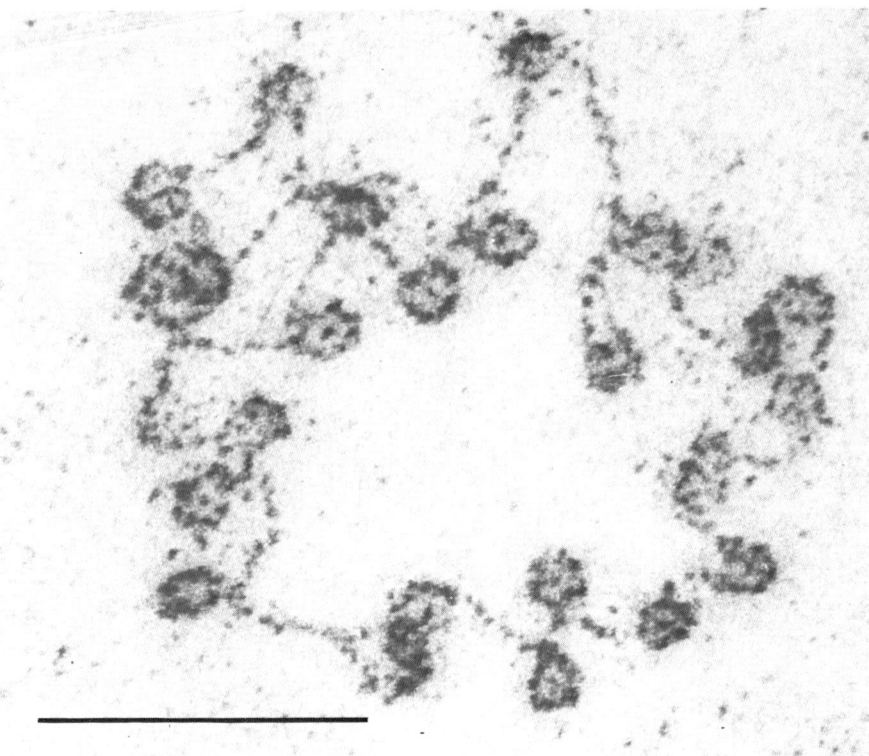

FIG. 2. Scanning transmission electron microscope (STEM) micrograph of spread SV40 chromatin. The virus chromatin prepared as described in Fig. 3 was observed in the high-resolution STEM of the European Molecular Biology Laboratory. Bar = 50 nm.

the "old" histone octamers segregate with the strand that is copied continuously (27), whereas the newly synthesized histones associate with the other DNA duplex (28; however, see also Seale, and Annunziato and Woodcock in this volume). This association is sequential; H3 and H4 associate first, followed by the assembly of H2A and H2B (20).

The extensive knowledge about the viral genome and its minichromosomes permits a more detailed analysis of the nature of the regulating elements governing the expression of the viral genome at the level of the DNA primary sequence and at the level of the chromatin. The following sections describe some of our recent contributions to the understanding of these matters.

I. Absence of Nucleosomes in a Fraction of SV40 Chromatin between the Origin of Replication and the Region Coding for the Late Leader RNA

Recent studies (29–31) reveal the presence of a nuclease-sensitive region on SV40 chromatin in proximity to the origin of replication. Such a region can be associated either with specific nonhistone proteins or with modified histones, or it can alternatively be deficient in histones. In an attempt to analyze the structure of this region, we prepared SV40 minichromosomes by different techniques, purified them by sedimentation, and adsorbed them to activated carbon grids (32). High-resolution, dark-field electron microscopy was employed to examine these samples. As shown in Fig. 3, whereas in most of the minichromosomes the nucleosomes are uniformly spread around the circular structure, 15–20% of the complexes revealed a gapped structure where a DNA segment appeared naked.

The thickness of the DNA in the gap was indistinguishable from that of naked DNA spread on the same grid. It is probable that the DNA in the gap is free from proteins, at least after the isolation of the minichromosomes on a sucrose gradient. To test whether the gap can be localized on the circular map of SV40, we digested the viral chromatin with several endonucleases that cleave SV40 DNA at unique but distinct sites. As shown in Fig. 3e, the enzyme $BglI$ that cleaves SV40 DNA at the origin of replication (0.67) generates linear molecules with a free DNA tail. The opposite end has no detectable free DNA beyond the last nucleosome. Digestion with $BamHI$, which cleaves SV40 DNA at 0.14 (Fig. 3d), and with $EcoRI$, which cleaves at 0.0, confirmed that the gap is located on the "late" side of the origin of replication. The mean length of the gap was 250 ± 13 base-pairs (33). Counting the number of nucleosomes in minichromosomes with or without the gap gave values of 23.5 ± 0.4 and 24 ± 0.17, respectively (these values are not significantly different, as shown by the application of Student's t test). Thus, we believe that the gap is not caused by the loss of one nucleosome. To check whether a specific conformation can be detected in the same region under more physiological conditions, we prepared infected nuclei under conditions that preserve the compactness of the chromatin (34) and digested them with DNase I. As shown in Fig. 4, a fraction of the viral DNA is converted into nicked circular (F-II) and linear (F-III) DNA. To locate the double-stranded cleavage site, the DNA preparation or the isolated linear molecules were further cleaved with $EcoRI$ or $BamHI$ restriction endonucleases. After $EcoRI$ treatment, we observed subgenomic frag-

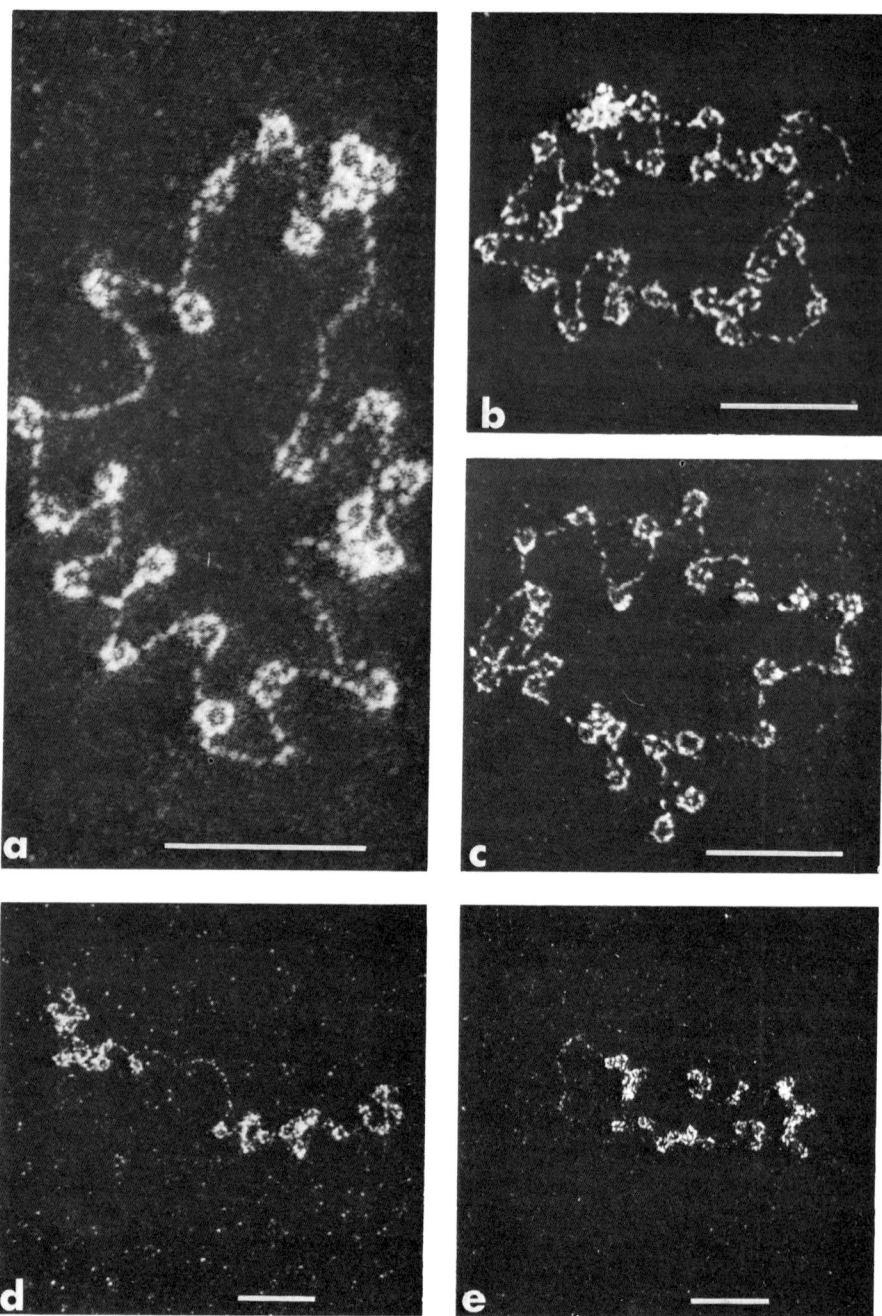

FIG. 3. Dark-field electron microscopy of spread SV40 chromatin. Virus chromatin was extracted from a nuclear pellet obtained from infected monkey kidney cultivated cells. Upon treatment in an isotonic medium in the presence of Triton X-100, the virus chromatin leaked out. It was purified on a 5 to 30% sucrose gradient. The minichromo-

ments with a length of 0.67 (SV40 unit length), 0.33 and 0.30 to 0.26 (a continuum), showing that in a fraction of the DNA the initial double-strand cleavage has occurred in the nuclei at the origin of replication and at certain specific sites between 0.67 and 0.74 of the physical map of SV40 (Fig. 5). The absence of cleavage between 0.68 and 0.69 on the viral DNA suggests that this site may be occupied *in vivo* by a nucleosome or other proteins. In fact, in several *Bgl*I linear minichromosomes, we observed by electron microscopy an unknown particle at the end of the free DNA segment (*33*).

The finding that the gapped region is asymmetric relative to the origin of replication, together with its presence throughout the late phase of infection (24–44 hours), tends to exclude its participation in the control of replication or in encapsidation. It is tempting to attribute this specific conformation of the viral chromatin to an element that participates in the control of transcription. In fact the major cap for late SV40 mRNA was mapped at position 0.72, a site that is exposed to DNase I. In addition, the following section provides evidence that the same region may participate in the control of early transcription.

To determine whether the specific configuration of the chromatin can be recognized by an RNA polymerase, we undertook *in vitro* transcription studies in collaboration with E. Jakobovits and Y. Aloni from the Weizmann Institute (*35*). *Escherichia coli* RNA polymerase recognizes a strong initiation site on *naked SV40 DNA* at position 0.17 of the map. The DNA sequence in this region is very similar to the canonical sequence of a prokaryotic promoter. The RNA transcribed from this site in an anticlockwise direction is homologous to the early viral strand. When we added *E. coli* RNA polymerase and ribonucleoside triphosphate to the SV40 minichromosome and hybridized the product to a blot of SV40 DNA restriction fragments, a different distribution emerged. After a brief synthesis, the major product came from the region between 0.67 and 0.76. Longer incubations increased the amount of hybridization to an adjacent fragment (0.76–1.0), and showed an increase in incorporation due to initiation at 0.17. Hybridization of a short synthesis product to a blot of separated stands

---

somes were diluted in twice-distilled water and spread on positively changed carbon-coated grids. The molecules were positively stained with uranyl acetate, which binds to the phosphate groups of the DNA. Thus the inner proteic part of the nucleosome appears unstained. The dark field was obtained by the beam-tilt method. Bar = 50 nm. (a) Example of a minichromosome displaying 24 nucleosomes and a naked region. (b, c) Molecules displaying a regular nucleosome spacing with no visible gap. (d) Minichromosome from a preparation digested with *Bam*HI; the naked region is located centrally. (e) Minichromosome from a preparation digested with *Bgl*I. Distal location of the gap.

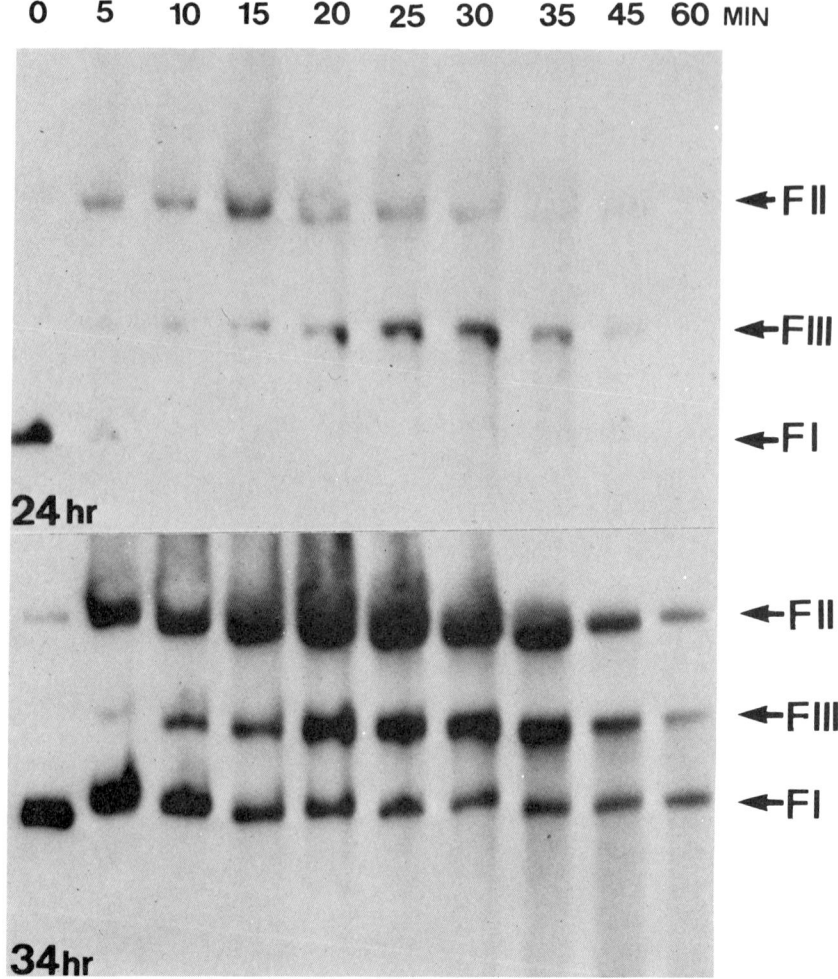

FIG. 4. Electrophoretic analysis of SV40 DNA after DNase I digestion of nuclei 24 hours and 34 hours after infection. Nuclei isolated from SV40-infected monkey cells (MA line) were incubated with DNase I (40 μg per 10^8 nuclei in 1 ml); aliquots were removed at various times, and the total DNA was isolated. The DNA samples were analyzed on 1% agarose gels and transferred to DBM paper; the viral DNA was detected by hybridization with nick-translated SV40 DNA (33). F-I, F-II, and F-III show the positions of superhelical, nicked circular, and linear SV40 DNA, respectively.

showed that the major synthesis in the region close to the origin comes from the late strand. Previous mapping of the minor *E. coli* binding sites on naked SV40 DNA did not detect any specific sites close to the origin of replication (36). The presence of naked DNA in

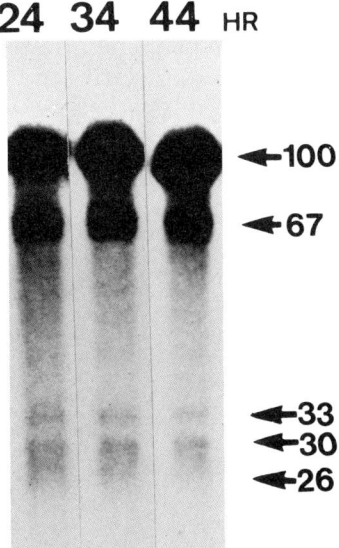

Fig. 5. Size of viral DNA fragments produced after digestion with DNase I followed by EcoRI cleavage. Aliquots of DNA from nuclei digested with DNase I for 30 minutes at 24, 34, and 44 hours after infection (Fig. 4) were further digested with EcoRI. Both supercoiled (F-I) and nicked circular DNA (F-II) give unit length linears (100%). The viral DNA previously cleaved around the origin give fragments of 67% and 33% of the total SV40 genome. Electrophoresis and detection of DNA after blotting was carried out as described in the legend to Fig. 4.

the gapped region or a more complex structure found in this region may explain the specificity of initiation *in vitro* observed by Jakobovits *et al.* (35).

II. Sequence Rearrangements in Polyoma Mutants That Grow in Embryonal Carcinoma Cells

Another approach in our studies of the regulation of gene expression of papova viruses led us to analyze the sequence of polyoma mutants in the region between the origin of replication and the beginning of the late mRNA, a region homologous to the gapped region in SV40 (Fig. 1). The natural host for polyoma virus is the mouse. Primary cells and many established mouse cell lines are permissive to this virus—permitting the expression of early and late functions and production of virions. One of the outstanding examples of mouse cells refractory to the infection by polyoma are embryonal carcinoma (EC) cells isolated from mouse teratomas (37). These cells resemble the inner mass cells of early blastocysts. They can undergo differentiation either *in vivo* after injection into mice or *in vitro*. The differentiated cells obtained under these conditions become permissive to the virus. Various lines of evidence prove that the viral DNA reaches the nuclei of the embryonal carcinoma cells, but does not express its early functions. The fusion of these infected cells with nonpermissive hamster cells (BHK21) permitted the production of virus in the heterokaryons (38). The hamster cells permit the synthesis of the early viral proteins (T antigens, Tag), and the mouse embryonal carcinoma cells supply the cellular function(s) required for viral DNA replication and late RNA and protein synthesis. Both these latter processes do not occur in the hamster cell by itself.

Polyoma mutants that overcome the block in the PCC4 aza line of embryonal carcinoma cells have been isolated (39). The mutants and the wild type are indistinguishable as to their growth in differentiated mouse cell lines. Molecular cloning in *E. coli* of the mutant DNA permitted the isolation of a unique DNA species that gave rise to virus that grew on the PCC4 cells. Restriction enzyme analysis indicated that the difference between the mutant and the wild-type DNA is located on the late side of the origin. The nucleotide sequence of the wild-type and two mutant DNAs is given in Fig. 6, and a schematic representation of the sequence rearrangements is given in Fig. 7. A deletion of 62 base-pairs in Py EC 204 and of 31 base-pairs in Py EC 97 is replaced by a duplication of a DNA fragment of 83 and 74 base-pairs, respectively, originating from a region found downstream in the noncoding late region.

This sequence rearrangement by itself now permits the synthesis of early RNA and proteins (Tag) in the infected PCC4 cells. The right-hand junction of the deletion duplication is identical in both mutants whereas the left-hand junction is different. The duplicated region carries some unique features that may explain its function: it contains an oligonucleotide very rich in A and T (position 201–214), which is bound to some of the heterogeneous late-capped T1 oligonucleotides (40). The leader-to-leader splice (41), probably occurring at position 216, is also included in the duplicated region. The sequence analogous to the canonical Hogness box for late transcription is found in position 140, included in the duplication of Py 204 but not in Py 97. The residues 106 to 121 can be organized in a very (G + C)-rich stem; this structure is conserved after the insertion due to the (C + G)-homology between the deleted and inserted segments. An (A + T)-rich stem and loop structure located in the wild type before the (G + C)-rich palindrome (residues 82–100) is replaced by another (A + T)-rich stem and

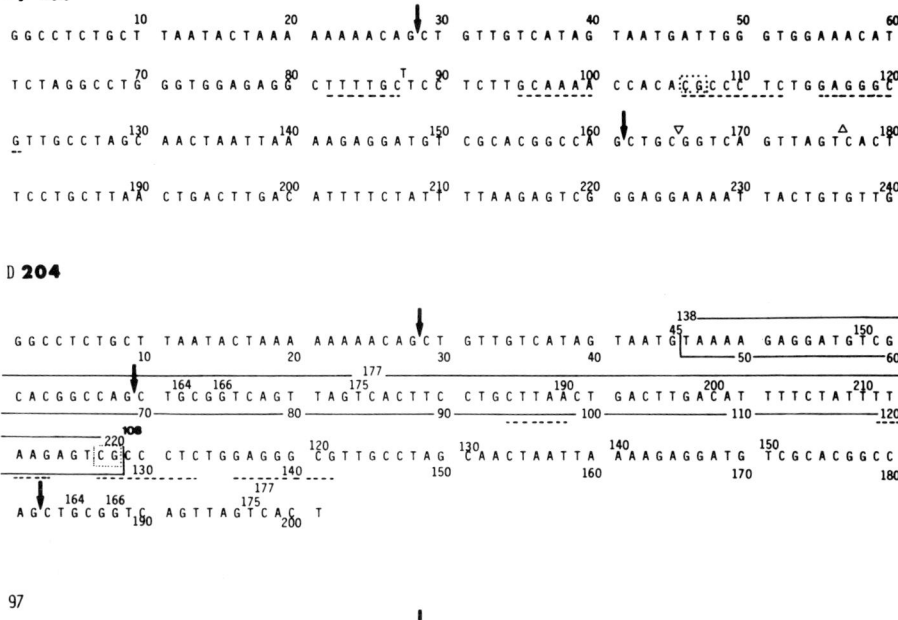

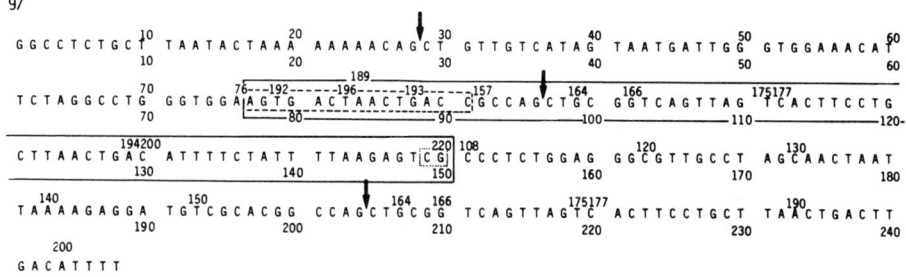

FIG. 6. The nucleotide sequence of wild-type and mutant polyoma DNA in the beginning of the late region. The sequence of the DNA strand homologous to the late coding RNA is given. The 5' end corresponds to the cleavage site of HpaII between fragments 5 and 3 (3). The numbers above the sequence are those corresponding to the wild-type A2 strain sequence (3). The numbers underneath assign each mutant running numbers. The duplicated sequences inserted into the deleted regions are enclosed with a solid line. The dashed line encloses a somewhat more complex duplication in the beginning of Py 97. The dotted line framing the C-G outlines the possible short homology between the two fragments that recombined. The sequences underlined with dashed lines indicate palindromes that may form stem and loop structures. The arrows indicate the position of the PvuII cleavage sites.

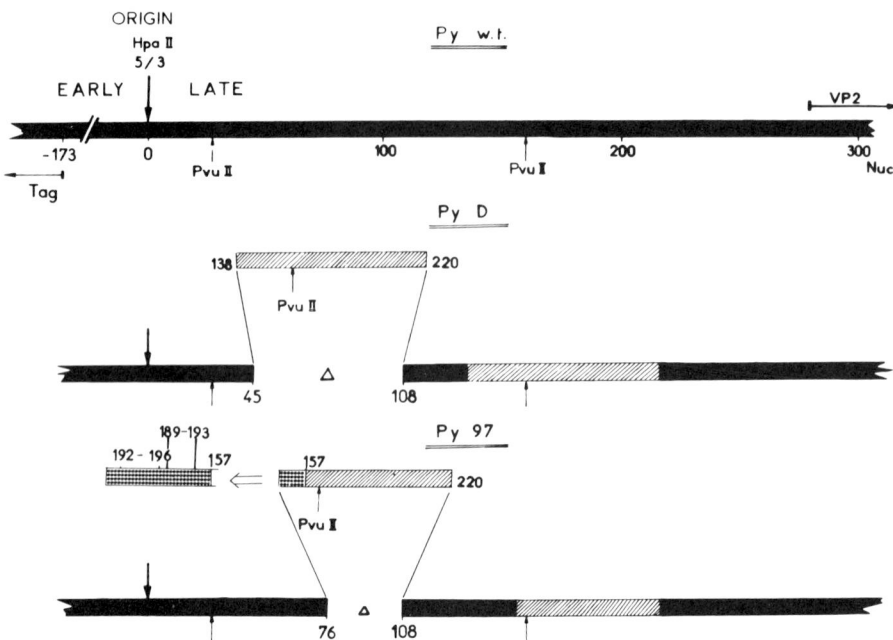

FIG. 7. Rearrangements in the late region of polyoma mutants. The figure at the top gives the position of the initiation sites for early proteins (Tag) at −173 from the origin of replication (the HpaII 5/3 junction) and for VP2 at 290. Below: The position of the deletions and their boundaries and of the duplicated region (marked in hatched areas).

loop (residues 186–215 in the duplicated region). This combination of an (A + T)-rich region next to a (G + C)-rich region is found frequently in the recognition site for RNA polymerase II, where the (A + T)-rich sequence resembles the canonical Hogness box. The sequence of the complementary strand (identical to early RNA) between positions 144 and 138 (TTAAAATA) may be the new recognition site for the RNA polymerase in the mutant virus.

These findings suggest that under certain physiological conditions, the nontranslated late region can be the initiation site for both late and early RNA. In fact, after deletion of a *BglI* site (82 nucleotides from the origin in the early region), Kamen (personal communication) observed the synthesis of early RNA that initiates on the late side of the origin. The specific conformation of the chromatin in this region, observed both in SV40 (33) and in polyoma (unpublished observations), may be part of the controlling element for the selection of the initiation sites for both late and early transcription. As far as the regulation of transcription in embryonal versus differentiated cells is concerned, we have to postulate that the factors participating in the initiation of RNA synthesis (42) are different in these two types of cells. Rearrangement of the wild-type viral sequence is required to permit recognition by the RNA polymerase II (B) in embryonal cells. Thus, regulation at the level of initiation of transcription may play an important role in the control of embryonal development (43).

An alternative hypothesis for the block of viral expression in EC cells was suggested by Segal et al. (44, 45), who showed that upon SV40 infection of another mouse EC cell line, F9 (this line is not permissive to our mutants), early viral RNA is synthesized in small amounts, but is not spliced, whereas this RNA is normally spliced in differentiated F9 cells. At present, we cannot exclude the possibility that early RNA made off the wild-type polyoma DNA in PCC4 cells cannot be processed and is rapidly degraded. The new 5' end of the mutant RNA will permit maturation of the early RNA. However, the fact that no unspliced early RNA was found in PCC4 cells infected with wild-type polyoma (F. Kelly and R. Kamen, personal communication) tends to support our hypothesis that the control occurs at the level of transcription.

III. Relation between Histone Acetylation and Gene Activity

Our previous studies showed that the four core histones associated with SV40 DNA are highly acetylated (46). Two parameters have been used to demonstrate this acetylation: (a) the presence of stained bands of mono-, di-, and triacetylated histones indicates the steady-state level of acetylation, which state depends on both the rates of acetylation and deacetylation; (b) the pulse-labeling of the histones with highly radioactive acetate reveals the rate of histone acetylation. In fact, whereas in the steady state mainly histones H4 and H3 contain acetylated forms, pulse-labeling reveals a high rate of incorporation in H2B, H3, H4, and H2A. Thus the acetate groups probably turn over faster in H2B and H2A than in H3 and H4. By both these criteria, the viral intranuclear chromatin is more acetylated than that of the chromatin of infected cells (46). Previous studies (47) suggest that in polyoma-infected cells the lack of small and middle T antigens in hrt (host range nontransforming) mutants causes a decrease in the level of acetylation of the viral histones. The capacity to acetylate the viral histones may be related to the transforming activity.

The viral chromatin in the infected cells undergoes several processes: replication, transcription, and encapsidation. To check the eventual role of the acetylation in any one of these steps, we pulse-labeled the infected cells and followed the distribution of the acetylated histones. Most of the recently acetylated histone is not incorporated into replicative intermediates that can be separated from the bulk of the chromatin (28). Furthermore, by labeling the newly synthesized histones with labeled amino acids, we could show that the histones bound to replicative intermediates were in a lower acetylated state than those bound to recently replicated chromatin. Thus, after sequential assembly of newly synthesized histones on replicating SV40 chromatin (20), and completion of replication, the core histones undergo further acetylation. To check whether this acetylation precedes immediately the encapsidation, we extracted the labeled nuclei under conditions that preserved the previrions and virions (48). A very low percentage of newly acetylated histones was found in these previrions and virions. To study the possible role of histone acetylation in the process of transcription, we undertook, in collaboration with P. Gariglio, the study of viral transcription complexes that permit RNA chain elongation *in vitro*. About 0.5% of the viral chromatin contains endogenously bound RNA polymerase II(B) (23). The nascent RNA chains that can be elongated *in vitro* increase the sedimentation constant of the DNA · histone transcription complexes. After labeling with

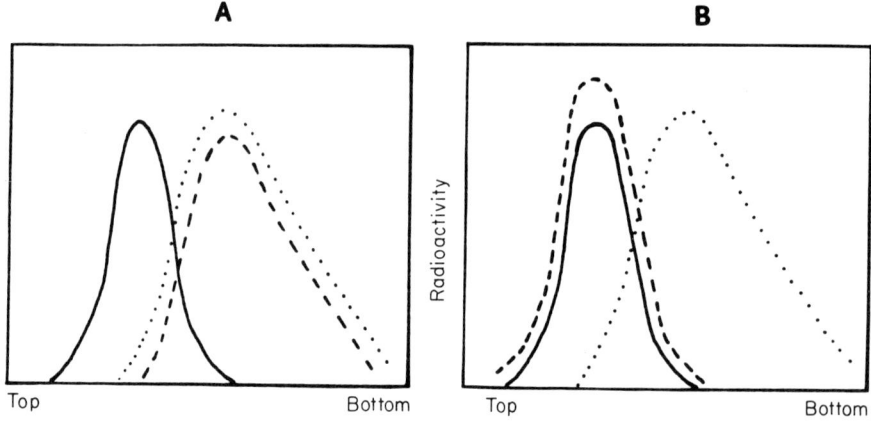

FIG. 8. Experimental approach to analysis of the role of acetylation in chromatin transcription. CV1 cells were infected with SV40, labeled with [^{14}C]thymidine between 24 and 40 hours after infection, and pulse-labeled for the last 10 minutes with [^{3}H]acetate in the absence or the presence of sodium butyrate. After extraction of the viral minichromosomes with 0.4 M NaCl, they were incubated for 60 minutes under conditions that permit elongation of nascent RNA chains, then submitted to sucrose gradient analysis. ——, [^{14}C]thymidine; ----, [^{3}H]acetate; . . . ., [α-^{32}P]CTP. Panel A: acetylated histones preferentially associated with transcriptional complexes. Panel B: acetylated histones associated with all viral chromatin.

[³H]acetate *in vivo*, in the presence or the absence of sodium butyrate to inhibit deacetylation, we extracted the viral chromatin and elongated the RNA *in vitro*. As shown in Fig. 8, two alternative results can be obtained, depending (*a*) on whether the transcribing complexes are more highly acetylated, (*b*) on the acetylation of a large fraction or the entire population of the viral chromatin. The results obtained did not show any massive displacement of the recently acetylated minichromosomes upon transcription. The fraction of viral chromatin that became acetylated in a 10-minute pulse greatly exceeds the small fraction of transcriptionally active minichromosomes (A. Chestier, M. Yaniv and P. Gariglio, unpublished). In spite of the absence of a direct correlation between the degree of acetylation and the number of transcription complexes, this modification may still be associated with gene activity. One of the possible models in the formation of transcription-competent minichromosomes by acetylation. A specific pattern of core histone acetylation may permit the binding of HMG14 and HMG17 proteins that are associated with active genes (49, 50). Only a small fraction of these potentially active complexes will in fact contain bound RNA polymerase.

IV. Conclusions

The DNA of the small circular DNA tumor viruses is associated with cellular histones in a structure that resembles that of the cellular chromatin fiber. Viral chromatin undergoes both replication and transcription in the nucleus. Regulatory mechanisms that are involved in chromatin replication, histone segregation, and assembly are probably very similar for the viral and cellular genomes. Our present studies have indicated that the region containing the initiation sites for late, and probably for early, transcription has a unique conformation in the viral minichromosomes. This region is devoid of nucleosomes in 5% to 20% of the viral chromosomes, as detected by electron microscopy and by DNase I digestion of nuclei isolated from infected cells. We are currently analyzing SV40-transformed rat cells in an attempt to probe the structure of the integrated viral genome and its regulatory regions in these cells. The absence of a strong association between the histones and the DNA sequence recognized by the eukaryotic RNA RNA polymerase may facilitate correct initiation. During embryonic development or during cell differentiation, new sequences may become available for transcription by displacement of histones from promoter regions. Inversely, genes may be shut down by the formation of nucleosomes containing the controlling DNA elements. Nonhistone proteins or extensive local modifications of the histones may govern the presence or the absence of histones on defined regions of the DNA.

It is clear, from both sequence comparison and transcription studies *in vitro*, that specific sequences are responsible for "promoter" recognition by eukaryotic RNA polymerases and their cofactors (discussed by Bogenhagen *et al.*, Roeder *et al.*, and Kurjan *et al.*, in this symposium). Our sequence studies of polyoma mutants that grow on embryonal carcinoma cells suggest that the promoters recognized in a differentiated cell may differ from those recognized by an embryonic cell. This may be analogous to the modifications of the transcription machinery at both the enzyme and DNA levels after phage SPO1 infection of *Bacillus subtilis* (discussed by Pero *et al.* in this symposium). Embryonic development and differentiation may thus be accompanied by changes in RNA polymerase II or its cofactors and by changes in the sequence of the initiation sites recognized on the cellular genome. The control of gene expression by both the primary DNA sequence and the structure of chromatin may permit more refined and tighter control of gene expression in eukaryotic cells. Further *in vitro* studies with reconstituted transcription systems should permit a test of these hypotheses.

Acknowledgments

We are grateful to J. Dubochet, R. Freeman, and M. Groom (deceased) (EM group, EMBL, Heidelberg) and O. Croissant (EM Laboratory, Department of Virology, Institut Pasteur, Paris) for helpful advice and the availability of their electron microscopy facilities. We thank M. Buckingham for valuable comments and C. Maczuka for help with the preparation of the manuscript. Gilles Moyne was supported by a short-term EMBO fellowship to perform the high-resolution observations in Heidelberg. This work was supported by grants from the CNRS (LA 270 and ATP Chromatine), INSERM (ATP 50-77-82 and 72-79-104), and the DGRST (MRM 129).

References

1. W. Fiers, R. Contreras, G. Haegeman, R. Rogiers, A. van de Voorde, M. van Heuwerswyn, J. van Herreweghe, G. Volckaert and M. Ysebaert, *Nature* **273**, 113 (1978).
2. V. B. Reddy, B. Thimmappaya, R. Dhar, K. N. Subramanian, S. Zain, J. Pan, P. J. Ghosh, M. L. Celma and S. M. Weissman, *Science* **200**, 494 (1978).
3. E. Soeda, J. R. Arrand, N. Smolar, J. E. Walsh and B. E. Griffin, *Nature* **283**, 445 (1980).
4. R. C. A. Yang and R. Wu, *Science* **206**, 456 (1979).
5. I. Seif, G. Khoury and R. Dhar, *Cell* **18**, 963 (1979).
6. J. Tooze (ed.), "The Molecular Biology of Tumor Viruses," Part 2, DNA Tumor Viruses. Cold Spring Harbor Laboratory, Cold Spring Harbor, New York, 1980.
7. N. H. Acheson, *Cell* **8**, 1 (1976).
8. P. M. Frearson and L. V. Crawford, *J. Gen. Virol.* **14**, 141 (1972).

9. S. R. Lake, S. Barban and N. P. Salzman, *BBRC* **54**, 640 (1973).
10. G. Fey and B. Hirt, *CSHSQB* **39**, 235 (1975).
11. A. L. Olins and D. E. Olins, *Science* **183**, 330 (1974).
12. J. D. Griffith, *Science* **187**, 1202 (1975).
13. C. Crémisi, P. F. Pignatti, O. Croissant and M. Yaniv, *J. Virol.* **17**, 204 (1976).
14. M. Bellard, P. Oudet, J. E. Germond and P. Chambon, *EJB* **70**, 543 (1976).
15. M. Favre, F. Breilburd, O. Croissant and G. Orth, *J. Virol.* **21**, 1205 (1977).
16. J. E. Germond, B. Hirt, P. Oudet, M. Gross-Bellard and P. Chambon, *PNAS* **72**, 1843 (1975).
17. J. T. Finch, L. C. Lutter, D. Rhodes, R. S. Brown, B. Rushton, M. Levitt and A. Klug, *Nature* **269**, 29 (1977).
18. W. Keller, *PNAS* **72**, 4876 (1975).
19. A. J. Varshavsky, V. V. Bakayev, P. M. Chumackov and G. P. Georgiev *NARes* **3**, 2101 (1976).
20. C. Crémisi and M. Yaniv, *BBRC* **92**, 1117 (1980).
21. R. T. Su and M. L. DePamphilis, *PNAS* **73**, 3456 (1976).
22. T. L. Brooks and M. H. Green, *NARes* **4**, 4261 (1977).
23. P. Gariglio, R. Llopis, P. Oudet and P. Chambon, *JMB* **131**, 75 (1979).
24. T. Hunter, B. Francke and L. Bacheler, *Cell* **12**, 1021 (1977).
25. D. Perlman and J. A. Huberman, *Cell* **12**, 1029 (1977).
26. G. Kaufmann, R. Bar-Shavit and M. L. DePamphilis, *NARes* **5**, 2535 (1978).
27. M. M. Seidman, A. J. Levine and H. Weintraub, *Cell* **18**, 439 (1979).
28. C. Crémisi, A. Chestier and M. Yaniv, *Cell* **12**, 947 (1977).
29. A. J. Varshavsky, O. Sundin and M. Bohn, *Cell* **16**, 453 (1979).
30. W. A. Scott and D. J. Wigmore, *Cell* **15**, 1511 (1978).
31. W. Waldeck, B. Fohring, K. Chowdhury, P. Gruss and G. Sauer, *PNAS* **75**, 5964 (1978).
32. J. Dubochet, M. Ducommun, M. Zollinger and E. Kellenberger, *J. Ultrastruct. Res.* **35**, 147 (1971).
33. S. Saragosti, G. Moyne and M. Yaniv, *Cell* **20**, 65 (1980).
34. A. Worcel, S. Han and M. L. Wong, *Cell* **15**, 969 (1978).
35. E. Jakobovits, S. Saragosti, M. Yaniv and Y. Aloni, *PNAS* **77**, 6556 (1980).
36. S. Saragosti, O. Croissant and M. Yaniv, *EJB* **106**, 25 (1980).
37. M. Boccara and F. Kelly, *Ann. Microbiol. (Inst. Pasteur)* **129A**, 227 (1978).
38. M. Boccara and F. Kelly, *Virology* **90**, 147 (1978).
39. M. Vasseur, C. Kress, N. Montreau and D. Blangy, *PNAS* **77**, 1068 (1980).
40. A. J. Flavell, A. Cowie, J. R. Arrand and R. Kamen, *J. Virol.* **33**, 902 (1980).
41. R. Kamen, J. Favaloro and J. Parker, *J. Virol.* **33**, 637 (1980).
42. P. A. Weil, D. S. Luse, J. Segall and R. G. Roeder, *Cell* **18**, 469 (1979).
43. M. Katinka, M. Yaniv, M. Vasseur and D. Blangy, *Cell* **20**, 393 (1980).
44. S. Segal, A. J. Levine and G. Khoury, *Nature* **280**, 335 (1979).
45. S. Segal and G. Khoury, *PNAS* **76**, 5611 (1979).
46. A. Chestier and M. Yaniv, *PNAS* **76**, 46 (1979).
47. B. S. Schaffhausen and T. L. Benjamin, *PNAS* **73**, 1092 (1976).
48. E. A. Garber, M. M. Seidman and A. J. Levine, *Virology* **90**, 305 (1978).
49. B. Levy, C. W. Wong and G. Dixon, *PNAS* **74**, 2810 (1977).
50. S. Weisbord and H. Weintraub, *PNAS* **76**, 630 (1979).

IV. Control of Transcription in Eukaryotes
Chairman and Summarizer: WILLIAM J. RUTTER

Summary   171
WILLIAM J. RUTTER

Summary

WILLIAM J. RUTTER

Department of Biochemistry and
Biophysics
University of California
San Francisco, California

This session emphasized new advances in the understanding of gene expression that have resulted from studies with structurally defined (usually cloned) DNA species. They illustrate the range of experimental tactics being employed to study the structural features of genes that are important for transcription, and the transcription and processing reactions themselves.

I. The Structure and Expression of Pancreatic Endocrine and Exocrine Genes[1]

A. Insulin[2]

Comparative studies on the sequence of cloned insulin cDNAs from rat human and anglerfish have shown the expected high degree of sequence conservation within the regions coding for the insulin B and A chains, and low degree of conservation within the connecting peptide region. Unexpectedly, there was also considerable conservation within the region coding for the signal peptide. Conservation of the hydrophobic amino acids of the signal peptide was expected, but conservation of specific amino-acid sequences was observed. This suggests that the specific sequence of the signal peptide is involved in its function. The length of the 3' nontranslated region varied considerably, however; in each of the cDNAs there existed a C-rich region close to the termination codon, and a

---

[1] By Graeme Bell, Leslie Rall, Orgad Laub, Peter Hobart, Robert Crawford, Lu-Ping Shen, Raymond MacDonald, William Swain, Michael Crerar, Raymond Pictet, and William J. Rutter (Department of Biochemistry and Biophysics, University of California, San Francisco, California) presented orally and summarized by William J. Rutter.

[2] Part of the studies on insulin were carried out in collaboration with Drs. Cordell and Goodman.

sequence including, but not limited to, the AAUAAA segment close to the site of poly(A) addition on the mRNA.

The complete DNA sequences of genomic fragments containing rat insulin genes I and II and the human insulin gene have been obtained. A similar fragment containing the chicken insulin gene has also been partially sequenced (Efstratiadis and colleagues). A comparison of the sequences of these fragments shows that all genes contain an intervening sequence in the 5' untranslated region of the transcribed portion of the gene. This intervening segment I is located at the same position with respect to the start of translation except in the human gene, where it is a few nucleotides displaced. The human rat II and chicken genes have a larger intervening sequence located at precisely the same position in the DNA coding for the connecting peptide. In contrast, the rat insulin I gene contains no intervening sequences in this region. This suggests that the primordial insulin gene contains two intervening sequences, and that the rat insulin I gene arose by a mechanism eliminating the second intervening sequence.

There is a sharp divergence in the nucleotide sequence and the length of both of the intervening sequences. However, there is some degree of sequence conservation at the junctions, particularly at the 3' junctions. This suggests that some aspect of this structure (splicing?) is required for biological function. In the 5' flanking region adjacent to the initiation of transcription, there is a region of conserved sequence that contains a TATAA sequence (Hogness box). In addition, approximately 40 nucleotides upstream (in the 5' direction), there is another area of sequence homology. There are no further detectable regions of sequence homology 5600 bases in the 5', or 11,000 bases in the 3', direction as determined by heteroduplex mapping or by limited sequence analysis; thus most of the structural features of the DNA associated with the insulin gene expression are contiguous with the 5' end of the gene. Other specific regions would have to be very short, or be present long distances from the gene itself.

A further search was made of the human insulin gene for repeated sequences, as it has been hypothesized that these are involved in gene expression. There were no detectable repeated sequences proximal to the gene, but a member of the "Alu^3 family" was found approximately 6000 bases from the 3' end. The sequence of the DNA of this region is bordered by 19 base-pair direct re-

---

[3] Endodeoxyribonuclease AluI, EC 3.1.23.1 [Ed.].

peats; it contains an 83 base-pair sequence that is repeated twice. Each of the regions contains a 14 base-pair sequence found in the area of the origin of DNA replication in SV40. Therefore, it is possible that this region is associated with DNA replication.

Two alleles of the human insulin gene differ in only two bases in the 3' untranslated region of the gene. In addition, a significant number of humans contain an insert of a few hundred to more than a thousand bases at a site approximately 700 nucleotides from the initiation of transcription of the gene. This structural heterogeneity may influence the expression of the gene and hence may be involved in a genetic predilection toward diabetes.

Attempts have been made to obtain expression of the human insulin gene when it is introduced together with the herpes thymidylate kinase gene in both covalently linked and independent configurations. Under conditions where the cells are selected for thymidylate kinase gene expression in mouse L cells, up to 100 copies of the insulin gene per cell have been introduced, without detectable expression of insulin or proinsulin (less than 200 molecules per cell). In three isolated cases, low (about 1500 molecules per cell) levels of a proinsulin-like molecule were detected. In separate experiments, the human insulin genomic fragment has been incorporated into the late region of the SV40 virus and grown in monkey kidney cells with a complementing defective helper virus. Low levels of expression have been observed. Attempts to obtain expression of human insulin gene on injection into the germinal vesicle of *Xenopus* oocytes have thus far been unsuccessful. Incubation of the human insulin gene fragment with partially purified extracts of HeLa cells under conditions where polymerase III is active has resulted in the formation of a single 700-nucleotide RNA. This species is probably related to the *Alu* family segment, but its biological role (replication or gene expression?) is not yet known. Preliminary experiments with endogenous RNA polymerase II in such extracts shows that an RNA is produced from a truncated gene. This RNA is currently being analyzed.

B. Somatostatins I and II

Progress has also been made in the isolation of cDNA and genomic DNA fragments containing the other endocrine genes of the pancreas. Because of the extraordinarily low levels of the cells making glucagon, somatostatin, and pancreatic polypeptides, the isolation of cDNA of these genes is difficult from mammalian organisms. In birds and fish, however, there is a higher proportion of these hormones relative to insulin, and in certain species the islets are present relatively unencumbered by the acinar tissue. "Shotgun" cloning of cDNA prepared against anglerfish RNA resulted in the anglerfish insulin cDNA clones (see Section I,A). In addition, two cDNA clones containing nucleotide sequences coding for two distinct somatostatin precursors 121 and 125 amino acids in length were identified and sequenced. The region coding for the 14 amino-acid somatostatin moieties was at the extreme 3' (carboxy terminal) portion of the molecule. Somatostatin I has an amino-acid composition identical to that of human somatostatin. However, somatostatin II differs in that a tyrosine replaces phenylalanine at position 7 and glycine replaces threonine at position 10. The somatostatin moiety is differentiated from the remainder of the precursor by two dibasic amino acids at the junctions.

The structure of the somatostatin precursor predicted from cDNA structure includes a signal peptide estimated to be approximately 24 residues (somewhat related to the signal peptide of anglerfish insulin) and an intervening peptide of approximately 83 and 87 residues, respectively. The sequence of the intervening peptide varies extensively in somatostatins I and II except in the region proximal to the somatostatin moiety (the last 28 amino acids). It is not known whether the intervening peptide codes for a distinct functional entity as in the case of the ACTH/β-endorphin precursor. Discovery of distinct somatostatins, however, raises the possibility that the alpha and beta cells are independently regulated by somatostatins I and II. Utilizing the cloned somatostatin cDNAs as probes, it has been possible to show (by Southern analysis) that there are a number of somatostatin-like gene segments in anglerfish, in humans, and in other vertebrates. Thus the various effects of somatostatin may be mediated by a number of separate molecules. Some of the genomic DNA fragments containing somatostatin genes have now been isolated.

C. Acinar Cell Genes

Progress has also been made in the study of the major genes expressed in pancreatic acinar cells. The cloned cDNAs have been identified by nucleotide sequence analysis and include amylase, carboxypeptidase A, proelastases I and II, chymotrypsinogen A. Four other cDNAs have not yet been identified. The amylase gene family has been studied more intensively. Full length cDNAs of both pancreatic and salivary gland amylases have been obtained. The amylase cDNA has been sequenced. Thus the complete amino acid sequence of this 58,000 amylase precursor (containing the signal peptide) has been obtained. The amylase precursor (containing the signal peptide) has been obtained. The amylase cDNA has been employed as a probe for detection of genomic DNA fragments containing the amylase gene; 39 such fragments were obtained and analyzed extensively via restriction enzyme mapping; 9 distinct families could be identified. Since the "library" was obtained from a single, presumably heterozygotic, animal, this implies that at least five amylase genelike regions are present in the rat. All these genes contain regions of homology. Heteroduplex mapping and sequence analysis show that the pancreas amylase gene is approximately 9000 nucleotides long and contains 8 intervening sequences. Studies of the expression of this gene in cellular and in *in vitro* systems are now being carried out.

II. The Control of Xenopus 5 S RNA Transcription[4]

Cloned repeating units of *Xenopus* 5 S DNA are faithfully transcribed *in vitro* in an extract of *Xenopus* oocyte nuclei. This provides an assay system for the identification of DNA sequences essential for accurate initiation and termination by RNA polymerase III. Recombinant DNA techniques were used to generate deletions approaching and entering the *Xenopus borealis* somatic (Xbs) 5 S RNA gene from either the 5' or 3' orientation. These deletion mutants were tested for their ability to support 5 S RNA transcription in an oocyte nuclear extract. Mutants lacking the entire 5' flanking region synthesized little or no 5 S size RNA. Thus a control region within the gene directs RNA polymerase III to initiate transcription approximately 50 nucleotides upstream (5' direction) from the 5' boundary of this region.

The 3' boundary has been defined by *in vitro* transcription of a series of deletions of cloned 5 S DNA whose end points approach and enter the gene from the 3' side. Since these deletions lack the normal termination site of 5 S RNA synthesis, an assay was developed for accurate initiation of transcription in the absence of correct termination. *In vitro* transcription in the presence of cordycepin triphosphate (3'-dATP) causes synthesis of a reproducible array of shortened RNA molecules separable by polacrylamide gel electro-

---

[4] By Daniel Bogenhagen, Shigeru Sakonju, and Donald D. Brown (Department of Embryology, Carnegie Institution of Washington, Baltimore, Maryland) presented orally and summarized by Daniel Bogenhagen.

phoresis. This transcription assay shows that 3' deletions leaving the first 83 or more 5' gene residues support accurate initiation of transcription. Larger deletions leaving 80 or fewer gene residues do not support 5 S RNA specific transcription initiation. The gene region from nucleotide 41 to 87, recloned with any additional 5 S DNA sequences, contains sufficient information to direct specific transcription initiation at a site in the upstream plasmid sequence. These workers propose that a protein factor is required for transcription initiation.

Similar *in vitro* recombinant DNA methods have been used to help identify sequences required for transcription termination. Termination of Xbs 5 S RNA synthesis does not require extensive regions of either the 3' spacer sequences or the intragenic sequences preceding the termination site. The experiments to date suggest that a cluster of four or more thymidine residues is required for transcription termination. The efficiency of termination is increased by high (G+C)-content of the DNA sequences immediately preceding and following the thymidine cluster.

III. Factors Involved in the Transcription of Purified Eukaryotic Genes by RNA Polymerases II and III[5]

The Roeder laboratory has previously developed reconstituted cell-free systems in which purified genes (or fragments thereof) are accurately transcribed by purified class II or class III RNA polymerase in the presence of crude cellular extracts. Using DNA templates with intact transcription units, accurate initiation and termination have been shown for the adenovirus 2 VA RNA and *Xenopus* 5 S RNA and tRNA genes (class III). Using truncated DNA templates, accurate initiation has been shown for (*a*) Ad 2 early, intermediate, and late transcription units; (*b*) mammalian globin genes; and (*c*) parvovirus Hl and AAV transcription units. (In the latter case the cell-free system has been used to identify promoters not previously predicted from *in vivo* studies.) The results of transcription studies with heterologous and homologous cell extracts indicate that the transcription seen in the cell-free system is mediated by general transcription factors (in the extracts) that are neither tissue- nor spe-

[5] By R. G. Roeder, D. Engelke, B. Honda, D. Lee, D. Luse, T. Matsui, S. Ng, J. Segall, B. Shastry, and P. Weil (Department of Biological Chemistry, Washington University School of Medicine, St. Louis, Missouri) presented orally and summarized by R. G. Roeder.

cies-specific and are by themselves insufficient for the *in vivo* regulation of various of the class II and class III genes.

To further understand the nature and mechanism of action of these transcription factors, and ultimately their potential modulation by other regulatory factors, Roeder and colleagues have begun to fractionate the extracts from human KB cells and from *Xenopus* oocytes. In human KB cells they have identified a minimum of four "factors" (designated TF-IIA, TF-IIB, TF-IIC, and TF-IID) required for active and accurate transcription initiation at the Ad 2 major late promoter. Although TF-IIA may not be absolutely required for accurate initiation, it greatly stimulates the overall rate of transcription in the presence of the other components. TF-IIC appears to act by suppressing random transcription at nonspecific template sites but could also have a direct role in specific initiation. TF-IIB and TF-IID neither stimulate nor repress general transcription, but are essential for accurate initiation. These factors appear to be distinct from those required for active transcription (initiation and termination) of class III genes in human cell extracts. The latter include at least two "factors" (designated TF-IIIB and TF-IIIC) necessary for transcription of tRNA and VA RNA genes. Another factor (designated TF-IIIA) is required along with fractions containing TF-IIIB and TF-IIIC for transcription of 5 S RNA genes. These studies indicate the generally complex nature of eukaryotic transcription events and point to the existence of both gene-specific and common factors; however, an accuragre definition of the multiplicity and specificity of these factors awaits their purification.

Class III factors analogous to those identified in human cell extracts have also been identified in *Xenopus* oocyte extracts; one of these (TF-IIIA) has been purified to homogeneity and shown to bind to an intragenic control region on the 5 S gene. Roeder *et al.* have previously shown that egg extracts are deficient in the functional equivalent of the oocyte TF-IIIA and do not transcribe 5 S genes unless supplied with oocyte TF-IIIA, but extracts derived from somatic cells very efficiently transcribe oocyte and somatic-type 5 S genes. Thus these results suggest that oocytes and somatic cells contain functionally equivalent, but structurally distinct, A factors and raise interesting questions regarding the role of the A factor(s) in 5 S gene regulation.

The further purification and analysis of the various class II and class III factors is underway. In addition the crude and partially purified systems are utilized for the detection of possible specific DNA sequences important for transcription.

IV. Mutations That Block the Expression of a Yeast tRNA[Tyr] Gene: Genetic Analysis, DNA Sequence Changes, and in Vitro Transcription[6]

The yeast *SUP4* tRNA[Tyr] gene makes an initial transcript containing an 11- or 12-nucleotide leader and a 14-nucleotide transcribed intervening sequence [E. DeRobertis and M. Olson, *Nature* **278**, 137 (1979)]. These workers have made an extensive genetic and physical analysis of this gene to determine whether DNA sequences within the leader, intervening sequence, and flanking 3' and 5' regions are essential for tRNA[Tyr] synthesis. Inactivation of ochre suppression by a *SUP4*-o allele was used as the basis for selecting mutants affecting many different aspects of *SUP4* tRNA[Tyr] synthesis and function.

The genetic map positions of 26 different mutations are colinear with the DNA sequence changes they produce. The mutations sequenced range in position from nucleotide 3 of mature tRNA[Tyr] to the last nucleotide, with one mutation in the intervening sequence. None of the 69 mutations genetically mapped appear to lie outside the DNA region specifying mature tRNA[Tyr]. This suggests that the sequences essential for initiation of transcription lie within the gene rather than in the leader and 5' flanking regions. A direct test of this possibility has been made by transcribing the cloned mutant *SUP4* genes with *Xenopus laevis* RNA polymerase III *in vitro*.

Seventeen of the mutant genes are transcribed normally and two yield no detectable pre-tRNA[Tyr] molecules. Mutation of the fourth base-pair within the intervening sequence appears to cause premature termination at the mutant site. The effect of this mutation and another premature terminator located nearby (a deletion of 2 As from TTAATTT) are explainable by the action of oligo(T) sequences in causing termination. Five mutations within the mature tRNA[Tyr] coding region cause minor size reductions in the *in vitro* transcription product. These mutant effects imply that sequences between the 30th and 90th nucleotides of the tRNA[Tyr] transcription unit specify the initiation and termination points.

Consideration of these results in relation to the functional effect of deletion and insertion mutations in the *Xenopus* 5 S RNA and tRNA[fMet] genes [S. Sakonju, D. Bogenhagen and D. Brown, *Cell* **19**, 13 (1980); A. Kressman, H. Hofstetter, E. Di Capua, R. Grosschedl, and M. Birnstiel, *NARes* **7**, 1749 (1979)] suggests that there is a common mechanism for polymerase recognition for all genes transcribed by eukaryotic RNA polymerase III.

[6] By Janet Kurjan, Benjamin D. Hall, Shirley Gillam, Michael Smith, Raymond A. Koski, and Stuart A. Clarkson (Department of Genetics, University of Washington, Seattle, Washington) presented orally and summarized by Benjamin D. Hall.

V. Mechanisms of DNA Repair
Chairman: RICHARD F. KIMBALL

Repair Replication Schemes in Bacteria and Human Cells 181
PHILIP C. HANAWALT, PRISCILLA K. COOPER,
AND CHARLES ALLEN SMITH

Recent Developments in the Enzymology of
  Excision Repair of DNA 197
ERROL C. FRIEDBERG, CORRIE T. M. ANDERSON,
THOMAS BONURA, RICHARD CONE,
ERIC H. RADANY, AND RICHARD J. REYNOLDS

Multiprotein Interactions in Strand Cleavage
  of DNA Damaged by UV and Chemicals 217
ERLING SEEBERG

In Vitro Packaging of Damaged Bacteriophage T7 DNA 227
WARREN E. MASKER, NANCY B. KUEMMERLE,
AND LORI A. DODSON

The Inducible Repair of Alkylated DNA 237
JOHN CAIRNS, PETER ROBINS, BARBARA SEDGWICK,
AND PHILLIPA TALMUD

Repair Replication Schemes in Bacteria and Human Cells

PHILIP C. HANAWALT
PRISCILLA K. COOPER AND
CHARLES ALLEN SMITH

Department of Biological Sciences
Stanford University
Stanford, California

The complementary double-stranded structure of DNA is important for both its replication and repair. When one of the two strands is damaged, the information encoded in the nucleotide sequence is still retained by the intact, complementary strand. This redundancy of information forms the basis for schemes that repair damage to one strand by excision and replacement with a stretch of normal nucleotides. Furthermore, it is an essential feature in mechanisms for tolerance of damage, such as strand exchange, that may enhance cellular survival in the presence of persisting lesions in DNA. It is important to distinguish clearly between those recovery schemes that effect removal of the lesions (i.e., repair) and those that facilitate tolerance of persisting damage in the genome. The same biological end point (e.g., survival) may result from the operation of either general scheme, but there are important conceptual differences. In principle, either scheme may include inducible as well as constitutive pathways, and either may have "error-prone" components that enhance mutagenesis in the surviving population of cells. In the present discussion, we focus on those schemes that remove damage from the DNA and attempt restoration of the original nucleotide sequence.

I. Repair Responses to Damaged DNA

The general scheme shown in Fig. 1 illustrates multiple enzymic pathways documented or proposed as mechanisms for repairing various classes of alterations in DNA. These are detailed in reviews (1, 2) and in the proceedings of a conference on DNA repair mechanisms (3). Several general features of these pathways are evident from the diagram. The repair sequence is initiated by enzymic recognition of the damage. The simplest case is that of a phosphodiester bond scission that can be closed directly by polynucleotide ligase. However,

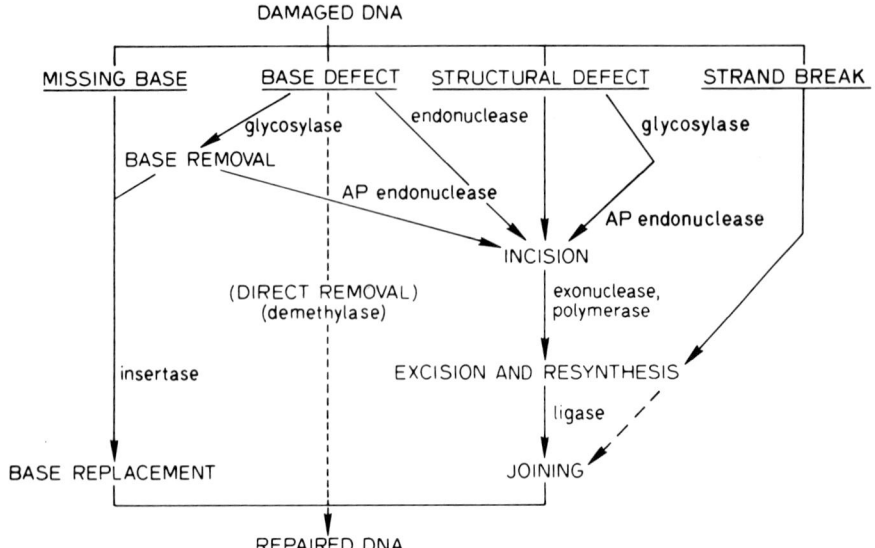

FIG. 1. Schemes for the enzymic processing of lesions in damaged DNA (1).

many strand scissions, such as those produced by ionizing radiations, cannot be directly ligated and require enzymic processing of the abutting strand ends followed by repair replication. Certain repair pathways may not even include strand breaks as intermediates; e.g., the direct removal of a methyl at the O^6 position of guanine can be accomplished in *Escherichia coli* apparently without disturbing the DNA structure (4). There is also evidence for "insertase"[1] activities that may directly replace proper purines at apurinic sites (5, 6).

The most general scheme for dealing with a variety of defects in DNA involves an enzymic process that nicks (cleaves) the damaged strand in preparation for subsequent excision of a stretch of DNA containing the damage, and repair resynthesis. This nicking may be effected by direct action of an endonuclease that recognizes damage, or it may be the result of sequential action of a glycosylase that cleaves a damaged base from its sugar, generating an AP (apurinic or apyrimidinic) site, followed by action of an AP endonuclease at a phosphodiester bond. Several glycosylases, highly specific for particular incorrect or altered bases, have been isolated from a variety of cell types (2). In addition, the small "UV endonucleases" of *Micrococcus luteus* and of

---

[1] See footnote 4 in chapter by Friedberg *et al.* in this volume, concerning "insertase" [Ed.].

E. coli infected by phage T4 possess glycosylase activities that cleave one pyrimidine (the one on the 5' side) of the dimer from its sugar (7–11). Whether those enzymes also possess AP endonuclease activity to complete the incision, or whether preparations of those enzymes have been contaminated with an AP endonuclease, is currently the subject of investigation in several laboratories. Most evidence suggests that those glycosylase activities are highly specific for pyrimidine dimers. The uvrABC system of E. coli, and the incision systems in mammalian cells, appear to act on a variety of different lesions; whether they incorporate a glycosylase activity in their mechanism is not clear. Some evidence suggests that the uvrABC system incises the dimer-containing strand directly (12).

There are several factors that may affect the steps in excision repair subsequent to incision. Obviously the chemical nature of the 3' end at the break is important. Repair replication may begin immediately at this point only if a 3' OH (rather than a 3' phosphate) is presented. In addition, if an AP site exists on the 3' end, some processing must be required before repair replications begins. Such processing would be expected following incision by T4 endonuclease V. A 3' exonuclease or a different AP endonuclease may cleave the strand at the 5' side of the AP site (13). Both E. coli and human cells can apparently perform such processing, as T4 endonuclease V has been shown to promote enhanced survival or repair synthesis when introduced into UV-irradiated repair-deficient cells (14–17).

The analysis of pyrimidine dimer excision and repair replication in mutant strains of E. coli has elucidated a number of alternative pathways for the intervening steps between incision and ligation (1). It is notable that no single mutation has been shown to eliminate either excision or repair resynthesis once incision has occurred. Mutants deficient in the polymerase (and 3' exonuclease) activities of DNA polymerase I are more sensitive to UV and more deficient in pyrimidine dimer excision than are mutants defective in the 5' exonuclease activity of this enzyme (18). Furthermore, double mutants deficient in both polymerases I and III are even more defective in dimer excision (19). Our current view is that excision and resynthesis are closely coupled processes, and that concurrent polymerization may "drive" the excision. Although the 5' exonuclease fragment of pol I can excise dimers *in vitro*, that activity is enhanced under conditions permitting DNA synthesis by the presence of the large proteolytic fragment containing the polymerase activity (20). Although each of the three known DNA polymerases in E. coli is capable of carrying out the repair resynthesis step, the relative importance of these enzymes for dealing with particular sites in the DNA has not been ascertained. The somewhat different substrate requirements of the three polymerases may require their coordination with particular exonuclease activities to effect repair. Thus, while pol I can bind to nicked DNA, the other two polymerases evidently require the expansion of the nick to a gap before binding and polymerization can be initiated (1). At least four exonuclease activities have been implicated in the excision of pyrimidine dimers. These include the 5' exonuclease of pol I, a weaker 5' exonuclease activity of pol III, exonuclease V (the *recBC* nuclease) and exonuclease VII, specific for single-strand DNA (1). It is not yet possible to assess the relative roles of these activities in the removal of lesions.

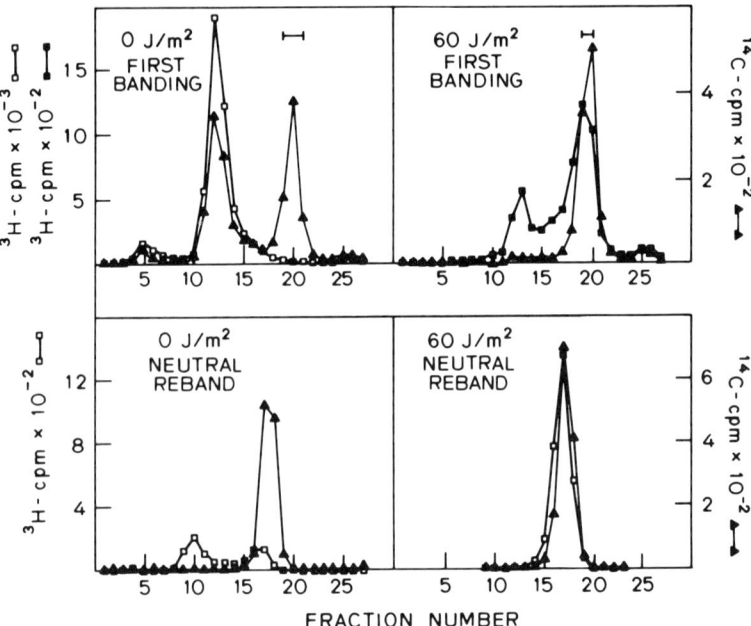

FIG. 2. Determination of repair replication in *Escherichia coli* (22). Exponentially growing cultures of strain W3110 thy^- were labeled with [^{14}C]thymine prior to UV irradiation in buffer (60 J/m² at 254 nm, or similarly treated control). Then the cells were incubated for 45 minutes in medium containing [^{3}H]bromouracil before lysis and CsCl density gradient sedimentation. Aliquots from the collected fractions were assayed for ^{3}H and ^{14}C in DNA. The peak parental density fractions (indicated in upper panels by the bars) were pooled and subjected to second CsCl gradient centrifugations, the results of which are shown in the lower panels. Density increases from right to left in these and all other gradients shown.

II. Repair Patch Size Distributions

The various experimental approaches for determining the repair resynthesis step in excision repair have been detailed elsewhere (1, 3). The 5-bromouracil density-labeling scheme originally utilized to demonstrate repair replication in *E. coli* permits the physical separation of DNA fragments containing repair patches from semiconservatively replicated portions of the genome (21). Furthermore, through controlled shearing and molecular-weight determination of the sheared fragments, an estimate for the patch size distribution can be obtained (22). Figure 2 and the top panel of Fig. 3 illustrate the application of this approach to UV-irradiated *E. coli*. The upper frames in Fig. 2 show both the inhibition of semiconservative DNA replication by 60 J/m² UV and the appearance at parental density of DNA synthe-

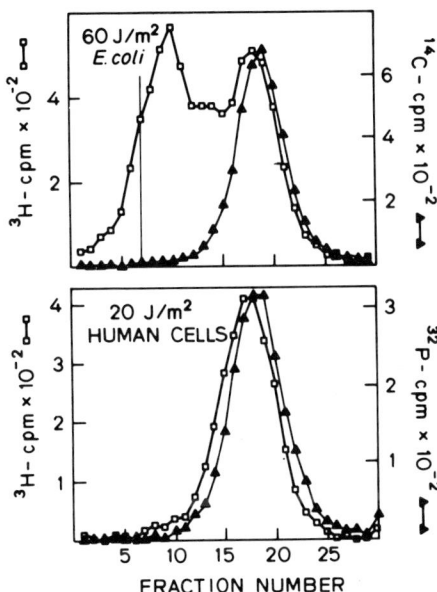

FIG. 3. Repair patch size analysis in *Escherichia coli* and in human cells (22). *Upper panel:* DNA in fractions 16–18 from the gradient presented in the lower right-hand panel of Fig. 2 was subjected to sonication to reduce the number average molecular weight of the DNA to about 250 nucleotides, as determined by alkaline sucrose gradient sedimentation. Alkaline CsCl density gradient sedimentation of this DNA produced the radioactivity profile shown. The vertical line indicates the density of fully bromouracil-substituted DNA strands. *Lower panel:* The analogous experiment with contact-inhibited T98G human cells is illustrated. ^{32}P-labeled cells were irradiated and then incubated for 23 hours in [^{3}H]BrdUrd. DNA of parental density was isolated from a neutral CsCl gradient and analyzed in the same way.

sized by repair replication. The parental density DNA is rebanded in neutral CsCl gradients to further purify the unreplicated DNA containing repaired regions, as shown in the lower frames of Fig. 2. The amount of repair synthesis is calculated from the ratio of [^{3}H]bromouracil repair label to [^{14}C]thymine label in the parental density region. This isolated DNA is then sheared by sonication and analyzed in alkaline CsCl equilibrium density gradients to determine the patch size distribution (Fig. 3, top panel). A bimodal distribution of the repair label is evident, with a fraction of the repaired DNA having only a slight density shift relative to the parental DNA, while the remainder is shifted to a density nearly that of fully bromouracil-substituted DNA. The former represents DNA repaired with short patches of approximately 20 nucleotides (as calculated from the density shift and the known fragment size), while the latter results from repair patches that are long relative to the size of the DNA fragments containing them. These long patches are at least 200 nucleotides in extent; some may be considerably longer (23; P. Cooper, unpublished). Although much of the repair label is contained in the long patches, only a small fraction of the damaged sites (10% or less) is repaired by this mode.

Similar analyses of the patch size distribution in cultured human cells after UV irradiation (Fig. 3, lower panel) reveal only a single class of patches (15–30 nucleotides), about the same size as the "short" patches in *E. coli*. No evidence for a significant frequency of much longer patches was observed in HeLa cells irradiated with 10–70 J/m^2 and allowed to repair for 4 hours (24). Similar results have been obtained for repair patches synthesized in WI38 cells, both at early (0–2 hours) and late (12–16 hours) times after irradiation with 10 J/m^2 UV (25) and also when repair was allowed to proceed for 40 hours after irradiation with 50 J/m^2 (C. A. Smith, unpublished). The same patch size distribution was found in WI38 cells exposed to UV or activated aflatoxin B$_1$(26) and in WI38 and T98G cells containing photoadducts of 8-methoxypsoralen or angelicin (27).

Using the bromouracil photolysis technique, other investigators have analyzed the patch size distribution in human cells irradiated with UV or treated with a variety of other DNA damaging agents (28, 29). These studies also provide no evidence for a class of repair patches in human cells comparable in size to the long patches of *E. coli*. They have, however, provided evidence for two different size classes of patches, which appear to be lesion specific. Patches resulting from repair of the major DNA products of ionizing radiation and some alkylating agents were reported to be only 3 or 4 nucleotides long and were termed "short." The average patch size reported following UV irradiation or exposure to a number of chemical agents ranged from 40 to 160 nucleotides in extent; these have been termed "long." Thus "long" and "short" patches have fundamentally different meanings when applied to human cells or to *E. coli*, respectively.

Although the results obtained by bromouracil photolysis are in agreement with those obtained from density-shift measurements with respect to the absence in human cells of long repair patches analogous to those in *E. coli*, there are some discrepancies that remain to be resolved. Unfortunately, the only damaging agent carefully investigated by both methods appears to be UV. Values obtained from density-shift experiments (15–30 nucleotides) appear to be significantly lower than those obtained by bromouracil photolysis (50–100 nucleotides). While the density-shift method has not revealed significant size differences between patches synthesized in response to UV, aflatoxin adducts, or psoralen adducts, the bromouracil photolysis method has indicated differences in average patch size for the repair of lesions produced by UV, N-acetoxyacetamidofluorene, ICR-170, and certain diol epoxides of benzo[*a*]pyrene, respectively.

III. Long-Patch Repair in *E. coli*

Excision repair in UV-irradiated *E. coli* is dependent upon the products of genes *uvrA*, *uvrB*, and *uvrC* to produce the incisions at pyrimidine dimers (12). Thus, no UV-stimulated repair resynthesis is observed in a *uvrA* mutant (30). Early studies (23) revealed the existence of two general classes of repair synthesis in *E. coli* resulting in short and long patches. The long-patch component was shown to be dependent on the rec^+ genotype, and repair synthesis was markedly enhanced in a *polA* strain (31). This led to the suggesting that at least two pathways for repair resynthesis exist, one mediated by DNA polymerase I and the other involving *recA*. We can now report a more complete characterization of the long-patch pathway and its genetic control.

A. Time Course

The size distribution of patches synthesized during various intervals after irradiation was investigated using the *dnaB266* mutant (strain PCH10, temperature sensitive for DNA replication) at the restrictive temperature. When repair label was present continuously for the entire 45 minutes needed to complete the repair process under those conditions, the typical distribution of long and short patches was obtained (Fig. 4). There was essentially no incorporation of repair

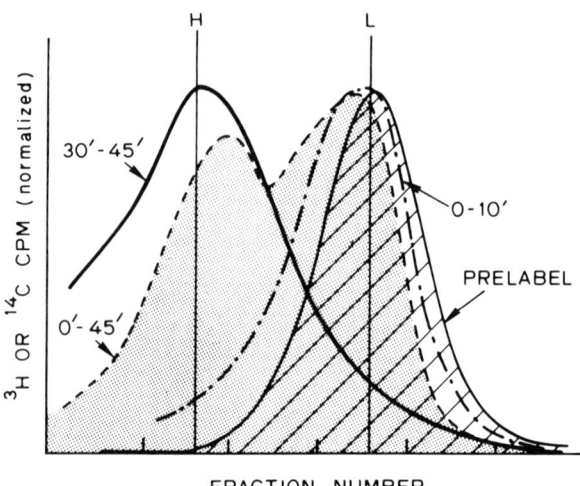

FIG. 4. Time course of long-patch repair replication in *Escherichia coli*. Radioactivity profiles of alkaline CsCl gradients are shown for sonicated DNA from the *dnaB266* mutant strain PCH10. Cultures were labeled with [^{14}C]thymine at 32°C, irradiated with 40 J/m^2 UV, and incubated in bromouracil-containing medium at 43°C to totally suppress semiconservative DNA synthesis. [^{3}H]Bromouracil was added to portions of the culture for the time intervals indicated and patch size analysis was carried out as previously described. The radioactivity profiles obtained have been superimposed, using the peak position of parental density DNA (L) for alignment. Each ^{3}H profile for the various times shown has been normalized by the radioactivity in its peak fraction. The position of fully bromouracil-substituted DNA is indicated by (H).

label in unirradiated cells, thus ruling out the possibility that long patches represent small stretches of semiconservative replication. After 10 minutes of postirradiation incubation, all the repair label was found in the short-patch position, and it was not subsequently chased into long patches when incubation was continued for an additional 35 minutes in medium containing nonradioactive bromouracil. Thus, only short patches are produced immediately after irradiation. Conversely, when the [^{3}H]bromouracil was added to the nonradioactive density label for the final 15 minutes of the 45-minute repair period, all the repair label appeared in long patches. At intermediate times during the excision-repair process, both long and short patches were synthesized. The delay in appearance is consistent with a requirement for induction of new proteins to perform the long-patch repair, a possibility also raised by the original demonstration that the $recA^+$ genotype is required for long-patch repair (31).

B. Inducibility and Genetic Control

Early studies showed that repair synthesis is reduced under conditions of amino-acid starvation in UV-irradiated *E. coli* strain TAU-bar (32), and it was subsequently shown that protein synthesis inhibition by chloramphenicol reduces repair synthesis in strain W3110 thy^- by preferentially inhibiting long-patch repair (19). These findings support the interpretation that long-patch repair is induced in response to DNA damage. We have examined the UV dose-response for repair synthesis in the presence and in the absence of chloramphenicol in W3110 and *dnaB266* strain PCH10 and find that the reduction in repair systhesis by inhibition of protein synthesis is slight at low doses and becomes maximal at 60–80 J/m^2, consistent with a requirement for induction. A similar dose dependence has been demonstrated for UV-induced Weigle reactivation of λ in uvr^+ bacteria (33). The repair patch size distribution in strain W3110 after 40 J/m^2 or 60 J/m^2 is shown in Fig. 5, together with the striking inhibition of the long-patch component by chloramphenicol. In other experiments, a similar inhibition of long-patch repair by rifampicin has been demonstrated.

Direct evidence that the long-patch mode of excision repair is induced by damage was obtained by the demonstration (19) that the inhibition of long-patch repair by chloramphenicol after 40 J/m^2 UV can be overcome if the cells were previously irradiated with a lower dose of UV (20 J/m^2) and allowed to carry out protein synthesis for 30 minutes. More recent studies (P. Cooper, unpublished) have shown that the long-patch pathway, like other "SOS functions,"[2] is induced by incubating a *tif-1* mutant at 43°C. Thus, long-patch repair was not reduced by rifampicin in the *tif* mutant when the UV irradiation was performed after 1 hour of incubation at 43°C. The final extent of repair synthesis was increased by *tif*-induction of the long-patch system prior to irradiation as compared to the usual case in which induction occurs subsequent to and as a result of DNA damage. However, this increase was much less than the (at least 10-fold greater) amount of repair synthesis expected if long patches were produced at each excision site once the system is induced. We interpret this result to mean that there must be a small defined class of sites reparable with long patches.

[2] SOS repair (or functions) may be defined as the error-prone DNA repair activity (or functions associated therewith) in *E. coli* that is dependent upon *recA* and *lexA* derepression (i.e., expression) and is brought about by UV irradiation or other agents having in common the ability to damage DNA or interrupt its synthesis in *E. coli* (After E. Witkin, Bacteriol. Rev. **40**, 869 (874) (1976). See also Witkin in this volume (p. 247). [Ed.].

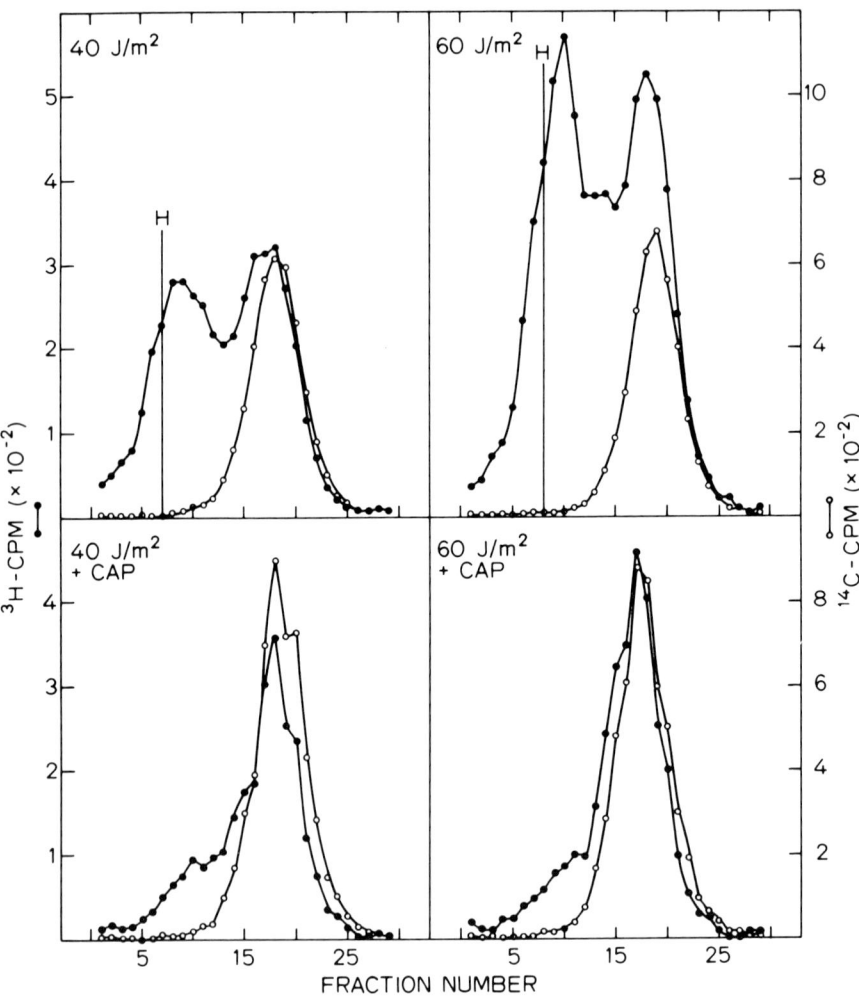

FIG. 5. Protein-synthesis dependence of long-patch repair replication. Radioactivity profiles of alkaline CsCl gradients of sonicated DNA are shown. *Escherichia coli* W3110 thy^- was labeled with [^{14}C]thymine at 32°C, irradiated in buffer with UV as noted, and incubated with [^{3}H]bromouracil for 45 minutes at 43°C. Chloramphenicol (CAP; 100 μg/ml) was added to portions of each culture for 15 minutes preceding irradiation and for the entire post-UV repair period. The vertical line indicates position of fully bromouracil substituted DNA.

We have used thermal induction of the long-patch system in *tif* mutants to evaluate the biological significance of this pathway. An appreciable enhancement of survival was obtained when *tif-1 sfiA11* cells were induced by incubation at 43°C prior to UV irradiation as compared to cells that had not been induced (P. Cooper, unpublished). This increased survival was largely, though not entirely, dependent on the uvr^+ genotype. Increased survival of UV-irradiated bacteriophage λ in the *tif* mutant incubated at 43°C prior to infection was identical to that of irradiated cells under the same conditions. The effect was similarly much more pronounced in *tif⁻ sfi⁻ uvr⁺* than in the *tif⁻ sfi⁻ uvrA6* strain. Thus, an inducible, uvr^+-dependent process, probably long-patch excision repair, is important for survival of UV-irradiated *E. coli* as well as for the enhanced survival (Weigle reactivation)[3] of UV-irradiated phage λ.

To determine which gene products are required for inducible long-patch repair, we have assayed repair synthesis in the presence and in the absence of chloramphenicol for a variety of mutant strains (Fig. 6). The inducible repair synthesis component is determined in these experiments as the difference between the value obtained with and without protein synthesis. Mutations in the *recB* and *recC* genes have no effect on inducible repair synthesis, thus ruling out the possibility that these are required for long-patch repair. This result also eliminates the possibility that the evident reduction in long-patch repair by chloramphenicol in wild-type cells is a consequence of increased postirradiation DNA degradation and the resulting dilution of repair label, since there is very little DNA degradation in the *recB recC* mutant. The very slight amount of inducible repair synthesis and the results of patch size analysis in the *lexA* mutant support a requirement for $lexA^+$ as well as $recA^+$ for long-patch repair. No one of the three DNA polymerases appears to be essential for the inducible repair component. When polymerase I is deficient, total repair synthesis is dramatically increased (*31*), but the proportion of that synthesis that is inducible is approximately the same as in wild-type cells. The most plausible interpretation of this result is that constitutive repair synthesis performed by polymerases II or III produces somewhat longer patches than the pol I-dependent 20-nucleotide patch, probably as a result of the intrinsic differences in properties of the enzymes, but

---

[3] Weigle (W-) reactivation may be defined as the increased survival and degree of mutagenesis of UV-irradiated λ phage brought about by a low level of UV irradiation of the host *E. coli* prior to infection. (After E. Witkin. *Bacteriol. Rev.* **40**, 869 (874) (1976). [Ed.].

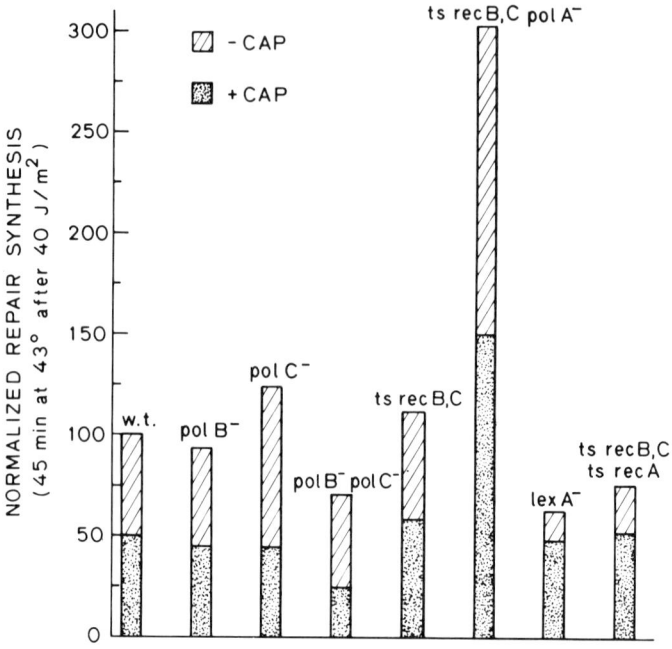

FIG. 6. Constitutive and inducible repair synthesis in different mutant strains of *Escherichia coli*. Repair synthesis determinations in the presence and in the absence of chloramphenicol (CAP) were carried out as described for Figs. 2 and 5, using a UV dose of 40 J/m^2, followed by postirradiation incubation for 45 minutes at 43°C. The repair synthesis value obtained for the appropriate wild-type strain without CAP in each experiment was used to normalize the value for the corresponding mutant strain without CAP. Strains used in order from left to right were as follows: W3110 (*thy*$^-$) or AB1157 (*thy*$^-$); HMS85 (*polB*1 *thy*$^-$ *lys*$^-$ *lac*Z); BT1026-1 (*dna*E1026); H10265/1 (*polB*1 *dna*E1026 *thy*$^-$ *endA strA*); *thy*$^-$ derivative of SK129 (AB1157 *recB*270 *recC*271); *thy*$^-$ derivative of SK211 (*polA*1 *recB*270 *recC*271); *thy*$^-$ derivative of DM822 (AB1157 *lexA*3); *thy*$^-$ derivative of JC11,391 (*recB*270 *recC*271 *recA*200).

that these are distinct from the much longer patches that are made at a minor class of sites when induction occurs. The reduced constitutive repair synthesis in the *polB polC* double mutant also implies that in wild-type cells some dimer sites are repaired by polymerases II and III with slightly longer patches in the absence of induction. Although it is difficult to be certain that no residual activity of polymerases III or II remains, the observation that inducible long-patch repair still occurs in this strain suggests that polymerase I may be capable of long-patch repair under inducing conditions.

C. Possible Mechanisms

There are several possible molecular mechanisms that may account for the dependence of the long-patch pathway upon induction of *recA* protein. This protein participates in a complex regulatory system in *E. coli* by possessing a protease activity that is activated when cellular DNA replication is inhibited (e.g., by damage) or when the altered protein in a *tif* mutant is incubated at 43°C. This protease activity cleaves the *lexA*-coded repressor, inducing the synthesis of more *recA* protein, and also results in the induction of a number of other proteins (34), either by cleaving their repressors or because they too are under *lexA* control. In addition, the *recA* protein participates directly in reactions with DNA, e.g., the promotion of strand assimilation (35). Thus induction of *recA* protein may enable long-patch repair by causing derepression of one or more essential proteins or by providing large amounts of the activity of the *recA* protein that interacts directly with DNA, or perhaps even by providing more of the *recA* protease activity.

Why does the inducible pathway result in repair synthesis in stretches longer than those synthesized by the constitutive system? It seems likely that the long-patch system operates on some special lesion or lesion configuration. Although possible, it seems unlikely that some minor photoproduct of UV irradiation can be repaired only by the long-patch pathway. Special configurations that might require the inducible pathway would include dimers in close proximity to other lesions, and dimers in or near specific regions of DNA in which repair by the constitutive pathway is blocked or prevented from completion. These might include actively transcribing regions, blocked replication forks, and possibly regions important for the maintenance of the ordered structure of the bacterial chromosome. In such instances, the increased size of the patch may reflect a necessity for more extensive excision and resynthesis brought about by the nature of the lesion-configuration being repaired. The size of the patch could also reflect intrinsic properties of the enzymic components used in this pathway that are modified or different from those of the constitutive pathway. If long-patch repair utilizes the direct interaction between *recA* protein and DNA, it is possible that the inducible system might not be analogous to the constitutive system, but might promote some recombinational event or at least some process involving steps similar to those involved in recombination. This process must be uvr^+ dependent, but might or might not result in the actual loss of pyrimidine dimers from DNA. The long patches might thus represent repair-type synthesis

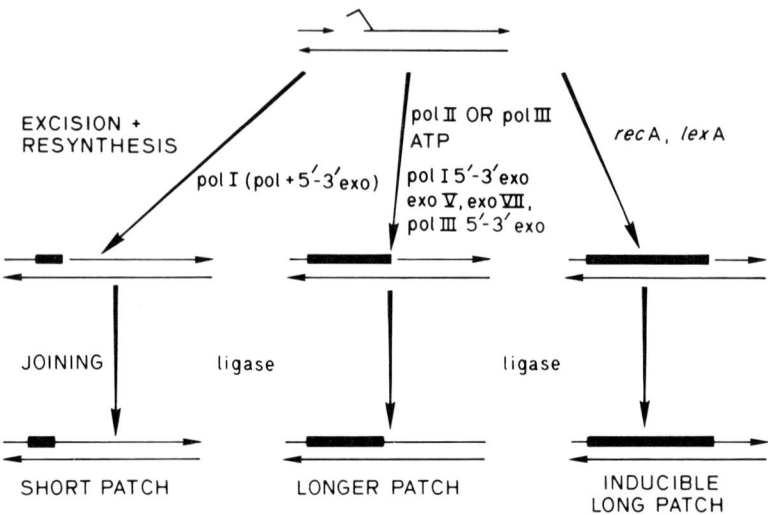

FIG. 7. Pathways for excision-repair in *Escherichia coli*, a working model.

necessary for the completion of this process, rather than resynthesis following excision of damage. Although our current knowledge of the details of long-patch repair bears on these possibilities, direct evidence concerning them is not yet available.

A model for the alternative pathways of excision repair in *E. coli* is presented in Fig. 7. It attempts to incorporate the results obtained with most of the mutants that have been studied.

IV. Summary

The excision-repair of damaged DNA must involve a coordinated sequence of steps beginning with recognition of the lesion. After phosphodiester bonds are broken, the integrity of the interrupted DNA strand is restored through the enzymic processes of degradation, repair resynthesis, and ligation. Pathways for excision-repair in *E. coli* have been elucidated through analysis of mutants deficient in various enzyme activities.

Three pathways of excision-repair can be distinguished in *E. coli*:

a pathway mediated by DNA polymerase I that results in short repair patches approximately 20 nucleotides in extent; a pathway that results in a somewhat larger average patch size, probably due to intrinsic properties of DNA polymerases II and III; and a pathway that is induced in response to DNA damage and is dependent upon protein synthesis, requires $recA^+$ and $lexA^+$, and leads to much longer patches. Short patches are initiated and completed earlier than the long patches produced by the inducible pathway. The inducible long-patch system appears to operate at a defined class of sites constituting a minority of the lesions, but the ability to perform such long-patch repair seems to be important for survival of irradiated cells. Thus, induction in uvr^+ cells prior to irradiation results in appreciably higher survival, while only a small increase in resistance is seen in uvr^- cells. Enhanced survival of UV irradiated phage λ by preinduction of the host in the same way is similarly much greater in uvr^+ than in $uvrA^-$ cells. Therefore, long-patch repair may be an important component of Weigle reactivation (see footnote 3).

In UV-irradiated human cells only a single class of repair patches, comparable in size to the short patches in *E. coli*, is synthesized at both early and late times after irradiation. The same patch size distribution also results from treatment of human cells by angelicin or 8-methoxypsoralen plus long-wavelength UV, or by activated aflatoxin B1.

Acknowledgment

This work was supported by a Contract No. DE-AC03-76SF00326-7 with the Department of Energy and a Grant No. GM-09901 from the National Institute of General Medical Sciences, DHEW.

References

1. P. C. Hanawalt, P. K. Cooper, A. K. Ganesan and C. A. Smith, *ARB* **48**, 783 (1979).
2. T. Lindahl, This Series **22**, (1979).
3. P. C. Hanawalt, E. C. Friedberg and C. F. Fox, eds., "DNA Repair Mechanisms." Academic Press, New York, 1978.
4. J. Cairns, P. Robins, B. Sedgwick and P. Talmud, this volume.
5. W. A. Deutsch and S. Linn, *PNAS* **76**, 141 (1979).
6. Z. Livneh, D. Elad and J. Sperling, *PNAS* **76**, 1089 (1979).
7. L. Grossman, S. Riazzudin, W. A. Haseltine and C. P. Lindan, *CSHSQB* **43**, 941 (1979).
8. W. A. Haseltine, L. K. Gordon, C. P. Lindan, R. H. Grafstrom, N. L. Shaper and L. Grossman, *Nature* **285**, 634 (1980).
9. E. H. Radany and E. C. Friedberg, *Nature* **286**, 182 (1980).
10. P. C. Seawell, C. A. Smith and A. K. Ganesan, *J. Virol.* **35**, 790 (1980).
11. E. C. Friedberg, C. T. M. Anderson, T. Bonura, R. Cone, E. H. Radany, and R. J. Reynolds, this volume.

12. E. Seeberg, this volume.
13. H. R. Warner, B. F. Demple, W. A. Deutsch, C. M. Kane and S. Linn, *PNAS* **77**, 4602 (1980).
14. A. Taketo, S. Yasuda and M. Sekiguchi, *JMB* **70**, 1 (1972).
15. K. Tanaka, H. Hayakawa, M. Sekiguchi and M. Okada, *PNAS* **74**, 2958 (1977).
16. G. Ciarrocchi and S. Linn, *PNAS* **75**, 1887 (1978).
17. C. A. Smith and P. C. Hanawalt, *PNAS* **75**, 2598 (1978).
18. P. Cooper, *Mol. Gen. Genet.* **150**, 1 (1977).
19. P. K. Cooper and J. G. Hunt, see p. 255 of reference 3.
20. E. C. Friedberg and I. R. Lehman, *BBRC* **58**, 132 (1974).
21. D. E. Pettijohn and P. C. Hanawalt, *JMB* **8**, 170 (1964).
22. C. A. Smith, P. K. Cooper and P. C. Hanawalt, *in* "DNA Repair: A Laboratory Manual of Research Procedures" (E. C. Friedberg and P. C. Hanawalt, eds.) in press, Dekker, New York, 1981.
23. P. K. Cooper and P. C. Hanawalt, *JMB* **67**, 1 (1972).
24. H. Edenberg and P. C. Hanawalt, *BBA* **272**, 361 (1972).
25. C. A. Smith, see p. 311 of reference 3.
26. A. Sarasin, C. A. Smith and P. C. Hanawalt, *Cancer Res.* **37**, 1786 (1977).
27. J. Kaye, C. A. Smith and P. C. Hanawalt, *Cancer Res.* **40**, 696 (1980).
28. J. D. Regan and R. B. Setlow, *Cancer Res.* **24**, 3318 (1974).
29. J. D. Regan, A. A. Francis, W. C. Dunn, D. Hernandez, H. Yagi and D. M. Jerina, *Chem. Biol. Interact.* **20**, 279 (1978).
30. P. K. Cooper and P. C. Hanawalt, *Photochem. Photobiol.* **13**, 83 (1971).
31. P. K. Cooper and P. C. Hanawalt, *PNAS* **69**, 1156 (1972).
32. P. K. Cooper, Ph.D. Thesis, Stanford University, Stanford, California, 1971.
33. M. Defais, P. Caillet-Fauquet, M. S. Fox and M. Radman, *Mol. Gen. Genet.* **148**, 125 (1976).
34. C. J. Kenyon and G. C. Walker, *PNAS* **77**, 2819 (1980).
35. K. McEntee, G. M. Weinstock and I. R. Lehman, this volume.

Recent Developments in the Enzymology of Excision Repair of DNA

> ERROL C. FRIEDBERG
> CORRIE T. M. ANDERSON
> THOMAS BONURA
> RICHARD CONE[1]
> ERIC H. RADANY AND
> RICHARD J. REYNOLDS[2]
>
> *Laboratory of Experimental*
>   *Oncology*
> *Department of Pathology*
> *Stanford University*
> *Stanford, California*

Recent years have witnessed an exciting explosion of new information on the enzymology of DNA repair, particularly that involving the excision of damaged bases, and the purpose of this article is to review, in a historical context, the progress in our understanding of the enzymology of excision repair from the time of the discovery of the phenomenon in the mid 1960s to the present. Current information suggests that the specificity of the enzymes involved in the excision of damaged or inappropriate bases from DNA resides with the very early enzymic events. Once the phosphodiester backbone of DNA is incised in proximity to sites of base damage, the enzymes involved in the subsequent excision, resynthesis, and ligation are not dependent on the presence of base damage in the DNA. Thus, for example, the excision of pyrimidine dimers by exonucleolytic degradation of DNA in the $5' \rightarrow 3'$ direction is dependent on the availability of single-strand breaks (nicks), which are recognized as substrate sites by specific exonuclease activities, irrespective of the presence of dimerized pyrimidines on the $5'$ side of the nicks. Similarly, repair synthesis of DNA is primarily dependent on available primer-templates, and DNA ligation on suitably juxtaposed $3'$ OH and $5'$ P termini. In this article, much of our attention is focused on the molecular events associated

---

[1] Present address: Office of Student Affairs, School of Medicine, University of Cincinnati, Cincinnati, Ohio 45221.

[2] Present address: Radiobiology Laboratories, Department of Physiology, Harvard School of Public Health, Boston, Massachusetts 02115.

with the *incision* of DNA rather than the enzymology of excision, resynthesis, and ligation. Except where necessary and appropriate in our view, a detailed description of most of the enzymes discussed has been avoided. In this regard we would point out that Lindahl has provided a comprehensive and authoritative review of DNA glycosylases and apurinic/apyrimidinic endonucleases in an earlier volume of this series (1). The interested reader should also consult the proceedings of the 1978 ICN–UCLA Symposium on DNA Repair (2).

I. Excision Repair of Pyrimidine Dimers in DNA—The Original Model

An early, and until recently widely quoted, model of the excision of damaged bases from DNA evolved in the mid-1960s as a result of two fundamentally important experimental observations. The first, reported by Setlow and Carrier (3), by Boyce and Howard-Flanders (4), and by Riklis (5), was that, when wild-type strains of *Escherichia coli* were exposed to ultraviolet (UV) radiation and then incubated for varying periods of time, thymine-containing pyrimidine dimers disappeared progressively from the acid-insoluble fraction of the DNA and appeared in the acid-soluble fraction. Certain strains known to have an abnormal sensitivity to UV radiation failed to demonstrate this response. The second observation, by Pettijohn and Hanawalt (6), was that, during post-UV incubation, wild-type strains of *E. coli* undergo a non-semiconservative mode of DNA synthesis (now designated "repair synthesis"). These two observations suggest that pyrimidine dimers are enzymically excised from DNA, leaving single-stranded gaps that are filled by repair synthesis. Given the information on nucleic acid enzymology existing at that time, it was totally rational to postulate that the initial enzymic event in this process was the incision of the DNA duplex catalyzed by a specific endonuclease that recognized sites of DNA damage in general, or pyrimidine dimers specifically (3–6).

A more detailed model of the molecular mechanism of the excision of pyrimidine dimers from DNA emerged as the result of studies from a number of laboratories on the characterization of specific enzymes believed to play a role in this process. Studies in this laboratory demonstrated that extracts of T4-infected cells catalyzed the selective incision of UV-irradiated DNA, whereas extracts of *E. coli* that had been infected with the UV-sensitive mutant $T4_{v1}$ failed to do so (7). We, as well as others, isolated and purified an activity designated as "T4 UV endonuclease" or "endonuclease V" of phage T4 (8–12). This activity

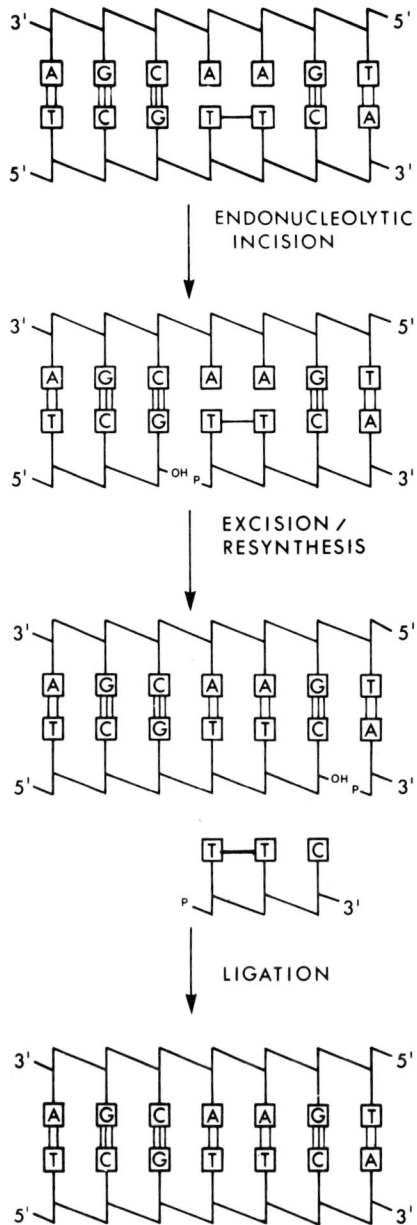

FIG. 1. Conventional model of the molecular mechanism of pyrimidine dimer excision from DNA. Hydrolysis of a phosphodiester bond on the 5' side of the dimer is postulated to occur by endonucleolytic incision as a primary enzymic event. Dimer excision and repair synthesis are shown as coordinate rather than sequential enzymic events.

catalyzes the production of single-strand breaks in duplex UV-irradiated DNA, creating 3' OH and 5' P termini at nicks located 5' with respect to pyrimidine dimers (12, 13). Properties identical to those of the phage T4 activity mentioned above were demonstrated with two DNA incising activities purified from extracts of *Micrococcus luteus* ("correndonucleases I and II") (14, 15) and with an activity isolated from *E. coli* (16).

The observation that the incision of UV-irradiated DNA occurs 5' with respect to pyrimidine dimers indicated that the excision of dimers *in vivo* must be effected by exonucleoytic degradation of DNA in the 5' → 3' direction. A number of enzymes with such defined polarity have been identified and characterized from *E. coli*. The 5' → 3' exonuclease activities of both DNA polymerases I and III can catalyze the excision of pyrimidine dimers *in vitro*, as part of an oligonucleotide sequence, from UV-irradiated DNA previously incised with dimer-incising activity from T4-infected cells (17) or from *M. luteus* (18), respectively. Similar results have been reported with exonuclease VII of *E. coli* (19) and with exonuclease V (20). Similarly, the presence of an available 3' OH group indicates the obvious potential for productive binding of a DNA polymerase for repair synthesis. On the basis of these observations, plus the discovery and characterization of the enzyme polynucleotide ligase, the now familiar model shown in Fig. 1 evolved (see references *1*, and *21-23* and references cited therein). Endonuclease-catalyzed hydrolysis of a phosphodiester bond 5' to the dimer was postulated as an initial event. The 3' OH terminus created allows for binding of an enzyme such as DNA polymerase I, which by virtue of its associated 5' → 3' exonuclease activity, can catalyze coordinate repair synthesis and dimer excision. After the replacement of all excised nucleotides, the last newly inserted nucleotide is ligated to the extant DNA in covalent phosphodiester linkage to complete the excision-repair process.

II. DNA Glycosylases and Base-Excision Repair

The discovery of a new class of enzymes that catalyze the excision of damaged or inappropriate bases from DNA as the free base was a significant landmark in the enzymology of DNA repair (24). Such enzymes are termed DNA glycosylases (25) because they catalyze the hydrolysis of the glycosylic bond linking the N-9 of purines or the N-1 of pyrimidines to the C-1 of the deoxyribose moieties in DNA. To date, four such enzyme activities have been demonstrated in extracts of both prokaryote and eukaryote cells. Each is apparently highly specific for a particular form of monoadduct base damage. The properties of most of these enzymes have recently been thoroughly reviewed by Lindahl (1; also see 26–29). All four activities isolated from extracts of E. coli are monomeric proteins of molecular weight < 30,000. None has any known cofactor requirement, and all are specific for deoxyribonucleic acid or deoxyribopolymers. (See Note Added in Proof, #1.)

We have investigated the properties of dUrd(DNA)[3] glycosylase from *Bacillus subtilis* (30) as well as the enzyme present in human KB cells (31). Our interest in the *B. subtilis* enzyme stems from the recognition that this organism is the natural host for the phages PBS1 and PBS2, both of which naturally contain uracil instead of thymine in their DNA (32). The enzyme from *B. subtilis* has properties very similar to those of *E. coli* (25). It is a monomeric protein of MW 24,000 with a Stokes' radius of 19.3 Å. The enzyme is fully active in the presence of 10 mM EDTA and has no cofactor requirement. Activity is inhibited in the presence of Co^{2+}, Fe^{2+}, or Zn^{2+}. The enzyme is specific for deoxyuridine in single- and double-stranded DNA or in deoxyribopolymers and does not release free uracil from RNA or from poly(rU) · poly(dA). In addition, neither dUrd, dUMP, nor dUTP are recognized as substrates. The K_m for poly(dU)$_{\overline{200}}$ is $1 \cdot 1 \times 10^{-9}$ M. The enzyme catalyzes uracil release from small oligo(dU)s, but the minimum size recognized as substrate is p(dU)$_4$.

Within approximately 6 minutes after the addition of phage PBS2 to a logarithmically growing culture of *B. subtilis*, dUrd(DNA) glycosylase activity is no longer detectable in extracts of harvested cells (33). This phenomenon apparently requires protein synthesis because, when chloramphenicol is added to the culture immediately after phage infection, normal levels of this glycosylase activity are observed. (33). These results strongly suggest that early after infection phage PBS2 expresses an inhibitor of the glycosylase activity. We have recently purified this inhibitor to near physical homogeneity (33a). It is a protein, as judged by its sensitivity to proteolysis and insensitivity to digestion with DNase and RNase. In the early stages of its purification, it shows a remarkable stability. It is insensitive to boiling for at least 5 minutes, and even when highly purified, the activity is recoverable from sodium dodecyl sulfate (SDS) or SDS–urea polyacrylamide gels.

[3] dUrd(DNA) = "deoxyuridine (in DNA)," the substrate for the glycosylase. Since uracil has no glycosyl bond, the name, "Ura-DNA glycosylase," sometimes used, is incorrect and should be abandoned. The same criticism may be leveled at "DNA(hypoxanthine) glycohydrolase" [EC 3.2.2.15; See *EJB* **104**, 1(1980)]; the substrate here is deoxyinosine (in DNA), not hypoxanthine. [Ed.]

The inhibitor has a molecular weight, estimated by gel filtration, of ~18,000 and a relative s value of 1.44 S, determined by velocity sedimentation in glycerol gradients. In dodecylsulfate/polyacrylamide gels the inhibitor behaves anomalously relative to a number of other proteins. Specifically, its electrophoretic mobility does not vary predictably with the acrylamide concentration of the gel. Thus it has not been possible to obtain a molecular weight measurement by this procedure. This anomalous behavior suggests that the inhibitor might be a glycoprotein; however, it fails to bind to a variety of lectin affinity columns. The inhibitor is fully active in the presence of EDTA and has no requirement for any known cofactor. Preliminary experiments indicate that it acts by binding directly to dUrd(DNA) glycosylase (R. Cone and E. C. Friedberg, unpublished). This binding is apparently highly specific, since the inhibitor has no activity against other DNA glycosylases from *E. coli* (P. Karran, R. Cone, and E. C. Friedberg, unpublished), but does inhibit the glycosylase from a number of biological sources (P. Karran, R. Cone and E. C. Friedberg, unpublished). While it is clear that the PBS2-induced inhibitor is expressed early after infection, an intriguing question yet to be answered is how the PBS2 genome (which has a base composition of 33% as uracil) survives degradation between the time of injection of the genome and expression of active inhibitor. Studies by others have shown that an inhibitor of the glycosylase is also expressed following infection of *B. subtilis* with phage PBS1 [34].

The dUrd(DNA) glycosylase is a ubiquitous enzyme, and a number of laboratories have found it in extracts of mammalian cells, including human cells [28, 29, 35, 36]. In our hands, the enzyme activity is located in both nuclear and mitochondrial subcellular fractions of human KB cells [31], suggesting a functional association with both sites of intracellular DNA replication (see Section III,A). It is difficult to be absolutely certain that the activity associated with the mitochondrial fraction reflects an intramitochondrial location in living cells. However, when crude mitochondrion-derived enzyme is incubated with isolated intact mitochondria, no spurious binding of activity to the mitochondria is observed [31]. Furthermore, the specific activity of soluble enzyme increases significantly when mitochondria are lysed, relative to the activity detected in mitochondria assayed under isosmotic conditions [31]. The enzyme from both subcellular fractions is currently being purified and characterized in our laboratory in the hope of detecting biophysical or biochemical differences suggestive of distinct genes coding for each. In this respect, it is of interest that Sirover [28] has recently reported the presence of two chromatographically distinct forms of dUrd(DNA) glycosylase in human peripheral blood lymphocytes stimulated by phytohemagglutinin, only one of which varies in activity as a function of the cell cycle.

Evidence that supports a role of DNA glycosylases in the excision of altered bases in living cells has emerged from a number of studies (37–43). Mutants of *E. coli* deficient in the glycosylase activity (ung^-) (38) are abnormally sensitive to treatment with nitrous acid (42) or with sodium bisulfite (43). Similarly, such mutants are deficient in the reactivation of T-odd phages treated with sodium bisulfite, but not in the reactivation of T-even phages so treated (43). This result is consistent with the observation that hydroxymethylcytosine (the form of cytosine present in the DNA of T-even phages) is resistant to deamination by sodium bisulfite (44). In addition, ung^- cells have an increased spontaneous frequency of mutation involving transition of G · C → A · T base-pairs (40). Mutants deficient in d(Urd)DNA glycosylase have also been isolated from *B. subtilis* (37) and have properties similar to those of ung^- strains of *E. coli* (39, 41). Mutants defective in 3-methyldeoxyadenosine(DNA) glycosylase activity (tag^-) have also been isolated (44a) and, as expected, are abnormally sensitive to treatment with alkylating agents such as methyl methanesulfonate.

The discovery of DNA glycosylases and their general mechanism of action led to the suggestion that two models of excision repair should be distinguished (45), i.e., nucleotide excision repair (best exemplified by the model of pyrimidine dimer excision quoted above) and base-excision repair (the excision of damaged bases as the free base). This distinction seemed reasonable on two counts. First, the form in which the damaged base was excised differed. Second, the enzymic mechanisms apparently operating were distinct, i.e., endonucleolytic incision of DNA as a primary event during nucleotide excision, and hydrolysis of N-glycosylic bonds as a primary event during base excision. As indicated in Section IV, very recent insights into the enzymology of pyrimidine dimer excision in at least two prokaryote systems suggest that, while the products of excision are indeed in the nucleotide form, the enzymes involved in incision are DNA glycosylases and apurinic/apyrimidinic endonucleases (see Section III).

III. Apurinic/Apyrimidinic Endonucleases

The enzyme-catalyzed excision of free bases leaves apurinic and/or apyrimidinic sites in DNA. Existing enzymologic information suggests two distinct mechanisms by which the repair of such sites can be completed.

A. The Degradative Pathway of Repair of Sites of Base Loss

Further degradation of the DNA can occur by the action of a specific nuclease of a class called apurinic/apyrimidinic (AP) endonucleases, which catalyze the hydrolysis of phosphodiester bonds specifically at sites of base loss (see ref. 1). This class of enzymes was first discovered in *E. coli* by Verly and his colleagues (46), who purified and characterized the quantitatively major AP endonuclease from this organism (47, 48). Since then, it has been established that this enzyme is the same protein as exonuclease III of *E. coli* coded by the *xth* gene (49, 50). The AP endonuclease activity is sometimes designated as endonuclease VI of *E. coli*, and is almost certainly the same enzyme as that originally called endonuclease II (see ref. 51 for a discussion of this nomenclatural confusion).

In addition to exonuclease III, several other AP endonuclease activities specific for sites of base loss in duplex DNA have been described in *E. coli*. These activities can be conveniently classified as enzymes with only AP endonuclease activity, and enzymes with other associated catalytic functions. The only known member of the former class is endonuclease IV, which has no associated exonuclease or phosphatase activities and, insofar as has been tested, recognizes no other substrates in DNA (52). On the other hand, endonucleases III (53) and V (54) have been shown to have a broad substrate specificity that includes sites of base loss in DNA. A question of obvious interest and importance is whether or not all the aforementioned AP endonucleases function as such in living cells. This is particularly relevant with respect to endonucleases III and V. The answer to this question can perhaps be most directly ascertained by the isolation of mutants defective in these activities. One rational basis for the existence of more than one AP endonuclease may relate to distinct K_m's *in vivo* for apyrimidinic and apurinic sites. Although it has been demonstrated qualitatively that some of the enzymes mentioned above recognize both sites in duplex DNA, no reliable quantitative comparisons have yet been reported.

None of the AP endonucleases of *E. coli* thus far discussed catalyze the hydrolysis of phosphodiester bonds at sites of base loss in single-stranded DNA. Recent studies in this laboratory have revealed the presence of an endonuclease activity in *E. coli* that does (T. Bonura and E. C. Friedberg, unpublished) (55). This enzyme is designated as endonuclease VII. It is a protein with an apparent molecular weight of about 60,000 as measured by sedimentation velocity in glycerol and by gel filtration. The single-stranded DNAs or deoxyribopolymers on which the enzyme is active include denatured phage PBS2 DNA, M13

DNA containing uracil [prepared from M13 phage grown in a strain of *E. coli* defective in dUrd(DNA) glycosylase and deoxyuridine triphosphatase activity], poly(dU), and poly(dT, dU). These all share in common a single-stranded conformation and the presence of deoxyuridine residues. Incubation with purified dUrd(DNA) glycosylase results in

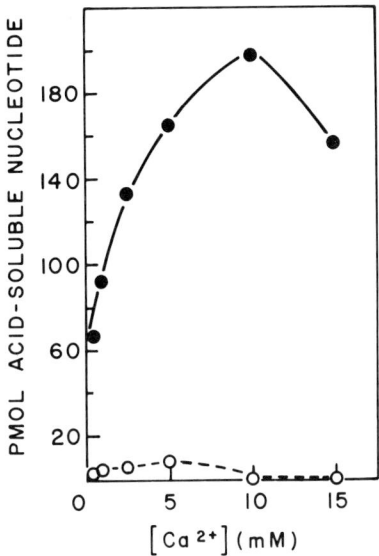

FIG. 2. Effect of CaCl$_2$ on the activity of endonuclease VII (55). A single-stranded deoxyribopolymer containing a random mixture of uracil and thymine [poly (dU · [^{3}H]dT)] was prepared with calf thymus terminal deoxynucleotidyl transferase. Reaction mixtures (0.2 ml) contained 40 mM KCl, 0.2 M cacodylate buffer (pH 7.2), 1 mM CoCl$_2$, 1 mM 2-mercaptoethanol, 11.3 nmol of dpT$_{34}$, 0.42 μmol each of dUTP and dTTP (Sigma), 0.5 mCi of [*methyl*-^{3}H]thymidine 5′-triphosphate (15.5 Ci/mmol, ICN), and 950 units of terminal transferase. Incubations were at 37°C for 20 hours, after which reactions were terminated by heating to 65°C for 10 minutes. Polymerized deoxynucleotides were isolated from the reaction mix by gel filtration through a Sephacryl S-200 column (0.5 × 15 cm) equilibrated in and eluted with 0.5 M KCl in 50 mM potassium phosphate buffer (pH 7.0). Fractions containing acid-precipitable material were pooled. The pooled material had an average size of about 200 nucleotides by sedimentation velocity on alkaline sucrose gradients. The (dU · [^{3}H]dT)$_{\overline{200}}$ thus obtained had a specific radioactivity of between 200 and 400 cpm per picomole of nucleotide. The endonuclease VII activity measures the release of acid-soluble radioactivity from depyrimidinated (dU · [^{3}H]dT)$_{\overline{200}}$ relative to that released from intact (nondepyrimidinated) polymer. Depyrimidination was effected enzymatically by incubation of the intact polymer with purified *Bacillus subtilis* uracil-DNA glycosylase (see text footnote 3). Substrate (500 pmol as nucleotide) was incubated with 5 mM TrisCl (pH 7.4), 200 μg of bovine serum albumin per milliliter, and varying amounts of CaCl$_2$ as shown. Incubations were at 37°C for 60 minutes. ●——●, Depyrimidinated substrate; ○---○, control substrate.

the selective excision of uracil, creating depyrimidinated sites. With such substrates the enzyme is fully active in the presence of EDTA, but is stimulated about twofold by 5 mM $CaCl_2$ (Fig. 2). No absolute requirement for any cofactor has been determined. The enzyme is an endonuclease, as evidenced by its ability to attack depyrimidinated circular single-stranded M13 DNA (Table I). Single-stranded DNA without depyrimidinated sites is not degraded.

At present the substrate specificity of this enzyme is not clear. Duplex PM2 DNA containing depurinated sites created by heat or by

TABLE I
ACTIVITY OF ENDONUCLEASE VII
ON DEPYRIMIDINATED M13 DNA GROWN
IN Escherichia coli BW313
($Hfr\ dut^-\ ung^-$)[a]

Conditions	Breaks/molecule
No additions	0.0 (0)[b]
NaOH, 0.5 M	0.36 (100)
Endonuclease VII	0.085 (24)

[a] Phage M13 DNA containing uracil was prepared by growing phage in a strain of *Escherichia coli* defective in both deoxyuridine triphosphatase activity (dut^-) and in dUrd(DNA) glycosylase activity (ung^-). Depyrimidinated M13 DNA was prepared by enzymic removal of incorporated uracil by incubation with dUrd(DNA) glycosylase. Reaction mixtures (0.1 ml) contain 23 μg of M13 DNA (3.5 × 10⁴ cpm/μg), 10 mM TrisCl (pH 7.5), 1 mM EDTA, and 375 units of dUrd(DNA) glycosylase (30). Incubations were at 37°C for 15 minutes. Reactions were terminated and deproteinized by the addition of two volumes of cold buffered (pH 7.9) phenol. The DNA was ether-extracted and extensively dialyzed against 10 mM potassium phosphate buffer (pH 7.0). Analysis by gel electrophoresis showed the presence of approximately 0.4 alkali-labile sites per M13 circular DNA molecule. M13 DNA not containing uracil was subjected to the same incubation and purification procedure and shown to contain no alkali-labile sites.

[b] Expressed as percentage of alkali-labile sites in depyrimidinated DNA.

acid treatment is not attacked by partially purified preparations of endonuclease VII. Duplex PBS2 DNA containing depyrimidinated sites is also not attacked by the enzyme. Further studies on this interesting activity are in progress. The enzyme is present at wild-type levels in mutants defective in exonuclease III activity (xth^-) and in extracts of uvr A, B, or C mutants. At present we have no data that address the biological function(s) of endonuclease VII.

B. The "Insertion" Pathway of Repair of Sites of Base Loss

The enzymic hydrolysis of phosphodiester bonds at sites of base loss generates nicks in DNA that can allow for the same sequence of degradative, resynthesis, and ligation events described earlier in relation to the excision repair of pyrimidine dimers. However, an alternative mechanism for the repair of sites of base loss (whether enzyme-catalyzed or arising spontaneously) is suggested by the discovery of "DNA purine insertase" activity.[4] Such enzymes have been found in extracts of human fibroblasts (56, 57) as well as in extracts of *E. coli* (58). They catalyze the direct addition of purines to apurinic sites in duplex DNA. We are not aware of any reports of enzyme activities that catalyze a similar addition of pyrimidines to apyrimidinic DNA.

IV. The Excision Repair of Pyrimidine Dimers in DNA—The Current Model

Grossman *et al.* (59) and Haseltine *et al.* (60) have suggested a highly intriguing model for the incision of UV-irradiated DNA at pyrimidine dimer sites by the DNA incising activities originally isolated from *M. luteus*. The model emerged from studies in which the *M.*

---

[4] This term ("purine insertase") in either the present form or as "purine insertion enzyme" (or "activity"), is doubly unfortunate, because it describes a phenomenon rather than a chemical reaction (see "Enzyme Nomenclature," 1968, p. 6), and because the phenomenon is not an "insertion": the chain is a poly(deoxyribose phosphate) with the bases appended to it as side chains. Rather than an insertion, the phenomenon is one of addition or replacement, hence "addition activity" and "replacement activity" seem appropriate as purely descriptive terms.

By the conventions established ("Enzyme Nomenclature," 1968, pp. 9–12), the enzyme can be classified as a *transferase* (if ATP hydrolysis were involved, it would be a ligase or synthetase), and named systematically, according to these conventions, *purine: apurinate purinetransferase*, and trivially (Recommended Name) *apurinate purinetransferase*. [Ed.]

luteus enzyme preparation was incubated with a fragment of UV-irradiated *E. coli* DNA of known nucleotide sequence. The object of using a sequenced DNA was to observe directly whether or not the enzyme creates nicks *immediately* to the 5' side of pyrimidine dimer sites, by comparing the lengths of single-stranded oligonucleotides generated after complete reaction of the DNA with enzyme to the lengths of oligonucleotides generated from the same unirradiated DNA by the Maxam–Gilbert DNA sequencing technique (*61*). The results of this analysis were surprising. The single-stranded oligonucleotides (generated by mild alkaline denaturation of the DNA after reaction with enzyme) were about one nucleotide longer than expected relative to the electrophoretic mobility of the analogous fragments generated by the sequencing technique. However, if, after incubation with enzyme, the UV-irradiated DNA was subject to extensive alkaline treatment prior to gel electrophoresis the anticipated size distribution of oligonucleotides was obtained. On the basis of these results, the authors proposed that two distinct enzymic events occurred during the incubation of the *M. luteus* enzymes with UV-irradiated DNA (Fig. 3). One was the hydrolysis of the N-glycosylic bond linking the 5' pyrimidine of a dimer to the deoxyribose moiety of the DNA backbone, thereby creating an alkali-labile, depyrimidinated site. The second was the hydrolysis of a phosphodiester bond on the 3' side of the depyrimidinated site by AP endonuclease activity either functionally associated with the DNA glycosylase or present in the enzyme preparation as a contaminant. An identical result was obtained with a preparation of so-called phage T4 "UV endonuclease" partially purified by the procedure of Friedberg *et al.* (*9*) (R. W. Haseltine, personal communication). Recent studies (*60*) have provided direct evidence for a pyrimidine-dimer-specific DNA glycosylase activity in *M. luteus*.

The properties of the *M. luteus* "correndonucleases I and II" (*14, 15*) are very similar to those of the T4 "UV endonuclease" (*12, 13*). We have therefore reexamined the T4 enzyme for evidence of the type of pyrimidine dimer, DNA-glycosylase activity implied by the model of Grossman *et al.* (*59*) and Haseltine *et al.* (*60*). Enzyme-catalyzed hydrolysis of either of the N-glycosylic bonds of a pyrimidine dimer should result in the release of *free* cytosine or thymine if the cyclobutane covalent linkage between the bases is disrupted. We prepared DNA containing $\sim 20\%$ of the thymine as dimers by irradiating *E. coli* DNA at 254 nm in the presence of the photosensitizer $AgNO_3$ (*62*). The DNA was subjected to complete reaction with a preparation of T4 "UV endonuclease," after which it was exposed to further prolonged

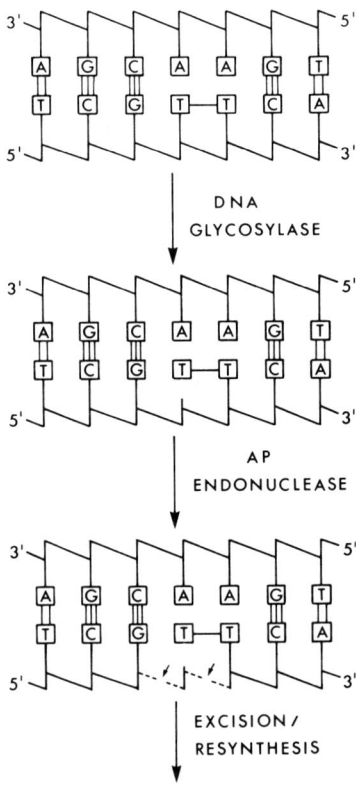

FIG. 3. New model of the enzymology of incision of UV-irradiated DNA containing pyrimidine dimers. Note that incision of the deoxyribose-phosphate backbone is postulated to result from the catalytic action of an apurinic/apyrimidinic (AP) endonuclease activity acting at a depyrimidinated site created by enzymic hydrolysis of the glycosylic bond on the 5′ side of the dimerized pyrimidines. Since it is not yet established that a unique AP endonuclease is involved in this process *in vivo*, strand breakage is shown as either 5′ or 3′ to the depyrimidinated site and the chemistry of the termini is not specified. In either event, chain incision occurs on the 5′ side with respect to the dimer.

irradiation at 254 nm. The second irradiation resulted in a net separation of pyrimidine dimers, since in DNA containing significantly more than 7% of thymine in dimers, the rate of dimer cleavage is greater than that of dimer formation; hence *net* dimer monomerization can be measured. This protocol resulted in the release of substantial amounts of free thymine measured by thin-layer chromatography of the acid-soluble fraction (63) (Fig. 4). Only background levels of free thymine were released by such photolysis of unnicked DNA, or incubation with enzyme without subsequent irradiation (Fig. 4). In further ex-

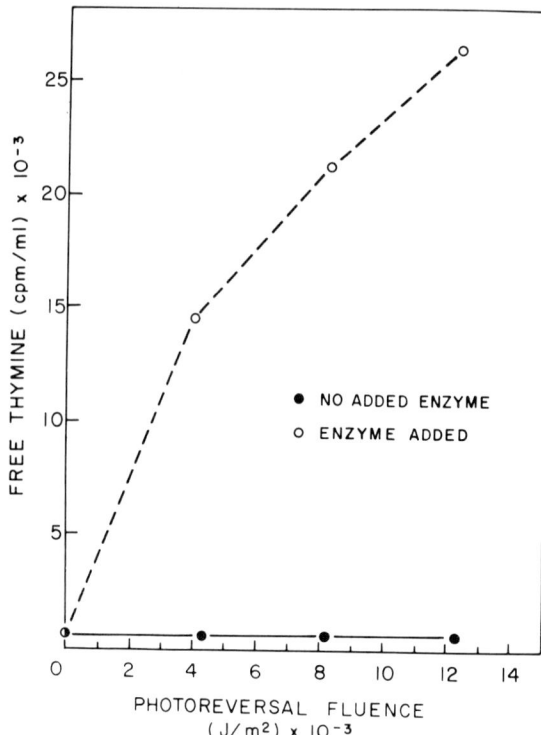

FIG. 4. Release of free thymine from DNA containing pyrimidine dimers is dependent both on incubation with phage T4 enzyme activity and on irradiation. [^{3}H]Thymine-labeled *E. coli* DNA (1.5×10^4 cpm/μg) containing 17.5% radioactivity as thymine-containing pyrimidine dimers was prepared by Ag$^+$-photosensitized UV-irradiation (62). The DNA (100 μg/ml) was incubated at 37°C with or without 9×10^4 units/ml of T4 UV-DNA incising activity (9) (generously provided by P. Seawell, Biology Department, Stanford University) in 10 mM TrisCl, 10 mM EDTA, 100 mM NaCl for 45 minutes. After incubation, the reactions were deproteinized by digestion with proteinase K and extraction with chloroform. Prior to direct photoreversal the ^{3}H-labeled DNA was mixed with 200 μg of [^{14}C]thymine-labeled *E. coli* DNA containing 21% thymine-containing pyrimidine dimers. The DNA mixtures were diluted in 10 mM TrisCl (pH 7.5), 1 mM EDTA to give a final A_{254} = 2.0 (5.0×10^5 ^{3}H cpm/ml). The ^{14}C-labeled DNA was added to monitor possible degradation of UV-irradiated DNA to acid-soluble nucleotide by photolytic irradiation. The DNA mixture was irradiated with a low-pressure germicidal lamp to the incident fluences indicated, after which duplicate 0.2-ml aliquots were withdrawn. [^{14}C]Thymine (9530 cpm) was added as a chromatographic standard to the aliquots, which were evaporated to dryness and from which the residue was extracted with ethanol. The extracts were spotted onto silica gel thin-layer plates and developed as described previously (64). ^{3}H and ^{14}C radioactivities were measured in individual chromatogram fractions. Total [^{3}H]thymine was determined by correction to 100% recovery of the [^{14}C]thymine marker. The aliquots contained < 10 cpm/ml [^{14}C]thymine prior to addition of the standard.

periments, we demonstrated an expected quantitative relationship between the amount of free thymine generated and the number of thymine-containing pyrimidine dimers lost (63a).

The release of free thymine following direct photolysis of thymine-containing pyrimidine dimers in DNA provides a convenient and specific assay for the pyrimidine-dimer, DNA-glycosylase activity present in extracts of E. coli infected with wild-type phage T4. We are currently attempting to purify this activity to physical homogeneity to determine whether the v gene product contains only the DNA glycosylase activity, or an associated AP endonuclease activity required to incise the DNA at the apyrimidinic site generated by the DNA glycosylase. (See Note Added in Proof, #2.)

A question of obvious interest is whether or not this molecular mechanism for the incision of DNA containing pyrimidine dimers is unique to M. luteus and bacteriophage T4 or is general in nature. If it can be established that most if not all organisms require a specific DNA glycosylase in physical association with a specific AP endonuclease, it might help resolve some of the apparent complexity of the biochemistry of the incision of UV-irradiated DNA in organisms such as E. coli. However, even the requirement for two distinct enzyme activities does not provide an obvious explanation for the apparent genetic complexity of the incision of UV-irradiated DNA in both lower and higher eukaryotes. Table II summarizes the results of recent studies we have carried out on a number of UV-sensitive and excision-defective strains of the yeast Saccharomyces cerevisiae (64). As noted in Table II, mutants defective at all tested alleles of four distinct loci (RAD 1, 2, 3, and 4) fail to generate measurable incision of their DNA in vivo during postirradiation incubation at 30°C. Furthermore, these strains fail to lose sites in their DNA (pyrimidine dimers) sensitive to the M. luteus UV DNA-incising activity [now thought to be a combination of a DNA glycosylase and one or more AP endonucleases (see above)]. Finally, none of these strains is defective in the enzyme activity that promotes the excision of pyrimidine dimers from preincised DNA in vitro (Table II). Thus, in S. cerevisiae, at least four and possibly more loci govern the incision of DNA during excision repair of pyrimidine dimers.

Analagous results have been obtained by others with studies of human cells in culture. Kohn et al. (65) have developed a technique for the measurement of small numbers of single-strand breaks in DNA by the gentle elution of DNA through filters, showing that during the incubation of normal human fibroblasts exposed to UV radiation, breaks in the DNA are detectable (66). When the same experiments

TABLE II
EXCISION REPAIR CAPACITY OF Saccharomyes cerevisiae[a]

Strain	Post-UV incubation-dependent single-strand-breaks (100 J m^{-2})	ESS removal (3 J m^{-2})	UV sensitivity[b]	Excision from incised DNA
WT	+++	+++	N	+++
rad1-2	−	−	VS	
rad1-4	−	−	SS	
rad1-11	−	−	VS	+++
rad2-2	−	−	VS	+++
rad2-3	−	−	VS	
rad2-4	−	±	SS	+++
rad3-1	−	−	VS	+++
rad4-2	−	−	VS	
rad4-3	−	−	VS	+++
rad7-1	+++	++	SS	+++
rad10-1			VS	+++
rad14-1	±	+	SS	+++
rad16-1				+++
rad18-1	+++	+++	VS	

[a] The table summarizes the results of a series of experiments with wild-type (WT) and mutant (rad) strains of the yeast S. cerevisiae (64); R. J. Reynolds and E. C. Friedberg, unpublished) in which the following parameters were measured: (a) the capacity of intact UV-irradiated cells to catalyze the formation of single-strand breaks in their DNA during incubation at 30°C (column 2); (b) the sensitivity of nuclear DNA isolated after incubation of UV-irradiated cells at 30°C, to a preparation of Micrococcus luteus UV-DNA incising activity that recognizes pyrimidine dimers [enzyme-sensitive sites (ESS)] in DNA (column 3); (c) sensitivity to killing by UV-irradiation (column 4); (d) the ability of extracts of yeast to catalyze the selective excision of thymine-containing pyrimidine dimers from UV-irradiated Escherichia coli DNA previously incised with M. luteus UV-DNA incising activity (column 5).

[b] N, Normal; VS, very sensitive; SS slightly sensitive.

were carried out with cells from patients suffering from the human disease xeroderma pigmentosum (in which there is a known defect in the ability to excise pyrimidine dimers) (see ref. 67 for a recent review), no measurable breaks were detected in four of the existing seven genetic complementation groups tested.

V. Conclusion

We seek in this review to emphasize the significant progress made in elucidating the details of the enzymic mechanisms of excision repair of damaged DNA. The recent discoveries of specific pyrimidine-dimer and DNA-glycosylase activities in *M. luteus* and phage T4-infected *E. coli* introduce some confusion about the current generality of classifying excision repair into distinct nucleotide (and base) excision-repair modes. However, it is quite likely that the incision of UV-irradiated DNA by an enzyme mechanism involving a specific DNA glycosylase is not universal (see article on *uvr A, B,* and *C* gene products by E. Seeberg in this volume). This argument nothwithstanding, even in the two biological systems just mentioned, evidence indicates that pyrimidine dimers are *excised* from DNA as part of a larger oligonucleotide, whereas in all known examples of base-excision repair, the products of excision are free bases. Finally, Livneh *et al.* (68) describe an activity in extracts of *M. luteus* that recognizes 8-(2-hydroxyisopropyl)purines in DNA. Preliminary evidence indicates that this activity is a true endonuclease without associated DNA glycosylase activity. Thus it is distinctly possible that the excision repair of cell damage by hydrolysis of *N*-glycosyl bonds involves a number of specific forms of base damage in which the degree of structural distortion of DNA secondary structure is limited. Very bulky DNA adducts such as those produced by the interaction of DNA with polycyclic aromatic hydrocarbons, acetamidofluorene, etc., may generate relatively nonspecific but significant distortion of the DNA helix that is recognized by a limited number of "general" endonucleases, such as that reported by Livneh *et al.* (68). Enzymes such as endonuclease V (54) of *E. coli* may also fall into this category.

Acknowledgments

The studies carried out in the authors' laboratory were supported by research grants from the United States Public Health Service (CA-12428), the American Cancer Society (NP-174), and the National Foundation—March of Dimes (1-672) as well as by contract No. EY-76-S-03-0326 with the United States Department of Energy. E. C. F. is the recipient of a Research Career Development Award (CA71005) from the United States Public Health Service. R. J. R. was supported by postdoctoral fellowships from the Bank of America—Gianinni Foundation and by Tumor Biology Training Grant No. 09151 from the United States Public Health Service. C. T. M. A. and E. H. R. are supported by a Medical Scientist Training Program grant (GM92210) from the United States Public Health Service.

References

1. T. Lindahl, This Series **22**, 135 (1979).
2. P. C. Hanawalt, E. C. Friedberg and C. F. Fox, eds., "DNA Repair Mechanisms." Academic Press, New York, 1978.
3. R. B. Setlow and W. L. Carrier, *PNAS* **51**, 226 (1964).
4. R. P. Boyce and P. Howard-Flanders, *PNAS* **51**, 293 (1964).
5. E. Riklis, *Can. J. Biochem.* **43**, 1207 (1965).
6. D. E. Pettijohn and P. Hanawalt, *JMB* **9**, 395 (1964).

7. E. C. Friedberg and J. J. King, *BBRC* **37**, 646 (1969).
8. E. C. Friedberg and J. J. King, *J. Bacteriol.* **106**, 500 (1971).
9. E. C. Friedberg, A. K. Ganesan and P. C. Seawell, *in* "Methods in Enzymology (L. Grossman and K. Moldave, eds.), Vol. 65, p. 191. Academic Press, New York, 1980.
10. P. C. Seawell, E. C. Friedberg, A. K. Ganesan and P. C. Hanawalt, *in* "DNA Repair: A Laboratory Manual of Research Procedures" (E. C. Friedberg and P. C. Hanawalt, eds.), Vol. I, Part A, p. 229. Dekker, New York, 1981.
11. S. Yasuda and M. Sekiguchi, *PNAS* **67**, 1839 (1970).
12. S. Yasuda and M. Sekiguchi, *BBA* **442**, 197 (1976).
13. K. Minton, M. Durphy, R. Taylor and E. C. Friedberg, *JBC* **250**, 2823 (1975).
14. S. Riazuddin and L. Grossman, *JBC* **252**, 6280 (1977).
15. S. Riazuddin and L. Grossman, *JBC* **252**, 6287 (1977).
16. A. G. Braun, M. Radman and L. Grossman, *Bchem* **15**, 4006 (1976).
17. E. C. Friedberg and I. R. Lehman, *BBRC* **58**, 132 (1974).
18. D. M. Livingston and C. C. Richardson, *JBC* **250**, 470 (1975).
19. J. W. Chase and C. C. Richardson, *JBC* **249**, 4553 (1975).
20. J.-I. Tanaka and M. Sekiguchi, *BBA* **383**, 178 (1975).
21. P. C. Hanawalt and R. B. Setlow, eds., "Molecular Mechanisms for Repair of DNA." Plenum, New York, 1975.
22. L. Grossman, *Adv. Rad. Biol.* **4**, 77 (1974).
23. P. C. Hanawalt, P. K. Cooper, A. K. Ganesan and C. A. Smith, *ARB* **48**, 783 (1979).
24. T. Lindahl, *Nature* **259**, 64 (1976).
25. T. Lindahl, S. Ljungquist, W. Siegert, B. Nyberg and B. Sperens, *JBC* **252**, 3286 (1977).
26. C. J. Chetsanga and T. Lindahl, *Bchem* **17**, 2110 (1978).
27. T. P. Brent, *Bchem* **18**, 911 (1979).
28. M. A. Sirover, *Cancer Res.* **39**, 2090 (1979).
29. M. Talport-Borle, L. Clerici and F. Campagnari, *JBC* **254**, 6387 (1979).
30. R. Cone, J. Duncan, L. Hamilton and E. C. Friedberg, *Bchem* **16**, 3194 (1977).
31. C. T. M. Anderson and E. C. Friedberg, *NARes* **8**, 875 (1980).
32. I. Takahashi and J. Marmur, *Nature* **197**, 794 (1963).
33. E. C. Friedberg, A. K. Ganesan and K. Minton, *J. Virol.* **16**, 315 (1975).
33a. R. Cone, T. Brown, and E. C. Friedberg, *JBC* **255**, 10354 (1980).
34. F. Tomita and I. Takahashi, *J. Virol.* **15**, 1073 (1975).
35. M. Sekiguchi, H. Hayakawa, F. Makino, K. Tanaka and Y. Okada, *BBRC* **73**, 293 (1976).
36. U. Kuhnlein, B. Lee and S. Linn, *NARes* **5**, 117 (1978).
37. F. Makino and N. Munakata, *J. Bact.* **131**, 438 (1977).
38. B. K. Duncan, P. A. Rockstroh and H. R. Warner, *J. Bact.* **134**, 1039 (1978).
39. H. Hayakawa and M. Sekiguchi, *BBRC* **83**, 1312 (1978).
40. B. K. Duncan and B. Weiss, *in* "DNA Repair Mechanisms" (P. C. Hanawalt, E. C. Friedberg and C. F. Fox, eds.), p. 183. Academic Press, New York, 1978.
41. H. Hayakawa, K. Kumura and M. Sekiguchi, *J. Biochem.* **84**, 1155 (1978).
42. R. DaRoza, E. C. Friedberg, B. K. Duncan and H. R. Warner, *Bchem* **16**, 4934 (1977).
43. R. R. Simmons and E. C. Friedberg, *J. Bact.* **137**, 1243 (1979).
44. H. Hayatsu and M. Shiragami, *Bchem* **18**, 632 (1979).
44a. P. Karran, T. Lindahl, I. Ofsteng, G. B. Evensen and E. Seeberg, *JMB* **140**, 101 (1980).
45. J. Duncan, L. Hamilton and E. C. Friedberg, *J. Virol.* **19**, 338 (1976).

46. W. G. Verly and Y. Paquette, *Can. J. Biochem.* **50,** 217 (1972).
47. W. G. Verly and E. Rassart, *JBC* **250,** 8214 (1975).
48. F. Gossard and W. G. Verly, *EJB* **82,** 321 (1978).
49. D. M. Yajko and B. Weiss, *PNAS* **72,** 688 (1975).
50. B. Weiss, *JBC* **251,** 1896 (1976).
51. W. G. Verly, *in* "DNA Repair Mechanisms" (P. C. Hanawalt, E. C. Friedberg and C. F. Fox, eds.) p. 187, Academic Press, New York, 1978.
52. S. Ljungquist, *JBC* **252,** 2808 (1977).
53. F. T. Gates, III and S. Linn, *JBC* **252,** 2802 (1977).
54. F. T. Gates, III and S. Linn, *JBC* **252,** 1647 (1977).
55. E. C. Friedberg, T. Bonura, R. Cone, R. Simmons and C. Anderson, *in* "DNA Repair Mechanisms" (P. C. Hanawalt, E. C. Friedberg and C. F. Fox, eds.), p. 163. Academic Press, New York, 1978.
56. S. Linn, U. Kuhnlein and W. A. Deutsch, *in* "DNA Repair Mechanisms" (P. C. Hanawalt, E. C. Friedberg and C. F. Fox, eds.), p. 199. Academic Press, New York, 1978.
57. W. A. Deutsch and S. Linn, *PNAS* **76,** 141 (1979).
58. Z. Livneh, D. Elad and J. Sperling, *PNAS* **76,** 1089 (1979).
59. L. Grossman, S. Riazzudin, W. A. Haseltine and C. P. Lindan, *CSHSQB* **43,** 947 (1978).
60. W. A. Haseltine, L. K. Gordon, C. P. Lindan, R. H. Grafstrom, N. L. Shaper and L. Grossman, *Nature* **285,** 634 (1980).
61. A. Maxam and W. Gilbert, *PNAS* **74,** 560 (1977).
62. R. O. Rahn and L. C. Landry, *Photochem. Photobiol.* **18,** 29 (1973).
63. E. H. Radany and E. C. Friedberg, *Nature* **286,** 182 (1980).
63a. E. H. Radany, J. D. Love and E. C. Friedberg, *in* "Chromosome Damage and Repair" (E. Seeberg and K. Kleppe, eds.). Plenum, New York (in press).
64. R. J. Reynolds and E. C. Friedberg, *in* "DNA Repair and Mutagenesis in Eukaryotes" (W. Generoso, M. D. Shelby and F. J. de Serres, eds.), p. 121. Plenum, New York, 1980.
65. K. W. Kohn, C. A. Friedman, R. A. G. Ewig and Z. M. Iqbal, *Bchem* **13,** 4134 (1974).
66. A. J. Fornace, K. W. Kohn and H. E. Kann, *PNAS* **73,** 39 (1976).
67. E. C. Friedberg, U. K. Ehmann and J. I. Williams, *Adv. Rad. Biol.* **8,** 85 (1979).
68. Z. Livneh, D. Elad and J. Sperling, *PNAS* **76,** 5500 (1979).

NOTE ADDED IN PROOF

1. There are now known to be at least 7 distinct DNA glycosylases in *E. coli*.

2. Recent evidence from our laboratory in collaboration with Dr. H. Edenberg, Indiana University, indicates that the T4v gene codes for both pyrimidine dimer-DNA glycosylase and a physically associated AP endonuclease activity (S. McMillan, H. Edenberg, E. H. Radany, R. C. Friedberg and E. C. Friedberg, unpublished data).

Multiprotein Interactions in Strand Cleavage of DNA Damaged by UV and Chemicals

ERLING SEEBERG

*Norwegian Defence Research
Establishment
Division for Toxicology
Kjeller, Norway*

Enzymes initiating excision of pyrimidine dimers from UV-damaged DNA have been isolated and extensively purified from *Micrococcus luteus* (1, 2) and bacteriophage T4-infected *Escherichia coli* (3, 4). These low-molecular-weight proteins were first characterized as endonucleases cleaving phosphodiester bonds on the 5' side of the dimers [e.g., endodeoxyribonuclease (pyrimidine dimer), EC 3.1.25.1]. However, it now appears that these enzymes catalyze the cleavage of the glycosyl bond between the pyrimidine at the 5' side of the dimer and the phosphodiester chain and hence are DNA glycosylases,[1] not classical endonucleases (5–7). The glycosylase action creates an AP site[2] in the DNA, and the strand break is made by an AP-endonuclease [endodeoxyribonuclease (apyrimidinic or apurinic), EC 3.1.25.2] that copurifies with the glycosylase. The AP-endonuclease may be a contaminant in the glycosylase preparation or an intrinsic property of the glycosylase itself (5, 8). This question remains to be settled.

Although the mechanism of strand incision at pyrimidine dimers now seems well established in the two cases described above, it is not clear that similar enzymes exist in other organisms. On the contrary, both genetic and biochemical studies of *Escherichia coli* and human cells indicate that strand incision at dimers in these organisms is much more complex than in *M. luteus* and T4-infected *E. coli*. First, the inci-

---

[1] The class names "glycohydrolase" and "deoxyribohydrolase" are used by the Enzyme Commission [see *EJB* **104**, 1 (1980)] in naming the enzyme (EC 3.2.2.15) that similarly removes hypoxanthine from deoxyinosine residues in DNA. "Nucleosidase" (or "deoxynucleosidase") would also be an appropriate class name. See also footnote 3 in paper by Friedberg *et al.* in this volume. [Ed.]

[2] "AP site" is short for "apyrimidinic site," which is equally appropriate for an "apurinic" site—in other words, for the site from which a base has been removed by the action of the glycosylase (glycohydrolase). [Ed.]

sion process requires several gene functions, i.e., the *uvrA*, *uvrB*, and *uvrC* functions in *E. coli* (9–11) and the XP functions A through E in human cells (12, 13). Second, incision in *E. coli* is ATP-dependent (14). Third, the incision enzyme in both *E. coli* and human cells appears to have a broad substrate specificity, since the same gene functions are responsible for recognition and incision of various types of DNA damage induced by UV or chemicals (reviewed in 15). The DNA glycosylases specific for pyrimidine dimers have no cofactor requirement and do not seem to act at any type of lesion other than pyrimidine dimers in DNA (2). In sum, studies at the cellular level indicate that dimer excision in both *E. coli* and human cells proceeds by means of a different type of incision enzyme than that studied in *M. luteus* and T4-infected *E. coli*.

The dimer-specific DNA glycosylases have been purified by selecting single proteins that produce strand breaks specifically in UV-irradiated DNA. This approach may not be feasible in the purification of an endonuclease activity that results from an interaction between several proteins, since these are likely to separate during purification. In our attempts to characterize the incision enzyme from *E. coli*, this problem was solved by using an *in vitro* complementation assay to purify separately the individual products required for the incision reaction (16). The assay is based on the observation that mutants known to be incision defective *in vivo* also are defective in the production of strand breaks in UV-irradiated DNA by extracts *in vitro* (17). However, the defective endonuclease activity in the mutant extracts can be restored by addition of a protein fraction containing the functional gene product that is defective in the mutant. The principal advantage of the *in vitro* complementation approach is that it allows purification of products of the *uvr*$^+$ genes without prior knowledge of their precise enzymic functions. This communication summarizes the characterization of the products of genes *uvrA*, *uvrB*, and *uvrC* purified by means of the complementation assay.

I. Properties of a Repair Endonuclease Reconstituted *in Vitro* from Partially Purified *uvr*$^+$ Gene Products

The *uvrA*$^+$, *uvrB*$^+$, and *uvrC*$^+$ gene products have been partially purified by ion-exchange chromatography of protein extracts from wild-type cells (16). The *uvrB*$^+$ and *uvrC*$^+$ gene products bind to DEAE-cellulose and elute in a single peak at an ionic strength of 0.3 M KCl. The *uvrA*$^+$ product has affinity for phosphocellulose and elutes at 0.2 M KCl. Neither the *uvrA*$^+$ nor the combination of *uvrB*$^+$

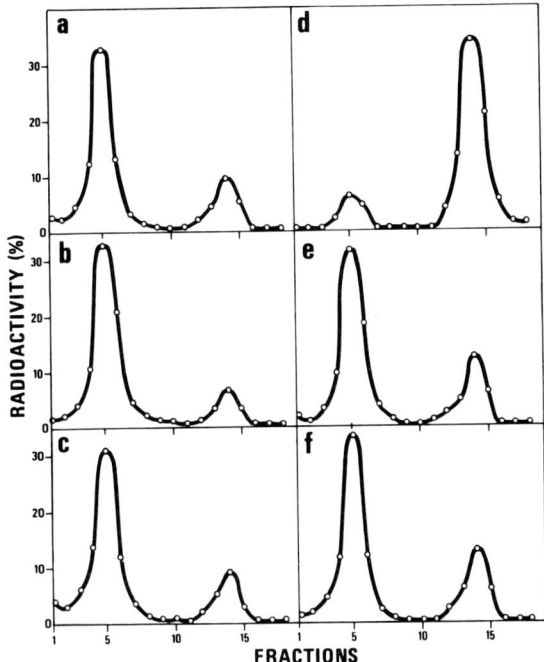

FIG. 1. Sedimentation profiles in alkali of UV-irradiated ColE1 DNA treated with products of *Escherichia coli* genes $uvrA^+$, $uvrB^+$, and $uvrC^+$. Covalently closed circular ^{3}H-labeled ColE1 DNA (20,000 cpm/µg) was irradiated with 18 J/m^2 of UV light (inducing about 2 pyrimidine dimers per molecule) and incubated with separated or combined $uvrA^+$, $uvrB^+$, $uvrC^+$ gene products in reaction mixtures containing 40 mM morpholinopropanesulfonic acid (pH 7.7), 100 mM KCl, 15 mM MgSO$_4$, 1 mM EDTA, 1 mM dithiothreitol, and 2.0 mM ATP, when included. After 20 minutes at 37°C, EDTA was added to a final concentration of 20 mM and the mixtures were layered onto preformed 5 to 20% alkaline sucrose gradients. The gradients were spun for 105 minutes at 40,000 rpm in a Spinco SW50.1 rotor. Fractions were collected directly into vials for scintillation counting. The amount of radioactivity in each fraction was calculated as percentage of total on the gradient. The amount of DNA used in each reaction was 0.1 µg. Enzyme addition to each mixture was (a) no enzyme; (b) 0.15 µg of uvrA protein; (c) 0.15 µg of uvrA protein and 0.3 µg of protein of a highly purified uvrB protein fraction; (d) 0.15 µg of uvrA protein and 20 µg of protein of the DEAE-cellulose fraction of $uvrB^+/uvrC^+$ products (16); (e) same as (d) without ATP; (f) 20 µg of protein of the DEAE-cellulose fraction of $uvrB^+/uvrC^+$ gene products (16).

and $uvrC^+$ gene products produce strand breaks in UV-irradiated DNA by itself. However, the combination of all three of these factors causes extensive cleavage of the UV-irradiated DNA (Fig. 1). This enzyme activity, referred to as the uvrABC endonuclease, is completely ATP-dependent (Fig. 1), has a pH optimum of 7.5–8.0 (data not

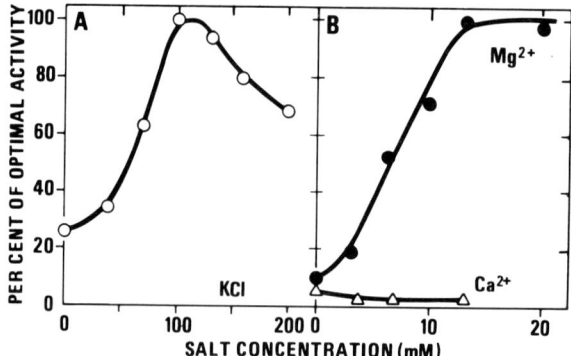

FIG. 2. Effect of ionic strength (A) and Mg^{2+} concentration (B) on the uvrABC endonuclease. Reaction conditions were as for panel (d) in Fig. 1 except for variations in the concentration of KCl (○——○), $MgCl_2$ (●——●), or $CaCl_2$ (△——△). Results are plotted as percentage of activity observed under optimal conditions for KCl and $MgCl_2$ and with respect to optimal $MgCl_2$ concentration for $CaCl_2$. Ultraviolet dose to the DNA was 18 J/m².

shown), a salt dependence of 100–130 mM KCl, and requires 12–20 mM Mg^{2+} for optimal activity (Fig. 2). Ca^{2+} cannot replace Mg^{2+} in the reaction. The endonuclease-sensitive sites in UV-irradiated DNA disappear upon photoreactivation (18). The UV-photoproducts recognized by the uvrABC endonuclease are hence positively identified as pyrimidine dimers. Table I shows the effect of various inhibitors and cofactors on the uvrABC endonuclease. Both N-ethylmaleimide and ethidium bromide are inhibitory. Neither of the ATP analogs[3] AMP-P[NH]P or ATP[γS], can replace ATP in the reaction, suggesting that an ATPase activity is required for the endonuclease reaction. The analogs are structurally very similar to ATP, but normally nonhydrolyzable by ATPases. Among the ATP hydrolysis products normally formed, AMP, phosphate, and pyrophosphate are without effect on the uvrABC endonuclease, but ADP causes a marked inhibition.

The uvrABC endonuclease is also active on chemically induced base adducts in DNA. Similar results to those presented above for UV-irradiated DNA are obtained by using DNA treated wtih psoralen plus light (18) or with N-acetoxy-2-acetamidofluorene (Fuchs and Seeberg, in preparation). In all cases, the DNA strand break formation depends on the presence of both the uvrA⁺ and the uvrB⁺/uvrC⁺ protein fractions and requires ATP. It thus appears that the uvrABC endonu-

---

[3] AMP-P[NH]P = adenosine 5'-[β,γ-imido]triphosphate; ATP[γS] = adenosine 5'-[γ-thio]triphosphate [see PNAS **74**, 2222 (1977)] [Ed.].

TABLE I

EFFECT OF ADDITIONS AND VARIATIONS IN
REACTION MIXTURES ON uvrABC ENDONUCLEASE
ACTIVITY[a]

Variations in reaction mixture	Relative activity
Standard	(100)
+ N-Ethylmaleimide, 20 mM	51
+ Ethidium bromide, 1 μg/ml	14
+ Novobiocin, 150 μg/ml	85
−ATP, +AMP-P[NH]P,[b] 1 mM	0
−ATP + ATP[γS],[b] 1 mM	5
+ ADP, 1 mM	31
+ AMP, 1 mM	95
+ P_i, 1 mM	103
+ PP_i, 1 mM	101
+ ATP[γS],[b] 0.5 mM	21

[a] The data are calculated as percentage of activity observed in standard reaction mixtures with 1 mM ATP. The UV-dose to the DNA was 9 J/m². The composition of standard mixtures and the procedure for break measurements were as previously described (16).

[b] See text footnote 3.

clease, as recovered by the complementation assay, has the broad substrate specificity and the complex nature indicated from studies of the uvr^+-dependent incision reaction in whole cells.

II. Molecular Properties of the uvrA⁺ Gene Product

The $uvrA^+$ gene product has been purified to apparent homogeneity from a strain carrying the uvrA gene on a multicopy plasmid (19). The uvrA plasmid was constructed by Sancar et al., (20) who identified the $uvrA^+$ product as a polypeptide of molecular weight 114,000 by specific radioactive labeling of plasmid-coded proteins by the maxicell method. This molecular weight corresponds to that observed in gel filtration experiments with active $uvrA^+$ product using the complementation assay (16) and to that observed by gel electrophoresis in dodecyl sulfate of the purified protein (19) (Fig. 3). In view of the possibility that ATP hydrolysis might be required for the uvrABC endonuclease reaction, the uvrA⁺ protein was assayed for an ATPase activity. It was found that the uvrA⁺ protein hydrolyzes ATP to ADP and phosphate (19). The K_m value for this reaction (2.3×10^{-4} mM ATP) corresponds to the K_m value of the ATP requirement of the

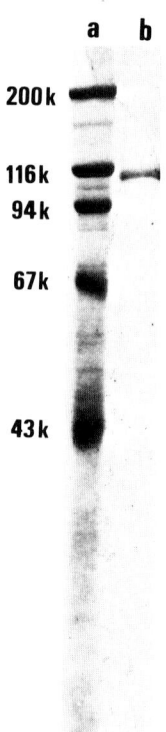

FIG. 3. Sodium dodecyl sulfate/polyacrylamide gel electrophoresis of purified uvrA⁺ protein. Lane a: Marker proteins (from top) myosin, β-galactosidase, phosphorylase, bovine serum albumin, ovalbumin. Lane b: Purified uvrA protein (1 µg). For purification procedure, see ref. (19). K = thousand daltons.

uvrABC endonuclease (19). Both ADP and ATP[γS] are competitive inhibitors of the ATPase activity of the uvrA⁺ protein. These compounds also inhibit the uvrABC endonuclease (Table I). It thus appears that the ATP requirement of the incision step of uvr^+-dependent excision repair in E. coli is determined by the ATPase function of the $uvrA^+$ gene product.

Studies of partially purified $uvrA^+$ product show that the uvrA⁺ protein is associated with a DNA binding activity with preferential affinity for UV-irradiated DNA (21). This binding property has been studied in more detail with the extensively purified protein (19). It appears that the uvrA⁺ protein binds preferentially to single-stranded rather than to double-stranded DNA, and has no affinity for damaged nucleotides per se. This has led to the speculation that the uvrA⁺ protein may be responsible for recognition of the damaged site by identifying regions of major distortions in the DNA helix. Such distortions are created by the bulky type of lesions acted upon by the uvrABC endonuclease.

The ATPase may be involved in the recognition process since the binding is stimulated by ATP and prevented by ADP. It is postulated that the uvrA$^+$ protein forms a complex with ATP and that this is a high-affinity DNA-binding complex. Conversion of ATP to ADP changes the binding property of the complex from high to low affinity. If the high-affinity complex initially binds to an arbitrary site on the DNA molecule, the cycle of the ATPase reaction and the ADP → ATP exchange may allow the protein to move along the DNA strand until the distortion is located. This model is supported by DNA binding experiments where ATP is replaced with ATP[γS], which greatly enhances binding of the uvrA$^+$ protein to DNA, but at the same time eliminates the preference for UV-irradiated or single-stranded versus nonirradiated DNA. These results would be expected if the high-affinity complex of the uvrA$^+$ protein with ATP[γS] is stuck at the initial binding site because it is not being hydrolyzed.

III. Properties of the uvrB$^+$ and uvrC$^+$ Gene Products

The *uvrB$^+$* gene product has also been extensively purified from a strain carrying the gene on a multicopy plasmid. Like the *uvrA$^+$* plasmid, the *uvrB* plasmid has been constructed by Sancar *et al.* (22), who identified the *uvrB$^+$* product as a polypeptide of M_r 84,000. This molecular weight corresponds to that observed for *uvrB*-complementing activity in gel filtration experiments (16) and to that observed for the major protein band in our most-purified fraction (unpublished data). Assays for DNA binding, ATPase or endonuclease activities in this fraction have yielded negative results. Addition of uvrA$^+$ protein to the uvrB$^+$ protein fraction does not restore any endonuclease activity on UV-irradiated DNA (Fig. 1), implying that the combination of *uvrA$^+$* and *uvrB$^+$* products alone is not sufficient for the uvrABC endonuclease reaction. Therefore, the *uvrC$^+$* product is an essential component of the uvrABC endonuclease.

The *uvrC*-complementing activity is very unstable in our hands, which has greatly complicated its purification. The *uvrC* gene has been cloned on a lambda phage vector, and this plaque-forming transducing phage is presently being used in attempts to amplify the intracellular level of the *uvrC$^+$* product. Two polypeptides of M_r 30,000 and M_r 70,000 are encoded by the cloned segment on the phage (23). A *uvrC* amber mutant derivative of the phage has been isolated to allow positive identification of the *uvrC* gene product. Van Sluis and Brandsma (24) have cloned the *uvrC* gene on a plasmid and their results indicate M_r 28,000 for the *uvrC* protein.

IV. On the Mechanism of Action of the uvrABC Endonuclease

The discovery that the so-called UV-endonucleases from *M. luteus*- and T4-infected cells are DNA glycosylases (see footnote 1), but not classical endonucleases, (see footnote 2) raises the question whether the uvrABC endonuclease also operates by a glycosylase mechanism or if it is a proper endonuclease. Studies of the action mechanism of the uvrABC endonuclease requires that all components be available in a highly purified form, so this question cannot be answered conclusively at the present time.

However, the evidence accumulated so far both from *in vivo* and *in vitro* studies indicates that the uvrABC endonuclease is indeed a proper endonuclease. First, the uvrABC endonuclease does not have the characteristics of a DNA glycosylase. All the DNA glycosylases so far characterized are small monomeric proteins with a high degree of specificity for a single type of base damage. The uvrABC endonuclease requires three different proteins for enzyme activity, of which at least two are of high molecular weight, and also recognizes a variety of different lesions including cross-links, monoadducts, and pyrimidine dimers in DNA (18). Still, the uvrABC endonuclease could be a DNA glycosylase of a novel type not previously recognized. If this were the case, one would expect that uvr^+-dependent repair of monoadduct damage in DNA should yield excision products as free bases. However, psoralen adducts in DNA are released *in vitro* as nucleotides, not as free bases (Seeberg, unpublished).

A glycosylase mechanism for the repair of cross-link damage implies that the cross-link should be released by the action of the glycosylase only. However, the release of DNA interstrand cross-links requires not only the functions of $uvrA^+$, $uvrB^+$, and $uvrC^+$, but also those of $polA^+$ and $uvrD^+$ (25). This is consistent with an endonuclease mechanism for the action of the uvrABC endonuclease, but not with a glycosylase mechanism. Therefore, it is only in the case of pyrimidine dimer damage that a glycosylase mode of action for the uvrABC endonuclease is consistent with the available data. However, the discrete pathways of dimer excision observed *in vivo* between UV-repair initiated by the uvrABC endonuclease and repair initiated by the "UV-endonuclease" from T4-infected cells suggest that these two physically different enzymes operate by different mechanisms

(26). It is possible that the process of multidamage recognition and incision is too complex to be effected by a simple enzyme of one subunit, but requires a complex enzyme with several subunits of specialized functions.

V. Concluding Remarks

It has long been assumed that the initial step of excision repair of UV-irradiated DNA in E. coli is similar to that characterized *in vitro* by the small dimer-specific glycosylases from M. luteus and T4-infected cells. This assumption has prevented major efforts to elucidate the uvr^+-dependent incision enzyme in E. coli. However, recent studies of the uvr^+ gene products *in vitro* show that these proteins are basically different from the small DNA glycosylases characterized so far. The interactions between the $uvrA^+$, $uvrB^+$, and $uvrC^+$ gene products resulting in the appearance of strand breaks in DNA damaged by UV and various chemical agents are now being subjected to extensive studies in several laboratories, and it is to be expected that we will soon have a more detailed knowledge about the excision mechanism in E. coli.

ACKNOWLEDGMENTS

The excellent technical assistance of Mrs. Anne-Lill Steinum is greatly acknowledged. This research was supported by the Royal Norwegian Council for Scientific and Industrial Research.

REFERENCES

1. W. L. Carrier and R. B. Setlow, *J. Bact.* **102**, 178 (1970).
2. S. Riazuddin and L. Grossman, *JBC* **252**, 6280 (1978).
3. S. Yasuda and M. Sekigūchi, *JMB* **47**, 243 (1970).
4. K. Minton, M. Durphy, R. Taylor and E. C. Friedberg, *JBC* **250**, 2823 (1975).
5. W. Haseltine, L. K. Gordon, C. P. Lindan, R. H. Grafstrom, N. L. Shaper and L. Grossman, *Nature* **285**, 634 (1980).
6. E. C. Friedberg, C. T. M. Anderson, T. Bonura, R. Cone, E. Radany and R. J. Reynolds, this volume.
7. B. Demple and S. Linn, *Nature* **287**, 203 (1980).
8. P. Seawell, C. A. Smith and A. K. Ganesan, *J. Virol.* **35**, 790 (1980).
9. P. Howard-Flanders, R. P. Boyce and L. Theriot, *Genetics* **53**, 1119 (1966).
10. R. B. Setlow and W. L. Carrier, *in* "Replication and Recombination of the Genetic Material (W. J. Peacock and R. D. Brock, eds.), p. 134. Australian Academy of Sciences, Canberra, 1968.
11. K. Shimada, H. Ogawa and J. Tomizawa, *Mol. Gen. Genet.* **101**, 245 (1968).
12. J. E. Cleaver and D. Bootsma, *Annu. Rev. Genet.* **9**, 19 (1975).
13. A. J. Fornace, K. W. Kohn and H. E. Kann, *PNAS* **73**, 39 (1976).
14. E. A. Waldstein, R. Sharon and R. Ben-Ishai, *PNAS* **71**, 2651 (1974).

15. P. C. Hanawalt, P. K. Cooper, A. K. Ganesan and C. A. Smith, *ARB* **48**, 783 (1979).
16. E. Seeberg, *PNAS* **75** 2569 (1978).
17. E. Seeberg, J. Nissen-Meyer and P. Strike, *Nature* **263**, 524 (1976).
18. E. Seeberg, *Mutat. Res.* (in press).
19. E. Seeberg and A.-L. Steinum (submitted).
20. A Sancar, R. P. Wharton, S. Seltzer, B. M. Kacinski, N. D. Clarke and W. D. Rupp, *JMB* (in press).
21. E. Seeberg, *in* "DNA Repair Mechanisms" (P. C. Hanawalt, E. C. Friedberg and C. F. Fox, eds.), p. 225. Academic Press, New York, 1978.
22. A. Sancar, N. D. Clarke, J. Griswold, W. J. Kennedy and W. D. Rupp, *JMB* (in press).
23. O. R. Blingsmo, E. Rivedal, A. L. Steinum and E. Seeberg, *in* "Chromosome Damage and Repair" (E. Seeberg and K. Kleppe, eds.). Plenum, New York, in press.
24. C. A. van Sluis and J. A. Brandsma, *in* "Chromosome Damage and Repair" (E. Seeberg and K. Kleppe, eds.). Plenum, New York, in press.
25. R. S. Cole, D. Levitan and R. R. Sinden, *JMB* **103**, 39 (1976).
26. E. Seeberg, A. L. Steinum, E. Rivedal and J. Nissen-Meyer, *in* "DNA Synthesis— Present and Future" (I. Molineux and M. Kohiyama, eds.), p. 967. Plenum, New York, 1978.

In Vitro Packaging of Damaged Bacteriophage T7 DNA

> WARREN E. MASKER
> NANCY B. KUEMMERLE AND
> LORI A. DODSON
>
> Biology Division
> Oak Ridge National Laboratory
> Oak Ridge, Tennessee

In vitro systems provide a powerful means for investigating various aspects of DNA metabolism. Use of cell-free systems allows introduction of exogenous substrates and enzymes and, in some cases, permits complementation assays that facilitate purification of important enzymic components of the reactions. Although in vitro characterizations have yielded considerable insight into the biochemistry of DNA replication, repair, and (most recently) recombination (1), there is occasional concern as to the degree to which in vitro reactions accurately mimic the in vivo situation. Use of biological activity as a criterion offers some reassurance that in vitro reactions proceed with sufficient fidelity to assure generation (or restoration) of DNA molecules that are genetically intact. We have made use of a cell-free system for packaging the DNA of bacteriophage T7 into phage heads so as to form viable phage particles, in order to monitor the biological consequences of in vitro DNA replication, DNA repair, and molecular recombination reactions. This article summarizes the results of these investigations.

Bacteriophage T7 is an excellent system for the study of DNA metabolism. The thorough genetic analysis of this phage, together with biochemical characterization of most of those enzymes important in DNA replication and recombination, affords a firm basis of knowledge for further investigation. The pattern of phage morphogenesis in T7 is relatively well understood, and, in particular, many features in the DNA replication and recombination processes of T7 have been carefully examined in vivo (2–4). Replication of T7's short (25×10^6 daltons) linear duplex molecule is discontinuous and bidirectional and proceeds primarily from an internal origin located near the genetic left end. Phage DNA synthesis depends upon the products of genes 2, 3, 4, 5, and 6 (2, 3). Late in infection, very complex, rapidly sedimenting DNA molecules, many times the size of an intact T7 genome, are observed (5–7). These molecules, some of which are concatemers, are matured and packaged by T7's late genes (genes 7 through 19) to produce viable phage (2).

I. *In Vitro* Packaging of Bacteriophage T7 DNA

Experiments of Sadowski *et al.* indicate that T7 DNA can be encapsulated into phage heads *in vitro* in a system that relies upon recombination between the exogenously added DNA and endogenous DNA in the extracts used to carry out the packaging (8–11). However, in order to examine biological activity of exogenous DNA, one needs a system as free as possible from endogenous DNA and unable to undergo genetic recombination. To approach this end, we prepared ex-

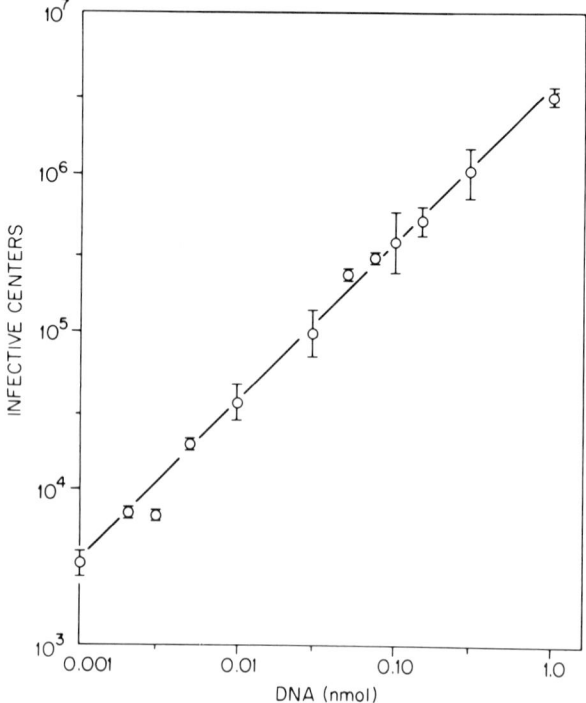

FIG. 1. *In vitro* packaging as a function of DNA concentration. The indicated amount of wild-type T7 DNA was incubated in *in vitro* packaging reactions with extracts from strain W3110 infected with T7 carrying amber mutations in genes 3, 5, and 6, and the resulting phage was plated on strain W3110. The average number of infective centers from several determinations are plotted as a function of DNA concentration. Reproduced from Fig. 1 of Kuemmerle and Masker (12).

tracts from suppressor-free *Escherichia coli* infected with T7 carrying amber mutations in three genes (genes 3, 5, and 6) essential for normal DNA synthesis and recombination, and then altered assay conditions in an effort to maximize the ability of these extracts to encapsulate purified, genome-length, wild-type T7 DNA (12). The use of suppressor-free indicator bacteria effectively eliminates expression of contaminating $T7_{3,5,6}$ phage particles remaining in the system. This *in vitro* system can encapsulate genome-length T7 DNA with an average efficiency of 0.03% phage production per genome equivalent of input DNA and, as seen in Fig. 1, shows a linear relationship between phage production and concentration of input DNA over a range exceeding three orders of magnitude (12). The packaging system is, by itself, unable to perform any significant level of genetic recombination (12, 13). This system is a useful means with which to examine biological consequences of manipulations performed upon T7 DNA in separate *in vitro* reactions. *In vitro* packaging has proved to be particularly valuable in monitoring DNA damage and repair.

II. DNA Repair *in Vitro*

Studies using *E. coli* cells made permeable with toluene or by sucrose plasmolysis have yielded several pieces of information concerning the excision-repair process (14, 15). However, these systems are limited by their impermeability to exogenous enzymes and DNA and by their relatively inefficient removal of pyrimidine dimers introduced by ultraviolet (UV) irradiation (16). *In vitro* excision repair of UV-induced damage appears to be limited by the incision step, and it has not yet been possible to achieve complete repair of UV-damaged DNA with *E. coli* lysates. However, if incision is completed by a separate endonuclease from *Micrococcus luteus*, the postincision steps of excision repair can be monitored *in vitro* (17). With this system, all the dimers introduced by UV doses up to 100 J/m² can be removed from T7 DNA, repair resynthesis with an average patch length of about 17 nucleotides is completed, and the DNA is restored to its original single-strand molecular weight (17). Thus, by several biophysical and biochemical criteria this system is able to complete the repair of UV-damaged DNA.

To examine the biological consequences of the *in vitro* repair process we employed the packaging system outlined above. First, UV-irradiated DNA was used as a substrate in an *in vitro* packaging reaction, and the product phage particles were plated on indicator strains that were wild-type, *polA*1, or *uvrA*. This experiment (12) demonstrated that the phage produced *in vitro* have host-cell reactivation characteristics very similar to those found *in vivo*. This result, together with other controls, indicates that damaged DNA is packaged *in vitro*, but that the resulting phage show reduced ability to form plaques, especially on an incision-deficient host (*17*). In additional experiments, irradiated DNA was repaired with the *in vitro* system described above, or left as an untreated control before being encapsulated into phage particles. These results (Fig. 2) demonstrate that as much as 80% of the original biological activity can be restored during the *in vitro* repair process. Collectively, these experiments indicate that the *in vitro* DNA repair system is reasonably accurate and that *in vitro* packaging is an acceptable means for determining the presence of UV-induced damage in DNA. It will be of considerable interest to extend these studies by attempting quantitative incision of thymine dimers using products of *E. coli* genes *uvrA*, *uvrB*, and *uvrC* described by Seeberg (*18*). Also, our preliminary results suggest that *in*

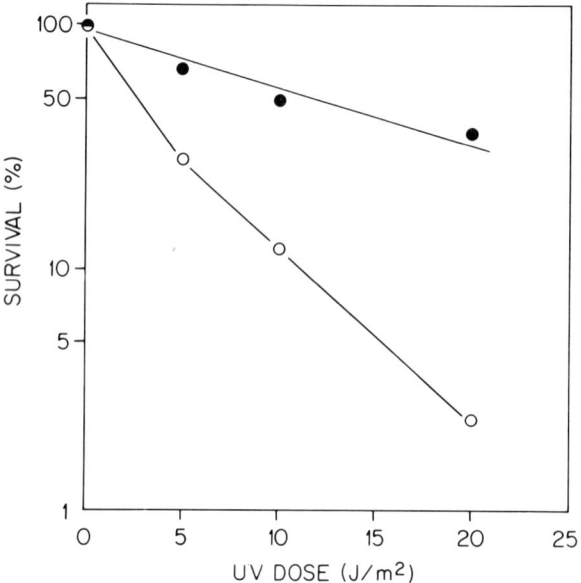

FIG. 2. Restoration of irradiated T7 DNA. T7 DNA was exposed to various doses of UV. A portion was retained as an untreated control, and the remainder was repaired *in vitro* by treatment first with a damage-specific endonuclease from *Micrococcus luteus*, then with an extract from *Escherichia coli* as described previously (*17*). The DNA was packaged *in vitro*, and the biological activity relative to unirradiated controls was determined. ○——○, Untreated control; ●——●, repaired DNA. Reproduced from Fig. 4 of Kuemmerle and Masker (*12*).

vitro packaging may be suitable for following the damage introduced by certain mutagenic chemicals, such as nitrosoguanidine, methyl methanesulfonate, and ethyl methanesulfonate, and may allow us to monitor the response of *in vitro* repair and *in vitro* replication systems to lesions introduced by these agents.

III. DNA Replication and Production of Phage *in Vitro*

T7 DNA can be replicated *in vitro* with good efficiency (19). The development of an *in vitro* system for copying this phage DNA has allowed separation of the synthesis system into its various components and provided insight into the replication of a linear duplex template (20). In its crude form, the *in vitro* DNA replication system appears to mimic the *in vivo* process quite well. The product of the reaction (produced in 3- to 10-fold excess over the amount of input T7 template) is synthesized semiconservatively, and has approximately

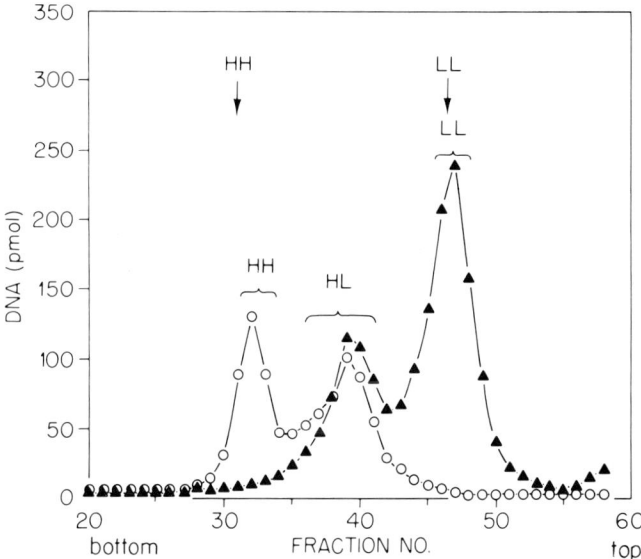

FIG. 3. Isopycnic analysis of DNA synthesized *in vitro*. An extract prepared from strain W3110 infected with T7 carrying amber mutations in genes 3 and 6 was used to replicate heavy-density (HH) T7 DNA *in vitro*. The product was labeled with normal-density (LL) precursors including [^{32}P]dATP. The DNA was sedimented to equilibrium in CsCl, and portions of each fraction were acid-precipitated to yield the profiles shown. Selected fractions, indicated by brackets, were dialyzed and subsequently packaged *in vitro*. ○──○, ^{3}H ^{13}C^{15}N (template) DNA; ▲──▲, ^{32}P product DNA. Reproduced from Fig. 2 of Masker *et al*. (13).

the same single-strand molecular weight as an intact T7 genome (21). Transfection assays have indicated that the product DNA is fully infective, but the low efficiency of this type of determination allowed only limited quantitative claims to be made regarding the biological activity of the newly synthesized DNA (22).

In view of these considerations, it was of interest to attempt synthesis of complete phage particles *in vitro* by first replicating T7 DNA *in vitro*, separating the product of two or more rounds of replication from the template, then encapsulating this DNA so as to form viable T7 phage. As seen in Fig. 3 it was possible to achieve good separation between product and template by using density-labeled DNA as template, replicating in the presence of normal (light) density precursors, and sedimenting the DNA to equilibrium in CsCl (13). DNA recovered from the "light-light" density region of this gradient and others like it could be packaged with an efficiency 40–95% of that of virion DNA used as a control in these experiments. Thus, it seems clear that most of the DNA produced *in vitro* is genetically intact and fully capable of producing phage (13).

The *in vitro* replication experiments outlined above were performed with extracts made from suppressor-free *E. coli* infected with T7 with amber mutations in genes 3 and 6. Apparently, normal levels of these gene products are not absolutely essential to *in vitro* DNA synthesis or phage production (13, 19). However, the presence of one or both of the enzymes coded for by gene 3 (an endonuclease) and gene 6 (an exonuclease) might be expected to alter quantitative aspects of the synthesis or to change the characteristics of the product DNA. In fact, extracts prepared from $T7_3$-infected cells (with normal levels of the gene 6 exonuclease) produce rapidly sedimenting DNA

TABLE I
BIOLOGICAL ACTIVITY OF T7 DNA SYNTHESIZED *in Vitro*[a]

Sample	Amount of DNA (pmol)	Number of infective centers ($\times 10^3$)	Efficiency relative to control
Control (virion)	10.0	5.0	1.0
$T7_{3,6}$ HH (template)	2.3	0.35	0.30
$T7_{3,6}$ HL (hybrid)	3.2	4.2	2.6
$T7_{3,6}$ LL (product)	1.1	0.52	0.95
$T7_3$	2.8	74	53

[a] DNA synthesis reactions were performed using extracts from strain W3110 infected with either $T7_3$ or $T7_{3,6}$, and the resulting DNAs were sedimented to equilibrium in CsCl as described in the legend to Fig. 3. After dialysis, the indicated amounts of DNA were incubated in an *in vitro* packaging reaction, and the resulting phage was plated on strain W3110. Data were taken from Table 4 of ref. 13.

(>400 S) that contains complex branched structures reminiscent of recombinant molecules and concatermeric DNA observed *in vivo* (*13*). The product of these reactions cannot be fully separated from density-labeled templates in neutral CsCl. Examination of DNA recovered from these reactions shows that it can be packaged *in vitro* with good efficiency. A comparison with virion DNA and DNA synthesized by extracts with reduced levels of the gene 6 product has shown that the DNA recovered from reactions carried out with normal levels of the gene 6 exonuclease is 10 to 50 times more efficient at phage production than intact genome length DNA (Table I). Although the reasons for this enhanced packaging efficiency are not clear, the data are compatible with Watson's model, which suggests a need for concatemer formation to facilitate normal phage development (*23*).

IV. *In Vitro* Recombination

The DNA structures observed after *in vitro* DNA synthesis include "H" forms that might have arisen by recombination. Also, the detection and characterization of T7 recombination *in vitro* by other workers (*9–11, 24, 25*) encouraged testing for genetic recombination during the *in vitro* DNA replication reactions. For this, we employed a mutant phage $T7_{ss}$ that, unlike wild-type T7, can grow on *Shigella sonnei*. This mutant has been used successfully by other workers to distinguish species of DNA present during recombination reactions (*10, 24, 25*). An experiment (shown in Fig. 4), designed to rescue an amber gene 3 mutation in DNA extracted from $T7_{3,ss}$ phage by *in vitro* recombination, demonstrated that recombination does indeed take place during the *in vitro* DNA replication reactions and that recombination frequencies depend on the distance between amber mutations employed in the crosses (*26*). Separate experiments demonstrated that reactions carried out by extracts from cells infected with T7 deficient in the gene 6 exonuclease showed markedly reduced levels of recombination (*25, 26*). Thus, deficiencies in the gene 6 product (*a*) reduce the formation of rapidly sedimenting DNA *in vitro*; (*b*) cause lower biological activity as determined by *in vitro* packaging; and (*c*) diminish the level of *in vitro* recombination.

V. Response to DNA Damage

Part of the rationale for initiating many of the experiments outlined above derives from our interest in the response of DNA replication enzymes to lesions encountered in their path. Thus, we have com-

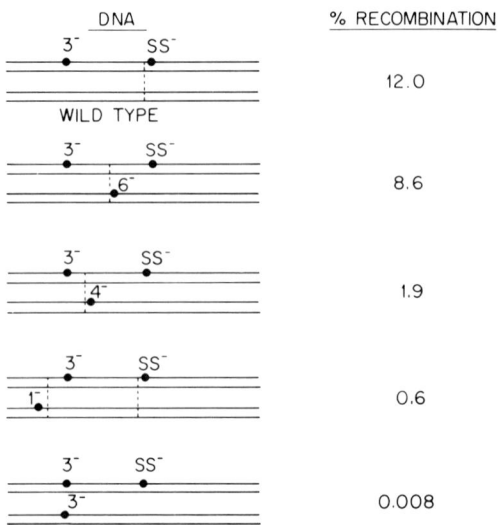

FIG. 4. *In vitro* recombination during the DNA replication reaction. An extract was prepared from strain W3110 infected with T7 carrying an amber mutation in gene 3 and used to perform *in vitro* DNA replication. The reaction mixtures included T7 DNA from phage with the ss mutation plus an amber mutation in gene 3 as well as an equal quantity of ss$^+$ T7 DNA with no mutations (wild type) or an amber mutation in either gene 1, 3, 4, or 6. After incubation the DNA was packaged *in vitro* and the resulting phage was plated on *Shigella sonnei* with and without an amber suppressor. Recombination frequencies, shown in the figure together with relative positions on the genetic map, were calculated from the ratios of infective centers found on the two indicators. Reproduced from Fig. 1 of Masker and Kuemmerle (26).

bined experiments with both irradiated DNA and chemically treated DNA into our investigations of *in vitro* DNA replication and recombination. Studies with UV-irradiated DNA showed that pyrimidine dimers constitute a formidable barrier to the *in vitro* DNA replication system. The principal product of these reactions are short duplex pieces of DNA about the length of the interdimer distance. Hybridization studies indicate that these fragments are copied from both the left and right halves of the T7 chromosome, suggesting that initiation events take place at alternate (secondary) sites on T7 DNA molecules containing UV-induced lesions (27). Recombination frequencies between irradiated DNA molecules are also compatible with this interpretation (26). Reduced levels of gene 6 exonuclease have little effect on most parameters of the DNA synthesis reaction. However, we have noted a higher proportion of biologically active DNA molecules produced from irradiated templates when the extracts used to perform DNA replication contain normal levels of the gene 6 product (26).

This may mean that recombination does play a role in producing damage-free DNA molecules *in vitro*. Studies on *in vitro* recombination of irradiated DNA molecules showed that portions of DNA molecules containing pyrimidine dimers are poor substrates for recombinational exchange. These observations prompt speculation that actively replicating molecules are preferred substrates for molecular exchange in this system. Recombinational events initiated at replication forks whose progress has been stalled by lesions encountered in their paths may eventually lead to generation of molecules relatively free of damage, and in this way may at least partially account for the increased yield of viable phage observed when irradiated DNA is replicated under conditions that permit recombination. *In vivo* studies (28) with irradiated T7 phage also show recombination between partial replicas.

Our characterization of the response of the DNA replication system to lesions introduced by chemicals is much less complete than what has been found using UV. Preliminary studies indicate that exposure of the *in vitro* DNA replication system to nitrosoguanidine reduces its ability to support DNA synthesis. Also, inclusion of this chemical in the DNA replication reaction mixtures reduces, but does not totally eliminate, production of viable phage particles after *in vitro* encapsulation. Further studies on *in vitro* repair of damage of this type and on the response of the replication-recombination system to chemicals may prove to be of special interest because of the potential for detecting *in vitro* fixation of mutation with our system.

VI. Summary

We have described experiments using *in vitro* packaging to monitor the biological activity of DNA recovered after *in vitro* repair, replication, and recombination reactions. The results suggest that the *in vitro* systems mimic the *in vivo* situation sufficiently well to allow generation (or restoration) of DNA molecules that can be encapsulated to form fully viable T7 phage particles. The *in vitro* packaging system has proved to be a convenient and relatively sensitive means for determining the amount of biological damage present in T7 DNA and for examining the response of various DNA metabolic systems to that damage.

Acknowledgments

We are pleased to acknowledge the partial financial support of this research by the Oak Ridge National Laboratory through the Exploratory Studies Program. Research was sponsored by the Office of Health and Environmental Research, U. S. Department of Energy, under contract W-7405-eng-26 with the Union Carbide Corporation.

References

1. A. Kornberg, "DNA Replication." Freeman, San Francisco, California, 1980.
2. F. W. Studier, *Science* **176**, 369 (1972).
3. R. Hausmann, *Curr. Top. Microbiol. Immunol.* **75**, 77 (1977).
4. R. C. Miller, *Annu. Rev. Genet.* **29**, 355 (1975).
5. T. J. Kelly and C. A. Thomas, Jr., *JMB* **44**, 459 (1969).
6. P. Serwer, *Virology* **59**, 70 (1974).
7. V. L. Paetkau, L. Langman, R. Bradley, D. Scraba and R. C. Miller, Jr., *J. Virol.* **22**, 130 (1977).
8. C. Kerr and P. D. Sadowski, *PNAS* **71**, 3545 (1974).
9. P. D. Sadowski and D. Vetter, *PNAS* **73**, 692 (1976).
10. P. D. Sadowski, *Virology* **78**, 192 (1977).
11. P. J. Vlachopoulou and P. D. Sadowski, *Virology* **78**, 203 (1977).
12. N. B. Kuemmerle and W. E. Masker, *J. Virol.* **23**, 509 (1977).
13. W. E. Masker, N. B. Kuemmerle and D. P. Allison, *J. Virol.* **27**, 149 (1978).
14. W. E. Masker, T. J. Simon and P. C. Hanawalt, *in* "Molecular Mechanism of DNA Repair" (P. C. Hanawalt and R. B. Setlow, eds.), p. 245. Plenum, New York, 1975.
15. E. Seeberg and P. Strike, *J. Bact.* **125**, 787 (1976).
16. W. A. Deutsch, J. W. Dorson and R. E. Moses, *J. Bact.* **125**, 220 (1976).
17. W. E. Masker, *J. Bact.* **129**, 1415 (1977).
18. E. Seeberg, this volume.
19. D. C. Hinkle and C. C. Richardson, *JBC* **249**, 2974 (1974).
20. C. C. Richardson, L. J. Romano, R. Kolodner, J. E. LeClerc, F. Tamanoi, M. J. Engler, F. B. Dean and D. S. Richardson, *CSHSQB* **43**, 427 (1979).
21. W. E. Masker and C. C. Richardson, *JMB* **100**, 543 (1976).
22. W. E. Masker and C. C. Richardson, *JMB* **100**, 557 (1976).
23. J. D. Watson, *Nature NB* **239**, 197 (1972).
24. L. Roberts, R. Sheldon and P. D. Sadowski, *Virology* **89**, 252 (1978).
25. G. S. Roeder and P. D. Sadowski, *CSHSQB* **43**, 1023 (1979).
26. W. E. Masker and N. B. Kuemmerle, *J. Virol.* **33**, 330 (1980).
27. W. E. Masker and N. B. Kuemmerle, *BBA* **609**, 61 (1980).
28. K. B. Burck, D. M. Taylor, H. W. Smith and R. C. Miller, Jr., *CSHSQB* **43**, 461 (1979).

The Inducible Repair of Alkylated DNA

> John Cairns
> Peter Robins
> Barbara Sedgwick and
> Phillipa Talmud
>
> *The Imperial Cancer Research Fund
> London, England*

For the last 3 years, we have been studying the inducible DNA repair pathway that protects *Escherichia coli* against mutagenesis by alkylating agents. It is too early for us to have accumulated the wealth of biochemical and genetic information that exists, for example, concerning the repair of pyrimidine dimers. However, it is already clear that the pathway we have discovered occurs in mammalian cells as well as in bacteria, provides a high level of protection against a very widespread class of mutagens, and seems to function by a most unusual mechanism. However, because the exact chemistry is still poorly understood and the protagonists of the reaction have not yet been purified, much of our information about mechanism comes from studies carried out *in vivo* and therefore may seem a little circumstantial.

Before reviewing the present status of the subject, it is worth considering how we discovered that *E. coli* has an inducible system that protects it against alkylating agents. For reasons that are not relevant in the present context but have been summarized elsewhere (1), we were measuring the mutation rate of *E. coli* continuously exposed to low levels of N-methyl-N'-nitro-N-nitrosoguanidine (MNNG). To our surprise, we found that, after the first 30 minutes, the bacteria became resistant to further mutagenesis (2) [actually, this phenomenon had already been observed by others but had not been investigated further (3)]. It soon became clear that we were dealing with some form of DNA repair induced by this mutagen. For example, we observed that it alkylated the DNA of induced and uninduced bacteria to about the same extent, which indicated that the resistant bacteria were still fully permeable to it (4). Also, the plating efficiency of alkylated plasmid DNA was raised if the plating bacteria had been exposed to MNNG (2). It may seem curious that the existence of an efficient form of repair, which protects against such a widely used mutagen, should have been discovered only recently, and then by accident. But most experiments with mutagens measure only the response to short exposures and high doses, even though most natural mutagenesis (and, perhaps, most natural carcinogenesis) is presumably the result of prolonged exposure to low levels of mutagens. For this reason, any inducible form of repair is easily missed. It is obvious now that there may be many forms of inducible repair awaiting discovery. If they exist, they would greatly influence our estimate of the hazards of our environment.

For the time being, we have called the repair pathway induced by methylnitronitrosoguanidine the "adaptive" response. However, as soon as the genetics and biochemistry of the pathway become better understood, it should be given a more specific name.

I. The Genetics of the Adaptive Response

As far as we can tell, none of the known DNA repair genes of *E. coli* is involved in the protection of MNNG-adapted bacteria against mutagenesis by MNNG (5, 6). In particular, the adaptive response appears to be distinct from the *rec-lex*-dependent form of repair, even though this too is inducible. This isolation of the adaptive response from other forms of DNA repair was confirmed when we showed that bacteria can be induced by, or protected against, almost any alkylating agent, but that there is no "cross-talk" between, for example, alkylating agents and ultraviolet light (5).

However, when one considers protection against the lethal (as opposed to mutagenic) lesion produced by MNNG, the situation is more complicated. Strains lacking polymerase I (*polA* mutants) or lacking 6-methyladenine in their DNA (*dam* mutants) can acquire resistance to mutagenesis by MNNG but not to its lethality (6). Conversely, we have isolated a mutant that can be induced to resist the lethality of MNNG but not its mutagenicity (7). Hence, we believe that the mutagenic and the lethal lesions produced by this agent are different and are repaired by separate pathways, both of which are inducible. Since little is known about the lethal lesion, the rest of this article is concerned almost entirely with protection against mutagenesis.

We have isolated two classes of mutants with abnormal adaptive responses. The members of the first class are unable to acquire resistance to mutagenesis by MNNG or other similar alkylating agents, and most of them also cannot acquire resistance to killing (7). These adaptation-defective (*ada*) mutations map at about 50 minutes. This region also contains two other genes involved in repair of lesions due to alkylated bases. The *alk* gene functions in a polymerase-I-dependent pathway for repair of certain lethal lesions, but is apparently not involved in the inducible pathways (8, 9). The *tag* gene codes for 3-methyldeoxyadenosine(in DNA) glycosylase,[1] the constitutive enzyme required for the repair of DNA containing the potentially lethal (but not highly mutagenic) alkylated base, 3-methyladenine (10).

The members of the other class of mutants we have isolated are constitutive for the adaptive response (11), but their mutations have not yet been mapped. Using the adaptation-defective mutants, we have measured the efficiency of the adaptive response, and from a study of the constitutive mutants we hope to learn something about the regulation and induction of the response (see Section III).

II. The Substrate for the Adaptive Response

Although the response provides strong protection against many alkylating agents, its spectrum shows certain interesting limitations. Methylating agents, such as methylnitronitrosoguanidine, methyl methanesulfonate (MeMS), and methylnitrosourea (MENU), are all powerful inducers of the response (5) although one of them, MeMS, is not a very powerful mutagen. By contrast, ethylating agents like EtMS and EtNU have little or no ability to act as inducers (7, 12), but *E. coli* becomes protected against them when induced by methylating agents (5, 7). It seems likely, therefore, that the signal for induction is not simply the presence in DNA of whatever mutagenic bases are the substrate for the response.

Many sites in DNA are attacked by alkylating agents (13, 14). The main site is the N-7 atom of guanine, but this does not affect base-pairing, and 7-alkylguanine is not perceptibly mutagenic. Of the other less readily alkylated sites, the most important is the O-6 atom of guanine (15) because O^6-alkylguanine has a high probability of being mistaken for adenine when it is being replicated or transcribed (16–18). Certainly, most of the mutations produced in *E. coli* by MNNG are $G \cdot C$ to $A \cdot T$ transitions (19); this implies that the most important lesions are those that affect guanine (or cytosine).

We were therefore pleased to find that MNNG produced the same amount of 7-methylguanine and 3-methyladenine in uninduced and induced bacteria, but much less O^6MeGua (20), for this meant that the

---

[1] This enzyme has been termed "3-methyladenine-DNA glycosylase" (Lindahl, This Series 22). However, as noted (loc. cit., footnote 2), the substrate is not the base released but the deoxynucleoside, and only when that nucleoside is in a DNA. Hence the name used here. See also footnote 3 in Friedberg *et al.*, this volume. [Ed.]

adaptive response was a reaction directed specifically against what was thought to be the main mutagenic lesion produced by methylating agents. Since the response induced by methylating agents also protects against ethylating agents, the gene product that interacts with DNA O^6MeGua presumably can also interact with O^6EtGua. In this context it is interesting that O^6EtGua, unlike O^6MeGua, can also be repaired by the *uvr* pathway. Hence, this pathway is normally the main protection against ethylating agents (12).

III. The Chemistry of the Adaptive Response

Although the chemical basis for the adaptive response seemed initially to be perfectly straightforward, the kinetics of the reaction proved to be unexpectedly complex. Bacteria that were adapted to a low concentration of MNNG could completely resist challenge with it up to a certain point (i.e., up to a certain dose or duration or exposure); below their limit, they did not accumulate many mutations (21) or significant levels of O^6MeGua (20), but beyond that point they would accumulate mutations and O^6MeGua as though they were no longer adapted at all. This suggested that the response could be saturated. The other peculiarity was the actual kinetics of removal of O^6MeGua. Immediately after challenge with labeled MNNG, adapted bacteria contained much less O^6MeGua than unadapted bacteria, but they could not be caught in the act of removing this missing O^6MeGua even when they had been cooled to 0°C before the challenge (20); yet their subsequent rate of removal of whatever O^6MeGua they did contain was not particularly rapid.

We gradually came to realize that all these features of the adaptive response would be accounted for if the action against O^6MeGua were conducted by a molecule that could act only once. The numerology seemed not unreasonable, because the rate of synthesis of this hypothetical molecule (i.e., the rate of removal of O^6MeGua following the challenge of unadapted bacteria) worked out to be roughly 100 molecules per bacterium per minute, and the maxium number of these molecules (i.e., the number of O^6MeGua required to saturate the response of maximally adapted bacteria) came to about 3000 per bacterium. These happen to be the same as the values for the induced synthesis of β-galactosidase. The test of this interpretation (22) was to show that any O^6MeGua present after the challenge of adapted bacteria (i.e., in excess of 3000 per bacterium) was stable for at least an hour in the presence of chloramphenicol, for this indicated that the molecule reacting with O^6MeGua has a turnover time of more than an hour.

Additional compelling support has now come from the *in vitro* studies carried out in Lindahl's laboratory. Extracts of MNNG-adapted cells have been shown to act *in vitro* on DNA containing O^6MeGua (23). The reaction is unusual, for it is not simply the operation of a demethylase or glycosylase (i.e., it does not liberate an acid-soluble product, such as methanol or methylguanine). Instead, the methyl group of O^6MeGua is apparently transferred to the reactant molecule, because it can be liberated by treatment with proteases (T. Lindahl, personal communication). It is not surprising, therefore, that the reactant molecule seems to be able to function only once.

As more is learned about this strange reaction, we may discover why it proceeds in such an extravagant manner. But it is already possible to hazard a guess. In terms of mutation (as opposed to lethality), the most dangerous changes to DNA are probably not lesions like chain breaks or pyrimidine dimers (which cannot act as template and therefore have to be repaired), but rather those where one base is, in effect, changed to another (which then can be copied and therefore may not be detectable by any form of mismatch repair). O^6MeGua is just such a case in point, because it tends to be copied in *in vitro* systems as if it were adenine (*16–18*), and in bacteria lacking the adaptive response it produces $G \cdot C$ to $A \cdot T$ transitions with an efficiency approaching 100% (unpublished calculations). We should therefore expect to find the evolution of some very rapid form of repair for O^6MeGua and any other similar miscoding bases.

The absolute efficiency of the adaptive response has recently been measured by comparing the on-going mutation rate of adapted bacteria growing in the presence of a low concentration of MNNG with the mutation rate produced in adaptation-defective mutants by the medium in which the adapted bacteria had been growing (*24*). The results show that the response lowers the mutation rate about 6000-fold (by comparison, the constitutive error-free *uvr* pathway lowers UV-induced mutation rates by about 10-fold (*25*), although it might conceivably do better than this at very low dose rates). The high efficiency of the adaptive response implies something about its mechanism. One three-thousandth of the bacterial chromosome is replicated every second. Therefore, the mutation rate would be lowered only 3000-fold if each O^6MeGua, formed in an adapted bacterium, were allowed one second in which to be the source of a mutation. It follows that each O^6MeGua is probably being restored to its normal coding properties (i.e., is demethylated) within about half a second of its creation. To achieve this level of protection, the cell may have to produce a catalytic reactant that provides an alternative site for the alkyl group.

So much for the reaction with O^6MeGua. There is, however, some indirect evidence that alkylating agents like MNNG produce other mutagenic lesions that are repaired by other branches of the adaptive response. For example, we now have several mutants that are constitutively protected against O^6MeGua (i.e., accumulate no O^6MeGua when challenged with MNNG) (11). Most of them are not significantly mutated by MNNG, but one of them, paradoxically, shows an appreciable mutation rate in low (but not high) concentrations of MNNG. Indeed, after prolonged exposures it has a higher mutation rate than its wild-type parent. Our interpretation is that this strain bears a mutation in the repressor of the adaptive response that allows the constitutive synthesis of the protein reacting with O^6MeGua; the mutation also interferes to some extent with the interaction between the repressor and whatever normally induces the response, and for this reason, the mutant is less ready than wild-type to switch on synthesis of the proteins that deal with the other mutagenic alkylated bases.

Obviously much more genetics and biochemistry are required before we have a clear picture of the adaptive response. In particular, we lack any positive information about the signals that induce the response.

IV. The Adaptive Response in Mammalian Cells

Preliminary experiments with Chinese hamster ovary cells growing *in vitro* suggested that animal cells could acquire some resistance to the lethality of methylnitronitrosoguanidine (26). However, definitive experiments by Montesano *et al.* (27) show that the livers of rats exposed to low levels of dimethylnitrosamine continue to accumulate 7-methylguanine in their DNA, but acquire the ability to avoid accumulating O^6-methylguanine. Furthermore, the kinetics of the reaction in adapted liver is much the same as the kinetics in *E. coli*.

It has long been known that the tissues containing the highest levels of O^6-alkylguanine, after whole body exposure to alkylating agents, are the ones that will subsequently show the highest incidence of cancer (28, 29). Hence the existence of an efficient inducible form of repair should have a considerable influence on the carcinogenicity of different schedules of exposure to nitrosamines. For example, occasional large doses should be more carcinogenic than a steady rate of exposure over the same period of time; large dose rates, above the capacity of the adaptive response, should be disproportionately more carcinogenic than smaller doses. At least the second of these predictions seems to be true, for there appears to be a marked threshold effect for several nitrosamines (30, 31).

Summary

Alkylating agents, which are probably the most widespread chemical carcinogens in our environment, induce, both in bacteria and mammalian cells, what appears to be the most efficient form of DNA repair yet described. In *E. coli* growing in the presence of low levels of N-methyl-N'-nitro-N-nitrosoguanidine, the induced repair pathway lowers the mutation rate by almost four orders of magnitude.

The main mutagenic lesion produced by such methylating agents is O^6-methylguanine. In induced *E. coli*, this is rapidly repaired by a protein that apparently acts both as a demethylase and as an acceptor of the liberated methyl group. Like certain restriction enzymes (32, 33), this protein seems to be capable of acting only once.

REFERENCES

1. J. Cairns, *Proc. R. Soc. London, Ser.* **B208**, 121 (1980).
2. L. Samson and J. Cairns, *Nature* **267**, 281 (1977).
3. A. Jimenez-Sanchez and E. Cerda-Olmedo, *Mutat. Res.* **28**, 337 (1975).
4. P. Jeggo, M. Defais, L. Samson and P. Schendel, *in* "DNA Synthesis—Present and Future" (I. Molineux and M. Kohiyama, eds.), p. 1011. Plenum, New York, 1978.
5. P. Jeggo, M. Defais, L. Samson and P. Schendel, *Mol. Gen. Genet.* **157**, 1 (1977).
6. P. Jeggo, M. Defais, L. Samson and P. Schendel, *Mol. Gen. Genet.* **162**, 299 (1978).
7. P. Jeggo, *J. Bact.* **139**, 783 (1979).
8. Y. Yamamoto, M. Katsuki, M. Sekiguchi and N. Otsuji, *J. Bact.* **135**, 144 (1978).
9. Y. Yamamoto and M. Sekiguchi, *Mol. Gen. Genet.* **171**, 251 (1979).
10. P. Karran, T. Lindahl, I. Øfsteṅg, G. B. Evenson and E. Seeberg, *JMB* **140**, 101 (1980).
11. B. Sedgwick and P. Robins, *Mol. Gen. Genet.* **180**, 85 (1980).
12. W. Warren and P. D. Lawley, *Carcinogenesis* **1**, 67 (1980).
13. P. D. Lawley, *in* "Screening Tests in Chemical Carcinogenesis" (R. Montesano, H. Bartsch and L. Tomatis, eds.), p. 181. IARC Sci. Publ. No. 12 (1976).
14. B. Singer, *Nature* **264**, 333 (1976).
15. A. Loveless, *Nature* **223**, 206 (1969).
16. L. L. Gerchman and D. B. Ludlum, *BBA* **308**, 310 (1973).
17. P. D. Lawley and C. N. Martin, *BJ* **145**, 85 (1975).
18. P. J. Abbott and R. Saffhill, *BBA* **562**, 51 (1979).
19. C. Coulondre and J. H. Miller, *JMB* **117**, 577 (1977).
20. P. F. Schendel and P. E. Robins, *PNAS* **75**, 6017 (1978).
21. P. F. Schendel, M. Defais, P. Jeggo, L. Samson and J. Cairns, *J. Bact.* **135**, 466 (1978).
22. P. Robins and J. Cairns, *Nature* **280**, 74 (1979).

23. P. Karran, T. Lindahl and B. Griffin, *Nature* **280,** 76 (1979).
24. J. Cairns, *Nature* **286,** 176 (1980).
25. A. Miura and J. Tomizawa, *Mol. Gen. Genet.* **103,** 1 (1968).
26. L. Samson, Ph.D. Thesis, Unviersity of London, 1978.
27. R. Montesano, H. Bresil and G. P. Margison, *Cancer. Res.* **39,** 1798 (1979).
28. R. Goth and M. F. Rajewsky, *PNAS* **71,** 639 (1974).
29. G. P. Margison and P. Kleihues, *BJ* **148,** 521 (1975).
30. B. Terracini, P. N. Magee and J. M. Barnes, *Br. J. Cancer* **21,** 559 (1967).
31. R. Preussmann, D. Schmähl and G. Eisenbrand, *Z. Krebsforsch. Klin. Oncol.* **90.** 161 (1977).
32. M. Meselson, R. Yuan and J. A. Heywood, *ARB* **41,** 447 (1972).
33. S. Linn, J. A. Lautenberger, B. Eskin and D. Lackey, *FP* **33,** 1128 (1974).

VI. Functions Induced by Damaged DNA
Chairman and Summarizer: EVELYN M. WITKIN

Summary EVELYN M. WITKIN	247
Inducible Error-Prone Repair and Induction of Prophage Lambda in *Escherichia coli* RAYMOND DEVORET	251
DNA and Nucleoside Triphosphate Binding Properties of recA Protein from *Escherichia coli* K. MCENTEE, G. M. WEINSTOCK, AND I. R. LEHMAN	265
Molecular Mechanism for the Induction of "SOS" Functions MICHIO OISHI, ROBERT M. IRBE, AND LEE M. E. MORIN	281
Induction and Enhanced Reactivation of Mammalian Viruses by Light LARRY E. BOCKSTAHLER	303
Comparative Induction Studies ERNEST C. POLLARD, D. J. FLUKE, AND DENO KAZANIS	315

Summary

EVELYN M. WITKIN

Department of Biological Sciences
Douglass College, Rutgers University
New Brunswick, New Jersey

The last session of this Symposium considered some aspects of the regulatory role of DNA damage. The *recA* protein of *Escherichia coli*, in addition to its role in promoting genetic recombination, is required for induction of "SOS functions"[1]: survival-promoting cellular activities that are repressed in intact calls but are expressed in response to certain types of DNA damage (*1*, review). Attention was directed to properties and activities of the *recA* protein (McEntee, Devoret), to the nature of the SOS "signal" generated by DNA damage (Oishi), and to evidence for inducible phenomena comparable to bacterial SOS functions in mammalian cells (Bockstahler). Additional contributions were made by Ernest Pollard, Graham Walker, and Alain Sarasin.

A central question in efforts to understand regulation of SOS functions in *E. coli* has been the role of the *recA* protein in this complex process. Induction of λ prophage, an SOS function that promotes phage survival rather than bacterial survival, is accompanied by proteolytic cleavage of λ repressor, and purified *recA* protein cleaves λ repressor *in vitro* (*2*). Current models of SOS regulation assume that *recA* protein is *activated*, i.e., that it acquires proteolytic activity, as a consequence of a process that begins with DNA damage, and that such activation of *recA* protein is a necessary condition for SOS induction. One of the products induced by DNA damage is *recA* protein itself, which is synthesized copiously in response to SOS-inducing treatments. Thus, DNA damage results in both the activation and the amplification of *recA* protein. Amplification of *recA* protein without its activation (e.g., by cloning the *recA*$^+$ gene on a plasmid or phage) does not result in SOS induction (*3*).

Devoret addressed himself to the question: Is amplification of *recA* protein *necessary* (albeit not sufficient) for SOS induction? In experiments designed to permit activation but not amplification of *recA* protein, the answer obtained was unequivocally negative. Two SOS functions (Weigle reactivation and λ prophage induction) were shown to be fully inducible under conditions preventing any significant in-

---

[1] Also defined in footnote 2 in Hanawalt *et al.*, this volume [Ed.].

crease in the rate of synthesis of *recA* protein following an SOS-inducing treatment. Efficient cleavage of λ repressor was also demonstrated under these conditions, indicating that the low constitutive level of *recA* protein present in uninduced cells is sufficient, given an appropriate inducing treatment, to promote the proteolytic cleavage of the prophage repressor. Devoret reviewed the evidence that *recA* protein has two distinct domains, one active in DNA binding and recombination and the other in the proteolysis required for SOS induction. His results support the conclusion that amplification of *recA* protein contributes to cell survival by increasing the efficiency of DNA repair and that activation of the proteolytic function of this remarkable protein, independently of its amplification, is both necessary and sufficient for SOS induction.

McEntee's contribution elucidated some aspects of the role of *recA* protein in genetic recombination, and presumably in recombinational repair of DNA. Reviewing recent advances, he described the ability of *recA* protein, in the presence of ATP or some of its analogs, to promote rapid reannealing of homologous single-stranded DNA chains, as well as association between single- and double-stranded, or between two double-stranded DNA species, not necessarily homologous. McEntee went on to define specific requirements for *recA*-dependent D-loop formation, for local unwinding of double-stranded DNA, and for the single-strand assimilation promoted by *recA* protein when sufficient homology exists between the duplex and the single strand associated in a D-loop. Strand assimilation proceeds well when *recA* protein is present in excess over the amount of single-stranded DNA, and is inhibited by excess single-strandedness. Masking excess single-strandedness by the addition of the single-strand-binding protein (SSB) increases the efficiency of strand assimilation. This suggests that recombination proficiency may be modulated *in vivo* by mutations affecting the ratio of *recA* protein to SSB produced constitutively and/or after DNA-damaging treatments. An altered ratio of these two DNA-binding proteins could explain the UV-resistance, proficient postreplication repair, and effective control of post-UV DNA degradation exhibited by mutant strains (*rnmA, rnmC*) that fail to amplify *recA* protein after SOS-inducing treatments (*4*). If this SSB or some other single-strand binding protein is overproduced in such mutants, strand assimilation and inhibition of DNA degradation may operate efficiently in spite of the presence of low levels of *recA* protein.

Oishi described recent developments in his efforts to characterize the early molecular events leading to SOS induction. Previous work in his laboratory had utilized a permeabilized cell system in which $\phi 80$ repressor controls synthesis of a bacterial enzyme (*5*). The ability of

SUMMARY

various oligonucleotides resulting from digestion of DNA to inactivate the phage repressor was monitored simply by assaying the enzyme. These studies led to the conclusion that certain breakdown products of DNA, for example d(AGGp), could promote phage repressor inactivation, relieving the requirement for exonuclease V activity that exists when induction is triggered by replication fork damage. More recent work has utilized a $\phi 80$ lysogen, permeabilized by osmotic shock, which remains capable of phage production following inactivation of the prophage repressor. Assaying induction by phage production, Oishi and co-workers have found that d(AGGp) is nearly as effective an inducer as UV, and that micrococcal nuclease is also an excellent inducer in strains possessing active exonuclease V (*recB/recC* gene product). Certain dinucleotides, including d(GG) have inducing activity at concentrations as low as 100 nM. Induction by d(GG) is abolished either by a *recA*$^-$ mutation in the host or by an *ind*$^-$ mutation in the gene coding the phage repressor, suggesting that the induction pathway initiated by the dinucleotide could be the same one that is initiated by genetic damage to the host. Some of Oishi's observations indicate a high degree of complexity and specificity in the early steps of SOS regulation. For example: d(GG) is an efficient inducer of $\phi 80$ prophage but a poor inducer of λ prophage. Furthermore, d(AGGp) fails to induce *recA* protein synthesis, nor does it promote cleavage of λ repressor *in vitro*.

Further evidence for complexity and specificity in SOS induction was provided by Pollard, who showed that various SOS functions in *E. coli* are expressed in the same irradiated population with quite different kinetics, and that their expression does not necessarily parallel the induced synthesis of *recA* protein. From his work on induced radioresistance, Pollard warned that extrapolation to low doses from data on carcinogenesis obtained at high radiation doses can be misleading, because populations exposed to high doses comprise only induced cells whereas those exposed to low doses comprise a mixture of induced and uninduced cells.

Walker described work from his laboratory (6) designed to detect DNA *d*amage-*in*ducible genes or operons ("*din*" loci) controlled by products of *E. coli* genes *recA* and *lexA*. A number of such presumptive SOS operons have been identified at various locations on the genetic map by the use of random insertions of a μ phage variant carrying the structural gene for β-galactosidase without its normal regulatory sequence. Clones carrying such insertions were screened to identify those in which β-galactosidase synthesis was inducible by DNA-damaging agents in *recA*$^+$ *lexA*$^+$ but not in *recA*$^-$ or *lexA*$^-$ strains. One of the *din* loci detected in this way proved to be the *uvrA* gene, not previously known to belong to the family of SOS-inducible genes. Characterization of the phenotypes of the other *din* insertion mutants, and the isolation of additional ones, can be expected to add much to our knowledge of the scope and significance of the SOS response.

Bockstahler reviewed evidence for radiation-inducible phenomena associated with mammalian viruses. He presented data showing induction of SV40 virus from SV40-transformed hamster kidney cells by photodynamic treatment, supporting an earlier hypothesis that clinical use of such treatment might result in the induction of latent tumor viruses. He reviewed evidence for "enhanced reactivation" of UV-irradiated mammalian viruses, i.e., for the increased survival of the virus that is observed when infectivity is assayed in radiation-damaged host cells rather than in undamaged host cells. Enhanced reactivation, similar in principle to Weigle reactivation[2] of bacteriophage, occurs in herpes simplex, SV40, and other nuclear-replicating viruses and in cells from many different normal and transformed cell lines treated with a variety of DNA-damaging agents. Whether or not the mechanism of this enhanced reactivation is similar to that of Weigle reactivation is not yet clear. Bockstahler then considered evidence that enhanced reactivation of animal viruses is or is not mutagenic. Citing conflicting experimental reports, he concluded that no unequivocal answer can yet be given to this question.

Sarasin's poster was directly related to the questions raised by Bockstahler. The data presented added to earlier work on induced reactivation of UV-irradiated SV40 virus in monkey kidney cells treated with carcinogens (7). The new data showed an increase in the yield of wild-type revertants of the virus associated with induced reactivation of UV-treated temperature-sensitive SV40 infecting cells pretreated with the carcinogen acetoxyacetamidofluorene (AAF). Sarasin concludes that an error-prone mode of DNA replication may be induced by DNA damage in mammalian cells.

REFERENCES

1. E. M. Witkin, *Bacteriol. Rev.* **40**, 869 (1976).
2. J. W. Roberts and C. W. Roberts, *PNAS* **72**, 147 (1975); J. W. Roberts, C. W. Roberts and N. L. Craig, *PNAS* **75**, 4714 (1978).
3. R. Brent and M. Ptashne, *PNAS* **77**, 1932 (1980).
4. M. R. Volkert, D. F. Spencer and A. J. Clark, *Mol. Gen. Genet.* **177**, 129 (1979).
5. M. Oishi and C. L. Smith, *PNAS* **75**, 3569 (1978).
6. C. J. Kenyon and G. C. Walker, *PNAS* **77**, 2819 (1980).
7. A. R. Sarasin and P. C. Hanawalt, *PNAS* **75**, 346 (1978).

[2] Defined in footnote 3 of Hanawalt *et al.*, this volume [Ed.].

Inducible Error-Prone Repair and Induction of Prophage Lambda in *Escherichia coli*

| RAYMOND DEVORET
| *Radiobiologie Cellulaire*
| *Laboratoire d'Enzymologie C.N.R.S.*
| *Gif-sur-Yvette, France*

Rapid progress was made in the molecular biology of repair of cellular damage when:

1. DNA was recognized as the main target of the 254 nm ultraviolet radiation produced by germicidal lamps, the use of which had become prevalent in the laboratory.
2. Pyrimidine dimers were identified as the main DNA lesions produced by UV light.
3. Repair was demonstrated to be enzymic and genetically controlled (*1, 2*).

This knowledge led to the characterization in *Escherichia coli* of two dark-repair processes of DNA damage: pyrimidine-dimer-excision repair (*3, 4;* see contributions by Friedberg *et al.* and by Hanawalt *et al.* in this volume), and postreplicative recombinational repair (*5, 6*).

A third dark-repair process, inducible and error-prone, was subsequently identified in prokaryotes (reviewed in *7* and *98*), and then in eukaryotes (see chapter by Bockstahler in this volume). Note that an inducible error-free repair pathway was discovered more recently (*8*) and characterized at the genetic and molecular levels (see chapter by Cairns *et al.* in this volume).

Let me state right away that the nature and the significance of inducible error-prone repair in *E. coli* (nicknamed "SOS" repair[1]), which the organizers of this meeting asked me to review, cannot be understood independently of another inducible cellular process, induction of prophage λ, which also results from the unscheduled arrest of DNA replication produced by DNA damage. Inducible error-prone repair is but one facet of a general cellular inducible mechanism. This is the gist of the unitary hypotheses postulated originally by Radman (*9*) and by Witkin (*10*).[2]

---

[1] Defined in footnote 2 of Hanawalt *et al.*, this volume [Ed.].
[2] See preceding paper.

This minireview is divided into three main parts. The first part (Section I) deals with the induction of prophage λ, which has been the paradigm of a complex inducible cellular pathway, and whose molecular mechanism has constantly served as a model for the other cellular functions induced by DNA damage. The second part (Section II) describes the experiments that led to the recognition of inducible error-prone repair. The regulation of cellular functions elicited by DNA damage (often called SOS functions[1]) are analyzed in the third part (Section III) of this review.

I. Induction of Prophage λ[3]

Most bacteria, if not all, are lysogenic: they carry in their genome the DNA of a virus (phage) in a dormant state (prophage) (*11*). Phage λ, whose host is *Escherichia coli* K12, has been one of the most studied viruses over the past 25 years (*12*).

A. Dormancy and Induction of Prophage λ

The dormant state of prophage λ is maintained by a repressor protein, which is the product of the prophage *cI* gene (*13*). The repressor prevents transcription of the prophage genes, other than the *cI* gene, by binding to two operators divided into six operator sites (*14*) flanking the *cI* gene.

When lysogenic *E. coli* bacteria (also called lysogens) are subjected to any treatment that halts DNA replication, prophage λ is induced (*11*): its DNA loops out of the bacterial DNA and directs the synthesis of proteins forming the virus particle, and a progeny of mature phage develops and bursts out of the host cell. Differing from other temperate prophages such as P2, prophage λ is inducible by agents or treatments that lead to carcinogenic transformation in mammalian cells (*15*).

In 1961, Jacob and Monod (*16, 17*) postulated that, following treatment of *E. coli* lysogens by DNA-damaging agents, a cellular inducer would inactivate the repressor of the prophage, and thereby would cause prophage induction. That induction of prophage λ indeed results from the inactivation of λ repressor was demonstrated (*18*) soon after λ repressor was isolated (*19*), and an easy assay for its titration in cell crude extracts was discovered (*20*). From the kinetics of disappearance of λ repressor, it is clear that as soon as there are no more than 10% of repressor molecules left in the cell (*21*) to block the pro-

---

[3] See also Oishi *et al.* in this volume, regarding the induction of prophage 80.

phage operator sites, prophage induction will occur. Transcription of the structural genes of the prophage begins (22), and within 60 minutes the host cell lyses and releases a progeny of about 100 phage particles.

B. DNA: The Primary Target of λ-Inducing Agents

The question to be asked first concerns the way λ-inducing agents such as UV light affect the cell to elicit the inactivation of λ repressor. Does UV light act directly on the repressor by a photochemical reaction that renders it unable to bind to operator regions? Or does UV light produce so much damage in the DNA of the prophage cI gene coding for λ repressor that transcription of this gene is prevented and, thus, the synthesis of λ repressor?

It has been demonstrated that the primary target of UV light is not the repressor protein or, more specifically, the DNA sequence coding for it. In effect, indirect induction of prophage λ occurs when the UV-damaged DNA of a replicon is introduced into a λ lysogen not itself exposed to UV light (23–26). This has been shown to occur with self-transmissible replicons such as F, R, and ColI (27). Except for phage P1 (28), other UV-damaged phages do not induce prophage λ.

Note that the part of the UV-damaged replicon, whose introduction into the cell elicits prophage induction, appears to be a small DNA fragment encompassing the origin of replication (oriF) of the F sex factor (Devoret, Couturier, and von Meyenburg, unpublished).

Transfer to a recipient λ lysogen of the UV-damaged DNA from an Hfr donor does not induce prophage λ, yet the introduction of such a damaged piece of nonreplicating DNA into a host cell will induce error-prone repair (29; see Section III).

What is then in a λ lysogen the target whose damage leads to lysogenic induction? We have argued that the bacterial chromosome being a replicon, damage at the replication fork disturbs the regulation of DNA replication, which in its turn provokes the induction of prophage λ (26). So much now for the primary inducing event.

C. Gene Products Involved in the Induction of Prophage λ

In a UV-irradiated E. coli lysogen, the pathway of induction of prophage λ begins with DNA lesions and ends with the disappearance of λ repressor. What are the successive biochemical steps involved in λ repressor inactivation?

This question was first approached by genetic methods. The identification of bacterial mutants in which λ was not inducible paved the way for the elucidation of the induction pathway. It was unexpectedly found that in bacterial mutants called $recA^-$, isolated as being unable to carry out genetic recombination (30), prophage λ is not inducible (31, 32). It was also observed that prophage λ is poorly inducible in $lexA^-$ mutants (33). Moreover, most mutations in either recA or lexA genes not only render the mutants sensitive to UV or X-ray irradiation (34, 35), but also affect several cellular functions induced by DNA damage: UV-reactivation (see below), mutagenesis by UV light, and filamentation (10).

What is the relationship between these two genes that regulate such diverse basic cellular functions (10, 35a)? It has been postulated that the lexA gene product is the repressor of the recA gene (36, 37). Indeed, the lexA protein binds to the operator region of the recA gene and inhibits recA transcription in vitro as has been demonstrated by Little et al. (personal communication). Furthermore, the lexA repressor appears to control its synthesis (38).

Other genes, such as recF (39, 40, 40a) and infA (41), are also involved in the induction pathway, but their role is still unknown. The infA mutants were isolated in an attempt to characterize various mutations that would render the bacteria unable to carry out one of the reactions required for prophage λ induction (42). Half of the isolated mutations were located in the recA region. The study of these mutants (43–48), coupled with that of others (49), led to the proposal that, in the recA gene, there are two clusters of mutations (lexB and recA proper) that may code, respectively, for two domains of recA protein (44). One domain may be involved in the binding of recA protein to DNA and control DNA degradation by the gene recBC product and thus, more specifically, recombination (as evidenced by recA mutations); the other domain would affect the cellular functions induced by DNA damage, such as prophage induction (as evidenced by the lexB and tif mutations).

What is the terminal reaction in the lysogenic induction pathway?

D. RecA Protein Cleaves λ Repressor

The terminal reaction in the pathway leading to induction of prophage λ has been elucidated by Roberts et al. (49–53), who demonstrated that (a) in vivo, inactivation of the λ repressor is due to proteolytic cleavage; (b) repressor cleavage can be produced in vitro; (c) cleavage is effected by the recA protein, and the reaction requires ATP and a polydeoxyribonucleotide chain of about 16–20 nucleotides; (d) prophage repressors other than λ, for instance P22 repressor, are also cleaved.

Note that lexB30 mutants, in which recombination but not induction of prophage λ can take place, are inducible for prophage 80 (Devoret and Pierre, unpublished). This indicates that the *lexB30* mutation may affect the specific binding of recA protein to a lambdoid repressor, not the catalytic site of the protease. Mutations in the *recA* gene that specifically affect the catalytic site of the protease have not yet been characterized.

When bacteria are submitted to a DNA-damaging treatment, transcription of the *recA* gene increases considerably. The amount of recA protein produced may amount to as much as 3% of the total of the cell proteins synthesized (54). The cloning of the *recA* gene by McEntee (55, 56) led many authors to identify recA protein as the formerly described protein X (57–60).

E. The recA Protein Cleaves the lexA Protein

The lexA protein, the repressor of the *recA* gene, is cleaved by the recA protein (61, 62). This appears to be the major mechanism whereby not only is the recA protein induced, but also the other cellular functions that follow DNA damage.

The recA protein cleaves the λ repressor and the *lexA* repressor, but carries basic functions in repair and recombination, the first steps of which are being elucidated (see contribution by McEntee *et al.* in this volume) (63, 64).

F. A Mechanism of Prophage Induction

The model presented here proposes in particular an explanation for the discrepancy between the elevated cell concentration of recA protein observed following DNA damage and its poor efficiency in cleaving λ repressor.

1. A stalled replication fork, which therefore contains gaps, is the substrate to which recA protein will bind. Bound recA molecules undergo an allosteric transformation into recA protease.
2. This transformation may occur at 42° in a *tif-1* bacterial mutant at Okazaki fragments in the absence of DNA lesions; Okazaki fragments may represent sites of "physiological lesions" that occur during the normal course of DNA replication; such occurrence would account rather simply for the activities of the recA-tif-modified protein in the absence of "real" DNA damage.
3. Since the lexA repressor is cleaved, recA transcription, normally at a low level, will be derepressed (64a). More recA protein is made, but the cell concentration in the activated form of recA protein depends on how many sites on the DNA are available to activate recA protein (62).

4. When DNA damage engenders sites (21) for the activation of recA protein, full induction of recA protein synthesis is not required for complete cleavage of λ repressor (65). The λ repressor is cleaved by the activated form of recA protein.

The model proposed implies that there are two forms of recA protein present in a cell, whose relative proportion is dependent upon the structural damage imposed on a replicon. This model is at variance with that of Oishi et al. (66; see paper by Oishi et al. in this volume).

Nature has certainly not invented the proteolytic form of recA protein to promote only induction of prophage λ. One of the roles of the recA protease is certainly to cleave first its repressor, the lexA protein. Phage λ adapted itself during the course of evolution to recognize when a cell is likely to die. Induction gives the prophage a chance to survive the doomed host cell. The progeny phage may be rescued from an impending death by parasitizing another bacterial population of phage-sensitive cells.

II. Inducible Error-Prone Repair

A. UV Reactivation

Weigle in 1953 (67) reported that, when *E. coli* K12 bacteria were exposed to a low UV dose and then infected with UV-irradiated phage λ, the survival of the phage was higher than in the control untreated host. Furthermore, the reactivated phage was heavily mutated (68). Weigle called this phenomenon "UV reactivation" (known also now as Weigle reactivation).

Has nature selected UV reactivation along with induction of prophage λ in order for the UV-damaged prophage to be repaired before being induced to leave a bacterium doomed to die? Or is UV reactivation a more general repair restoring also damage to chromosomal DNA?

B. A Third Repair Process

It has taken more than 20 years to elucidate the mechanism of UV reactivation. After the discovery of UV reactivation, two hypotheses were offered to account for this phenomenon. Weigle's hypothesis (69) was that UV reactivation resulted from illegitimate recombination between the DNAs of the infecting phage and the host chromosome. Harm's hypothesis (70) was that pretreatment of the host with UV light produced an enhancement of what is now known as excision repair. These two hypotheses have been proved incorrect.

In my laboratory, we provided evidence that UV reactivation was not the result of illegitimate recombination (71, 72); also, that UV-reactivation may be obtained in excision-repair-deficient bacteria (29, 73), although the efficiency of UV-reactivation can be low in some $uvrB^-$ mutants (73a). We reasoned (74) that UV-reactivation cannot result from postreplicative recombinational repair, since single-stranded phages can be UV-reactivated (75, 76). Neither could it result from a prereplicative recombinational process (74). Since the multiplicity of host infection by the single-stranded phage was low, UV reactivation is not due to multiplicity reactivation, a prereplicative recombinational process (74).

C. An Inducible Error-Prone Repair Type for UV-Damaged Phage

If the process of UV reactivation of UV-damaged phage λ is a unique repair mechanism, what is it? I have heard some people answer this question by saying "error-prone repair." Let me point out that this is but a tautological answer. UV reactivation is a highly mutagenic type of repair; to call it error prone is simply to acknowledge that it is mutagenic.

A real advance in the field occurred when it was established that this type of repair is inducible. The idea that UV reactivation of phage λ is inducible by the very presence of residual DNA lesions, or following the unscheduled arrest of DNA replication, was conceptually new.

The inducibility of UV reactivation of phage λ was demonstrated by two sets of experiments. On the one hand, it was shown that, similarly to lysogenic induction, UV reactivation can be produced by introducing a piece of damaged DNA by conjugal transfer into a host cell subsequently infected with UV-damaged phage λ (29). Note that the UV-damaged DNA can be either that of a transmissible replicon or not. On the other hand, Defais et al. (77) showed that the efficiency of UV reactivation of UV-damaged phage λ reached a peak after 30 minutes of incubation of the UV-irradiated host cells before infection; chloramphenicol treatment of the UV-irradiated host cells that stopped protein synthesis before phage infection prevented the occurrence of repair and mutagenesis of the phage.

More recently, it has been demonstrated that, upon UV reactivation of phage ϕX174, a single-stranded phage, DNA replication is not prevented by the presence of pyrimidine dimers; this was observed for a substantial fraction of UV-damaged single-stranded DNA molecules (78). The fraction of DNA molecules replicated corresponded to the fraction of repaired phage. Replication can occur on UV-damaged DNA templates. It has not yet been shown, at the molecular level, how replication can occur on a template containing pyrimidine dimers. We know from genetic experiments that it does occur. The solution to this question is the first step toward the understanding of the molecular mechanism of mutagenesis.

To account for the molecular mechanism of inducible error-prone repair, Villani et al. postulated that an inhibitor of the 3'—5' exonucleolytic activity of polymerase I is induced following DNA damage (79). In effect, when polymerase I makes a mistake, the error is immediately corrected; in the case of a coding abnormality this results in the idling of DNA polymerase (79). Polymerase idling must consequently be lifted by an inhibitor of the correcting mechanism (80). This hypothesis can account for the fact that DNA replication can proceed opposite pyrimidine dimers at the cost of a lower fidelity.

If there exists such a mechanism, the present likely candidate for the inhibitor of the correcting mechanism might be the product of the *uvm* or *umuC* gene (81–84a). This assumption is based upon the fact that $umuC^-$ or uvm^- mutations prevent UV-reactivation from occurring in $recA^+$ and $lexA^+$ bacteria. In any case, in UV-irradiated hosts in which the cellular concentration of recA protein was 50 times more than in the wild type, the time course of UV-reactivation of phage λ was as found by Defais et al. (77); the last experiment indicates that a protein other than recA is induced (Quillardet et al., unpublished). Is this the gene *umuC* product? Moreover, is the *umuC* gene under the control of the lexA repressor?

D. Inducible Error-Prone Repair Operates Also on Bacterial DNA

Since most studies on inducible error-prone repair have been mostly conducted on phages, the question arises whether this type of repair is restricted to phage DNA or operates also on chromosomal DNA.

A bacterial mutant whose recA protein is altered, this being due to a mutation called *tif-1* located in the *recA* gene (85–87) behaves, when shifted to 42°C, almost as though it has been exposed to UV light. The temperature shift induces prophage λ and, in a nonlysogen, cell mutagenesis, phage reactivation, and cell filamentation. In other words, the temperature shift mimics in this mutant (86) the effects of UV irradiation.

The behavior of the *recA tif-1* mutant has been taken as an argument in favor of the notion that inducible error-prone repair must operate on bacterial DNA. This idea has been challenged (88). Inducible error-prone repair in *E. coli* has certainly not been selected in the course of evolution for the repair of phage or prophage DNA, but mainly for the repair of cellular DNA (9). Based upon this assumption, UV-damaged viruses have been used to probe the existence of such a repair process in prokaryotes and in eukaryotes as well (see contribution by Bockstahler in this volume).

E. Targeted versus Nontargeted Mutagenesis

Since mutations can be revealed only in bacteria that have survived DNA damage, it is manifest that DNA replication can go to completion on damaged and, therefore, noncoding templates. This question breeds a second question: Do mutations arise necessarily in the newly formed DNA strand opposite the lesions residing in the old strand? An intact phage can mutate upon infection of a UV-damaged host, and the rate of mutagenesis can be as high as 10–40 times over the background (89, 90; P. Quillardet and R. Devoret, unpublished); this phenomenon, called *indirect mutagenesis*, can also take place in hosts treated with nitrosoguanidine (91). Moreover, Witkin reported that nontargeted mutagenesis occurs in bacterial DNA (92).

III. The Regulation of Cellular Functions Induced by DNA Damage

A. The SOS Repair Hypothesis

Radman proposed that UV reactivation of phage λ was the manifestation of an inducible error-prone repair system acting not only on phage DNA, but also on chromosomal DNA. DNA damage induces a process for its repair. This repair being mutagenic may result in a better adaptation of cells exposed to detrimental conditions. Mutagenesis permits survival. Hence, the choice of "SOS repair" to designate this cellular process (9).

The SOS repair hypothesis has provided a theoretical framework for the mechanism of mutagenesis. This is not the least of its merits.

B. A Generalized Mechanism for the Cellular Functions Induced by DNA Damage

Witkin, who had postulated a correlation between induction of prophage λ and cell filamentation (93), extended the SOS concept to the other cellular functions induced by DNA damage (10). Most importantly, she proposed that all these "SOS functions" result from a coordinate derepression of various operons, each controlling a given cellular function. In the absence of DNA damage, all these operons are turned off by specific repressors (10). She also postulated that derepression of SOS operons would result from the inactivation of the various repressors by proteolytic cleavage, as has been demonstrated for prophage induction.

C. The lexA Repressor Has a Major Regulatory Role

Witkin's concept (10) of a coordinate expression of cellular functions induced by DNA damage has received major support from the studies concerning the role of the lexA repressor. It may no longer be necessary to postulate that many specific repressors regulate the so-called SOS operons. The lexA repressor appears to play a central role in controlling most of the cellular functions induced by DNA damage, the list of which (10) may include also excision repair (94).

IV. Conclusion

The unitary hypotheses of SOS repair and SOS functions have provoked much experimental work. The cellular functions induced by DNA damage appear to be *lexA*-controlled functions. Note that experimental data now indicate that, apart from being under the probable control of a common repressor, all cellular functions induced by DNA damage have different molecular mechanisms.

Mutagenesis induced by DNA damage has been considered as one of the basic mechanisms of carcinogenesis. Some researchers in this field have considered of general value the fact that there exists in bacteria an inducible error-prone repair mechanism that can account for the appearance of most mutations. It must be stressed that epigenetic mechanisms such as induction of dormant viruses are also observed in bacteria.

The fact that there are strong analogies between prokaryotic and eukaryotic cells in their response to DNA damage amply justifies the research devoted to analyzing the regulation of cellular functions (genetic as well as epigenetic) by repressors, whose inactivation by proteolytic cleavage is irreversible. Commitment to cell differentiation implies such an irreversible mechanism. Studies conducted with the help of prokaryotic organisms should provide some valuable contribution to the models postulated for cell transformation in mammalian cells.

Acknowledgments

Mélanie Pierre is gratefully thanked for her help with the references. This work has benefited from grants from Fondation pour la Recherche Médicale, Ligue Française contre le Cancer, and Association pour le Dévelopment de la Recherche sur le Cancer.

References

1. A. Rörsch, C. Van den Kamp and J. Adema, *BBA* **80**, 346 (1964).
2. W. Harm, Z. *Vererbungslehre* **94**, 67 (1963).
3. R. B. Setlow and W. L. Carrier, *PNAS* **51**, 226 (1964).
4. R. P. Boyce and P. Howard-Flanders, *PNAS* **51**, 293 (1964).
5. W. D. Rupp, C. E. Wilde, D. L. Reno and P. Howard-Flanders, *JMB* **61**, 25 (1971).
6. P. Howard-Flanders, *Br. Med. Bull.* **29**, 226 (1973).
7. R. Devoret, *Biochimie* **60**, 1135 (1978).
8. L. Samson and J. Cairns, *Nature* **267**, 281 (1977).
9. M. Radman, *in* "Molecular Mechanisms for Repair of DNA (P. C. Hanawalt and R. B. Setlow, eds.), p. 355. Plenum, New York (1975).
10. E. M. Witkin, *Bact. Rev.* **40**, 869 (1976).
11. A. Lwoff, *in* "Phage and the Origins of Molecular Biology" (J. Cairns, G. S. Stent and J. D. Watson, eds.), p. 88. Cold Spring Harbor Laboratory, Cold Spring Harbor, New York, 1966.
12. A. D. Hershey and W. Dove, *in* "The Bacteriophage Lambda" (A. D. Hershey, ed.), p. 3. Cold Spring Harbor Laboratory, Cold Spring Harbor, New York, 1971.
13. M. Ptashne, *in* "The Bacteriophage Lambda" (A. D. Hershey, ed.) p. 221. Cold Spring Harbor Laboratory, Cold Spring Harbor, New York, 1971.
14. M. Ptashne, K. Backman, M. Z. Humayun, A. Jeffrey, R. Maurer, B. Meyer, and R. T. Sauer, *Science* **194**, 156 (1976).
15. P. Moreau and R. Devoret, *Origins of Human Cancer, Cold Spring Harbor Conf. Cell Proliferation,* Book **C 4**, 1451 (1977).
16. F. Jacob and J. Monod, *JMB* **3**, 318 (1961).
17. F. Jacob and J. Monod, *CSHSQB* **26**, 193 (1961).
18. H. Shinagawa and T. Itoh, *Mol. Gen. Genet.* **126**, 103 (1973).
19. M. Ptashne, *PNAS* **57**, 306 (1967).
20. A. D. Riggs, H. Suzuki and S. Bourgeois, *JMB* **48**, 67 (1970).
21. A. Bailone, A. Levine and R. Devoret, *JMB* **131**, 553 (1979).
22. P. Chadwick, V. Pirrotta, R. Steinberg, N. Hopkins and M. Ptashne, *CSHSQB* **35**, 283 (1970).
23. E. Borek and A. Ryan, *PNAS* **44**, 374 (1958).
24. E. Borek and A. Ryan, This Series **13**, 249 (1973).
25. R. Devoret and J. George, *Mutat. Res.* **4**, 713 (1967).
26. J. George and R. Devoret, *Mol. Gen. Genet.* **111**, 103 (1971).
27. R. Devoret, M. Monk and J. George, *Zentralbl. Bakteriol. Parasitenk. Infektionskr. Abt.* 1 **196**, 193 (1965).
28. J. L. Rosner, L. Kass and M. B. Yarmolinsky, *CSHSQB* **33**, 785 (1968).
29. J. George, R. Devoret and M. Radman, *PNAS* **71**, 144 (1974).

30. A. J. Clark, *J. Cell. Physiol.* **70,** Suppl. 1, 165 (1967).
31. K. Brooks and A. Clark, *Virology* **1,** 283 (1967).
32. I. Hertman and S. Luria, *JMB* **23,** 117 (1967).
33. J. J. Donch and J. Greenberg, *Mol. Gen. Genet.* **128,** 277 (1974).
34. D. W. Mount, K. B. Low and S. J. Edmiston, *J. Bact.* **112,** 886 (1972).
35. P. Howard-Flanders and R. P. Boyce, *Radiat. Res. Suppl.* **6,** 156 (1966).
35a. M. R. Volkert, D. F. Spencer and A. J. Clark, *Mol. Gen. Genet.* **177,** 129 (1979).
36. D. W. Mount, *PNAS* **74,** 300 (1977).
37. L. Z. Pacelli, S. H. Edmiston and D. W. Mount, *J. Bact.* **137,** 568 (1979).
38. R. Brent and M. Ptashne, *PNAS* **77,** 1932 (1980).
39. Z. I. Horii and A. J. Clark, *JMB* **80,** 327 (1973).
40. M. E. Armengod and M. Blanco, *Mutat. Res.* **52,** 37 (1978).
40a. A. J. Clark, M. R. Volkert and L. J. Margossian, *CSHSQB* **43**(2), 887 (1978).
41. A. Bailone, M. Blanco and R. Devoret, *Mol. Gen. Genet.* **136,** 291 (1975).
42. R. Devoret and M. Blanco, *Mol. Gen. Genet.* **107,** 272 (1970).
43. P. Morand, M. Blanco and R. Devoret, *J. Bact.* **131,** 572 (1977).
44. P. Morand, A. Goze and R. Devoret, *Mol. Gen. Genet.* **157,** 69 (1977).
45. K. McEntee, J. E. Hesse and W. Epstein, *PNAS* **73,** 3979 (1976).
46. K. McEntee, G. M. Weinstock and I. R. Lehman, *PNAS* **77,** 857 (1980).
47. M. Castellazzi, P. Morand, J. George and G. Buttin, *Mol. Gen. Genet.* **153,** 297 (1977).
48. B. W. Glickman, N. Guijt and P. Morand, *Mol. Gen. Genet.* **157,** 83 (1977).
49. J. W. Roberts and C. W. Roberts, *PNAS* **72,** 147 (1975).
50. J. W. Roberts, C. W. Roberts and D. W. Mount, *PNAS* **74,** 2283 (1977).
51. J. W. Roberts, C. W. Roberts and N. L. Craig, *PNAS* **75,** 4714 (1978).
52. N. L. Craig and J. W. Roberts, *Nature* **283,** 26 (1980).
53. J. W. Roberts, C. W. Roberts, N. L. Craig and E. M. Phizicky, *CSHSQB* **43** (2), 917 (1978).
54. L. J. Gudas, *JMB* **104,** 567 (1976).
55. K. McEntee, *Virology* **70,** 221 (1976).
56. K. McEntee and W. Epstein, *Virology* **77,** 306 (1977).
57. L. J. Gudas and A. B. Pardee, *PNAS* **72,** 2330 (1975).
58. K. McEntee, *PNAS* **74,** 5275 (1977).
59. P. T. Emmerson and S. C. West, *Mol. Gen. Genet.* **155,** 77 (1977).
60. L. J. Gudas and D. W. Mount, *PNAS* **74,** 5280 (1977).
61. J. W. Little and J. E. Harper, *PNAS* **76,** 6147 (1979).
62. J. W. Little, S. H. Edmiston, L. Z. Pacelli and D. W. Mount, *PNAS* **77,** 3225 (1980).
63. C. M. Radding, "UCLA-ICN Symposium on Replication and Recombination," Academic Press, (in press).
64. S. C. West, E. Cassuto, J. Mursalim and P. Howard-Flanders, *PNAS* **77,** 2569 (1980).
64a. A. McPartland, L. Green and H. Echols, *Cell* **20,** 731 (1980).
65. P. L. Moreau, M. Fanica and R. Devoret, *Biochimie* **62,** 687 (1980).
66. M. Oishi, C. L. Smith and B. Friefeld, *CSHSQB* **43** (2), 897 (1978).
67. J. J. Weigle, *PNAS* **39,** 628 (1953).
68. G. Kellenberger and J. Weigle, *BBA* **30,** 112 (1958).
69. J. Weigle, "Phage and the Origins of Molecular Biology (J. Cairns, G. S. Stent and J. D. Watson, eds.), p. 226. Cold Spring Harbor Laboratory, Cold Spring Harbor, New York, 1966.
70. W. Harm, "Repair from Genetic Radiation," p. 107. Pergamon, New York, 1963.
71. M. Radman and R. Devoret, *Virology* **43,** 504 (1971).

72. M. Blanco and R. Devoret, *Mutat. Res.* **17**, 293 (1973).
73. M. Defais, P. Fauquet, M. Radman, and M. Errera, *Virology* **43**, 495 (1971).
73a. R. H. Rothman, L. J. Margossian and A. J. Clark, *Mol. Gen. Genet.* **169**, 279 (1979).
74. R. Devoret, M. Blanco, J. George and M. Radman, *in* "Molecular Mechanisms for Repair of DNA" (P. C. Hanawalt and R. B. Setlow, eds.), p. 155. Plenum, New York, 1975.
75. J. Ono and Y. Shimazu, *Virology* **29**, 295 (1966).
76. E. S. Tessman and T. Ozaki, *Virology* **12**, 431 (1960).
77. M. Defais, P. Caillet-Fauquet, M. S. Fox and M. Radman, *Mol. Gen. Genet.* **148**, 125 (1976).
78. P. Caillet-Fauquet, M. Defais, M. Radman and J. J. Cornelis, *JMB* **117**, 95 (1977).
79. G. Villani, S. Boiteux and M. Radman, *PNAS* **75**, 3037 (1978).
80. M. Radman, G. Villani, S. Boiteux, A. R. Kinsella, B. W. Glickman and S. Spadari, *CSHSQB* **43**(2), 937 (1978).
81. G. Steinborn, *Mol. Gen. Genet.* **165**, 87 (1978).
82. G. Steinborn, *Mol. Gen. Genet.* **175**, 203 (1979).
83. T. Kato and Y. Shinoura, *Mol. Gen. Genet.* **156**, 121 (1977).
84. G. C. Walker, *CSHSQB* **43**(2), 893 (1978).
84a. G. C. Walker and P. P. Dobson, *Mol. Gen. Genet.* **172**, 17 (1979).
85. M. Castellazzi, J. George and G. Buttin, *Mol. Gen. Genet.* **119**, 139 (1972).
86. M. Castellazzi, J. George and G. Buttin, *Mol. Gen. Genet.* **119**, 153 (1972).
87. J. George, M. Castellazzi and G. Buttin, *Mol. Gen. Genet.* **140**, 309 (1975).
88. E. Salaj-Smic, D. Petranovic, M. Petranovic and Z. Trgocevic, *Mol. Gen. Genet.* **177**, 91 (1979).
89. F. Jacob, *C. R. Acad. Sci.* **238**, 732 (1954).
90. R. Devoret, *C. R. Acad. Sci.* **260**, 1510 (1965).
91. S. Kondo and H. Ichikawa, *Mol. Gen. Genet.* **126**, 319 (1973).
92. E. M. Witkin and I. E. Wermundsen, *CSHSQB* **43**(2) 881 (1978).
93. E. M. Witkin, *CNAS* **57**, 1275 (1967).
94. C. J. Kenyon and G. C. Walker, *PNAS* **77**, 2819 (1980).
95. J. D. Hall and D. W. Mount, This Series **25**, 53.

DNA and Nucleoside Triphosphate Binding Properties of recA Protein from *Escherichia coli*

K. McEntee

G. M. Weinstock and

I. R. Lehman

Department of Biochemistry
Stanford University School of
Medicine
Stanford, California

The product of the *recA* gene of *Escherichia coli*, a protein with a molecular weight of approximately 40,000, utilizes ATP to alter the conformation of DNA. Specifically, the recA protein catalyzes the ATP-dependent formation of duplex DNA from complementary single-stranded DNA chains ("annealing") (1) as well as the formation of joint molecules by annealing single-stranded DNA molecules with homologous regions of duplex DNAs ("strand assimilation") (2, 3). Several lines of evidence suggest that these enzymic activities reflect the function of recA protein *in vivo*, where this protein is required for homologous genetic recombination and for postreplication repair of several types of DNA damage.

In $recA^-$ cells, an early step in recombination is blocked, a result consistent with a defect in strand transfer between homologous chromosomal regions. Radding *et al.* have implicated recA protein in the formation of joint molecules during ϕX174 recombination. Using a transfection assay, they showed that joint ϕX molecules prepared *in vitro* yield recombinants in $recA^-$ cells where normal ϕX174 recombination is blocked (4). The defect in postreplication repair in $recA^-$ mutants appears to be the failure to fill "gaps" that result from blockage of DNA replication through a region containing pyrimidine dimers. These gaps are removed in $recA^+$ cells by strand transfer from an undamaged homologous chromosomal segment (5). The most direct evidence relating *in vitro* activities with *in vivo* functions of recA protein is derived from experiments with a conditional *recA* mutant. The recA protein purified from a cold-sensitive $recA^-$ strain possesses cold-labile strand-annealing activity (1) as well as cold-labile pairing of duplex and single-stranded DNA molecules (unpublished observation).

The recA protein possesses both ssDNA-dependent ATPase and dsDNA-dependent ATPase activities (1, 3, 6). In strand-annealing and strand-assimilation reactions, recA protein hydrolyzes ATP to ADP and P_i. Using nonutilized nucleoside triphosphates (such as GTP or TTP) or nonhydrolyzable ATP analogs, we have demonstrated that strand assimilation and strand annealing are activities usually coupled to ATP hydrolysis. A particularly useful analog for these studies is adenosine 5'-[γ-thio]triphosphate (ATPγS), which binds extremely tightly to recA protein and inhibits both ATPase and DNA annealing activities (1, 7). In the presence of ATPγS, recA protein binds tightly to single-stranded DNA and to duplex DNA (3, 7). Since this analog is not hydrolyzed by recA protein, DNA binding can be studied in the absence of ATP hydrolysis.

We have investigated both the DNA binding and nucleoside triphosphate binding properties of purified recA protein in order to elucidate the coupling of ATP hydrolysis to its strand-annealing and assimilation activities. In this paper we discuss several aspects of ssDNA and dsDNA binding by recA protein and the effect of nucleoside triphosphates on the interactions between recA protein and DNA. Additionally, we have used photoaffinity labeling to examine the effects of different nucleoside triphosphates on the binding of ATP to recA protein.

I. Characterization of Complexes between ssDNA and recA Protein

The recA protein binds ssDNA in the absence of a nucleoside triphosphate, a property that has been useful in its purification (8). A simple filter-binding assay has been used to characterize binding of recA protein to ssDNA. In this assay, we measure retention of ^{3}H-labeled M13 ssDNA on nitrocellulose filters (Millipore type HAWP) briefly treated with alkali (3). This treatment prevents retention of ssDNA in the presence of high salt, but protein–ssDNA complexes are efficiently retained. The retention of M13 ssDNA increases linearly with recA protein until an average stoichiometry of 1 recA protein monomer per 20–30 nucleotides of ssDNA is reached. As shown in Table I the amount of ssDNA retained depends upon the presence of a nucleoside triphosphate or analog in the reaction, and the conditions for washing these complexes during filtration. In the absence of NTP, 74% of the M13 DNA can be retained when the complexes are washed with low salt (50 mM NaCl) following incubation. Washing the complexes with 1 M NaCl reduces the amount of DNA retained to

TABLE I
Effects of Nucleoside Triphosphates on recA Protein Binding to ssDNA[a]

Nucleoside triphosphate or analog	ssDNA retained (%)	
	50 mM NaCl wash	1 M NaCl wash
—	74	27
ATP	53	30
ATPγS	89	95
UTP	43	19
UTPγS	ND[b]	100
GTP	53	25
GTPγS	ND	100

[a] Reaction mixtures containing 4.62 μM ³H-labeled M13 ssDNA (42,000 cpm), recA protein (216 nM), nucleoside triphosphate (1 mM), or analog (100 μM) were incubated for 30 minutes at 30°C, filtered through alkali-treated Millipore filters (type HAWP, 45 nm), and washed with buffer containing the indicated NaCl concentration. UTPγS (uridine 5'-[γ-thio]triphosphate) was a generous gift of Dr. Fritz Eckstein. GTPγS (guanosine 5'-[γ-thio]triphosphate) was obtained from Boehringer-Mannheim.

[b] ND, not done.

27.4%. In the presence of 1 mM ATP, only 52.7% of the input DNA is retained, and this is further reduced to 29.5% by treatment with 1 M NaCl. Incubation in the presence of ATPγS results in retention of 88.5% of the input DNA. Unlike complexes formed in the presence of other nucleoside triphosphates, these complexes are completely resistant to treatment with 1 M NaCl.

Similar results are obtained with the UTP analog UTPγS. This [γ-thio]triphosphate analog is also a potent inhibitor of the ATPase activity of recA protein (unpublished observation). The reduced DNA binding in the presence of ATP or UTP may result from the hydrolysis of these nucleoside triphosphates by recA protein. However, using GTP, which is poorly hydrolyzed by recA protein, the same reduction in DNA binding is detected. In the presence of GTPγS, all of the DNA is bound in salt-resistant complexes. From these and other data we conclude that the binding of recA protein to ssDNA is qualitatively changed by the binding of the [γ-thio]triphosphate analogs, possibly by inducing a different conformation of recA protein that binds tightly to ssDNA and is not dissociated from the DNA by high salt.

We have carried out a series of experiments to investigate the nature of the bound DNA and to determine whether ATPγS is bound in these complexes. These investigations yielded the following results.

1. The half-time for dissociation of ssDNA from recA protein–ssDNA complexes is approximately 20 minutes at 37°C. In the presence of ATPγS, the half-life of the complex increases to more than 2 hours. This half-life is unchanged even if ATP is present in a hundredfold excess over ATPγS.
2. Using [^{35}S]ATPγS, we detect tight binding of the nucleoside triphosphate analog to recA protein in the presence of ssDNA. Like the bound DNA, the ATPγS is nonexchangeable in these complexes.
3. No covalent recA protein-DNA or recA protein–ATPγS intermediate has been detected in these complexes.

We suppose that ATPγS binds to the same site as ATP on the recA protein, an idea supported by kinetic experiments (7, and unpublished) and by photoaffinity labeling experiments. Although this model predicts that ATPγS is a competitive inhibitor of the ATPase activity of recA protein, the slow dissociation of this analog from recA protein · ssDNA complexes effectively "traps" recA protein in a form unable to hydrolyze ATP. This trapping by ATPγS has been confirmed for both ATP and UTP hydrolysis activities of recA protein (unpublished results).

We have also investigated the binding of recA protein to a variety of DNA and RNA homopolymers by competitive binding in the presence of ATPγS. As shown in Table II, recA protein binds both polyribo- and polydeoxyribonucleotides with varying efficiency. The most efficient competitor of recA protein binding to natural DNA is poly(dT), which is approximately 6–7 times more efficient than φX ssDNA. A complementary experiment using labeled poly(dT) and unlabeled P22 ssDNA confirms this result. Poly(dC) also competes efficiently for the binding of recA protein to P22 ssDNA, although not as well as the polythymidylate. Polymers less efficient than φX174 DNA or P22 ssDNA included poly(dA), poly(dG), and poly(rA). These results indicate that binding of recA protein to DNA is not sequence-specific, since it binds extremely well to polymers containing only T or C residues. Furthermore, recA protein binds poly(rU) [and poly(rC)] indicating that the deoxyribose moiety is not obligatory for binding. The efficient binding of recA protein to poly(dT) might indicate a preference for base composition or it might reflect the fact that poly(dT) contains little secondary structure. This latter explanation would be consistent with the efficient binding of recA protein to

TABLE II
COMPETITION FOR recA PROTEIN BINDING
BY VARIOUS POLYNUCLEOTIDES[a]

Competing unlabeled polynucleotide	Concentration (μM) of unlabeled polynucleotide reducing binding to ^{3}H-labeled P22 ssDNA by 50%
P22 ssDNA	5
Poly(dA)	>12.5
Poly(dG)	≫7.5
(dT)$_{1000}$	0.8
Poly(dC)	1.37
(dT)$_{12}$	≫15
Poly(rA)	>7.5
Poly(rU)	3.75

[a] Reaction mixtures containing 5.25 μM ^{3}H-labeled heat-denatured P22 DNA, 100 μM ATPγS, and 0.216 μM recA protein were incubated with various concentrations of unlabeled competing polynucleotide for 15 minutes at 30°C, filtered, and assayed for radioactivity.

poly(dC), another polymer with a small amount of secondary structure.

Although poly(dT) is an excellent substrate for recA protein binding, (dT)$_{12}$ fails to compete for binding of recA protein when present in excess over P22 ssDNA. We have also failed to detect direct binding of recA protein to ^{3}H-labeled (dT)$_{12}$ in a nitrocellulose filter-binding assay. Moreover, although this oligonucleotide stimulates the ATPase activity of rep protein (9, and unpublished results), it does not serve as a cofactor for the ssDNA dependent ATPase of recA protein. We conclude that there is a minimum size of polynucleotide required for binding of recA protein and for stimulating ATP hydrolysis by recA protein. We do not know, however, whether the size requirements for binding and hydrolysis are identical.

Does the nucleoside triphosphate alter the polynucleotide preference of recA protein binding? Although we have not investigated this possibility exhaustively, we observe the same preferential binding of recA protein to poly(dT) in the absence of ATPγS. Furthermore, the homopolymers that stimulate ATPase also stimulate the UTPase activity of recA protein, indicating little or no effect of the nucleoside triphosphate on the selection of a polynucleotide cofactor (unpublished results).

Although the filter binding assay provides considerable information on the nature of the recA protein · ssDNA complexes, it fails to provide any information on changes in DNA structure upon recA protein binding. A useful assay for investigating changes in DNA secondary structure is ethidium bromide fluorescence. When ethidium bromide binds and intercalates into duplex regions of DNA, it displays a large (approximately 25-fold) enhancement of fluorescence. Proteins such as DNA binding proteins, which eliminate or reduce secondary structure in DNA, quench the fluorescence enhancement factor. Figure 1A shows that, when recA protein binds to ϕX174 ssDNA in the presence of ATPγS, there is significant quenching of the ethidium bromide fluorescence. By increasing the recA concentration, a saturation point can be reached beyond which there is no further reduction in fluorescence. From the data of Fig. 1 we calculate that this saturation occurs at a ratio of 1 recA monomer per 4–5 nucleotides. This stoichiometry is consistent with the amounts required for maximal strand assimilation (with respect to ssDNA nucleotides), and for recA-protein-dependent cleavage of phage λ repressor (10). Moreover, a kinetic analysis of the ssDNA-dependent ATPase activity of recA protein indicates that saturation is achieved at a ratio of 1 recA protein monomer per 4–5 nucleotides (unpublished results).

The results shown in Fig. 1 indicate that quenching of ethidium fluorescence by recA protein binding to ϕX174 ssDNA differs in two ways from fluorescence quenching caused by the helix-destabilizing or single-strand-binding (SSB) protein to this DNA.

1. In the absence of ATPγS, recA protein binding to ϕX ssDNA only slightly affects enhanced ethidium fluorescence. In order to remove secondary structure from ssDNA molecules, recA protein must also bind ATPγS. ATP does not replace ATPγS as a cofactor for quenching ethidium fluorescence, perhaps because of hydrolysis of the ATP during the binding reaction. These results support the idea that binding of ATPγS to recA protein significantly alters the way this protein interacts with DNA. The SSB protein removes DNA secondary structure in the absence of any nucleoside triphosphate cofactor (Fig. 1B).

2. RecA protein removes approximately 50% of the enhanced ethidium fluorescence when the protein saturates the DNA, indicating that a considerable amount of secondary structure is unaffected by tight binding of recA protein. Nearly complete denaturation of ss ϕX174 DNA duplex regions occurs in the presence of SSB protein. The ability of this protein to remove hairpins in ssDNA of small phages has been suggested as an important role of this protein for the replication of these phage DNAs *in vitro* and *in vivo*. RecA protein

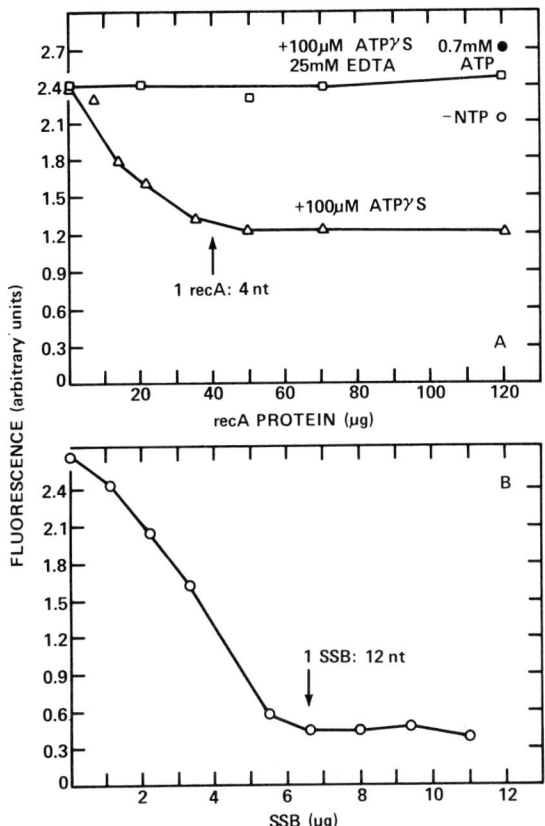

FIG. 1. The recA protein removes DNA secondary structure from φX174 ssDNA. Fluorescence measurements were made in a Turner Model 430 spectrofluorometer (γ^{ex} = 525 nm, γ^{em} = 590 nm). Reactions (180 µl) contained 4200 pmol of φX174 ssDNA, 20 mM Tris (pH 7.5), 10 mM MgCl$_2$, 0.1 mM EDTA, 1 mM dithiothreitol, 20 mM NaCl, and either recA protein (plus the indicated NTP or analog) or single-strand-binding (SSB) protein. Incubations were performed at 37°C for 5 minutes and stopped by addition of ethidium bromide (1 µg/ml). After fluorescence measurement, EDTA was added to each sample and the fluorescence was redetermined. nt = nucleotide.

cannot substitute for SSB protein in the replication of G4 ssDNA to the RF intermediate (data not shown).

We conclude that binding of recA protein to ssDNA in the presence of ATPγS results in an unwinding of duplex regions in the φX174 molecule. A considerable amount of DNA secondary structure is not disrupted by recA protein binding. Alternatively, binding of recA protein to φX174 ssDNA could completely remove existing secondary structure, but at the same time, the protein could impose a different conformation on the DNA that is stabilized by tight binding of the protein.

II. RecA Protein Binds and Unwinds Duplex DNA

RecA protein requires a nucleoside triphosphate for binding to duplex DNA (3). A complex between recA protein and duplex DNA could be detected by nitrocellulose filter binding when either ATP or ATPγS was present. Complexes formed with ATPγS are, like those formed between recA protein and ssDNA, extremely stable and resistant to high salt. Complexes formed between recA protein and duplex DNA in the presence of ATP are rapidly formed, and, as ATP is hydrolyzed in the reaction, the complexes dissociate (3). From these results, we suggested that recA protein unwinds duplex DNA when it binds cooperatively in the presence of ATP or ATPγS (3). Cunningham et al. demonstrated that under different conditions single-stranded DNA stimulates recA protein to unwind duplex DNA in the presence of ATPγS (11).

A more detailed analysis of duplex DNA binding by recA protein has shown this reaction to be extremely pH sensitive. Binding of recA protein to duplex DNA is optimal at pH 6.2–6.4. At pH 7.5–8.0, the rate of binding to duplex DNA is reduced by more than 50-fold but can be strongly stimulated by suitable amounts of ssDNA, whether heterologous or homologous. Over the entire pH range (6.2–8.5), recA protein binding to dsDNA requires a nucleoside triphosphate (ATP) or [γ-thio]triphosphate analog.

In Figure 2, recA protein-duplex DNA complexes (relaxed, covalently closed PM2 DNA) are shown. The clustering of recA protein on the DNA molecules confirms our earlier result that recA protein binding to duplex DNA is cooperative. The electron micrographs reveal two additional features of recA protein–duplex DNA interaction.

1. Using either circular PM2 DNA or linear P22 DNA, we have found no evidence for extensive DNA unwinding by bound recA protein. This result could mean that (a) recA protein binds to duplex DNA but does not cause unwinding of the helix; (b) recA protein binding causes only a local unwinding of a duplex region; or (c) recA protein causes extensive unwinding of duplex DNA, but bound recA protein holds the single strands in a tight conformation that permits only limited opening of these unwound regions.

2. RecA protein causes a lateral aggregation or association of duplex DNA molecules. In the case of circular PM2 DNA · recA protein

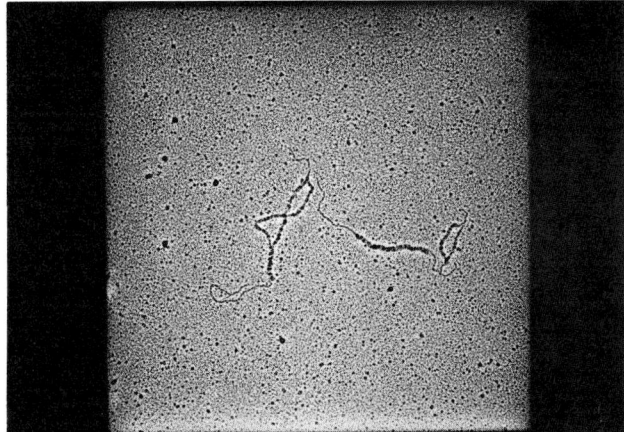

FIG. 2. Complexes of recA protein and duplex DNA were visualized without fixation by the method of Griffith (13). RecA protein (220 pmol) was incubated with relaxed, covalently closed PM2 DNA (700 pmol) in 20 mM maleate (pH 6.2), 10 mM MgCl$_2$, 30 mM NaCl, 0.5 mM dithiothreitol, and 100 μM ATPγS at 37°C for 20 minutes.

complexes, the duplex strands appear to be wound around each other. This conformation is most likely stabilized by protein–protein interactions in the complex.

We have tested directly whether recA protein binding promotes helix unwinding by treating the recA protein · PM2 DNA complexes with a "nicking–closing" enzyme from *Drosophila melanogaster* (12; T. Nelson, personal communication). In the relaxed, covalently closed PM2 DNA molecule, unwinding of a helical region due to recA protein binding results in the appearance of compensating supertwists in the DNA of the opposite sense (positive). These topological turns are removed by treatment with the *Drosophila* enzyme. The resulting molecules are covalently closed but contain a net reduction in the number of topological turns. Following removal of the protein (by detergent), the uncomplexed DNA will be negatively supertwisted. The results of this experiment are shown in Fig. 3.

The PM2 DNA in the recA protein · duplex DNA complexes can be converted to negatively supertwisted DNA having a mobility identical to that of supertwisted PM2 DNA from virions. Unwinding of duplex DNA by recA protein is also highly cooperative—the products of the unwinding reaction display a narrow distribution in superhelical density (based upon the mobility in agarose gels) reflecting the cooperative binding of recA protein to the DNA. RecA protein-dependent unwinding of duplex DNA is pH sensitive and can be stimulated significantly at pH 8.0 by the addition of heterologous ssDNA. This re-

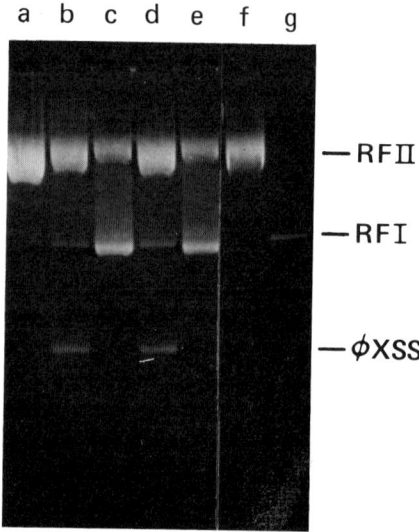

FIG. 3. RecA protein unwinds duplex DNA. Reactions containing either 20 mM maleate (pH 6.2) or Tris (pH 8.1) buffer, 10 mM MgCl$_2$, 9.1 nmol of relaxed covalently closed PM2 DNA, and the indicated amount of recA protein were incubated at 37°C for 30 minutes, treated with *Drosophila* "nicking–closing" enzyme at pH 8.1 for 30 minutes at 30°C, deproteinized with 1% Na dodecyl sulfate, and electrophoresed in a 0.8% agarose gel (Tris-borate). Lanes: (a) 865 pmol of recA, Tris, pH 8.1, 60 minutes; (b) 346 pmol of recA, Tris, pH 8.1, 60 minutes, 1 nmol of φX ssDNA; (c) 346 pmol of recA, maleate, pH 6.2, 60 minutes; (d) 346 pmol of recA, Tris, pH 8.1, 120 minutes, 1 nmol φX ssDNA; (e) 346 pmol of recA, maleate, pH 6.2, 120 minutes; (f) relaxed PM2 DNA; (g) supercoiled PM2 DNA (from virion).

sult is consistent with that of Cunningham *et al.* (*11*). RecA protein directly unwinds duplex DNA in the presence of ATPγS (ATP and UTPγS also serve as cofactors) at pH 6.2, and at pH 8.0 recA protein promoted unwinding of duplex DNA is stimulated by ssDNA. In both cases unwinding by recA protein does not require a free end or phosphodiester bond break in the DNA, as demonstrated here with PM2 DNA and circular φX174 ssDNA.

An alternative interpretation of this experiment and that of Cunningham *et al.* (*11*) is that recA protein does not cause unwinding of the duplex but wrapping of the duplex around the protein in a sense opposite to that of the helix (left-handed wrapping). Treatment of these complexes with nicking–closing enzyme, or, in the case of a nicked duplex circle, with DNA ligase, would produce negatively supertwisted DNA molecules following deproteinization. We favor the interpretation that recA protein causes unwinding of the DNA for three reasons. First, it explains the hydrolysis of ATP in the presence of a duplex DNA cofactor (creation of single-stranded regions in the DNA during unwinding). Second, we find no evidence for shortening of the contour length of the DNA when it is complexed with recA protein, which would be expected if extensive wrapping occurs. Third, unwinding of duplex DNA by recA protein is consistent with the unwinding of duplex regions of ϕX174 ssDNA determined by ethidium bromide fluorescence quenching.

We conclude that recA protein promotes duplex DNA unwinding directly at low pH (6.2) or in the presence of ssDNA at high pH (8.0). Unwinding of DNA by recA protein requires nucleoside triphosphate binding but not hydrolysis since either ATPγS or UTPγS serves as a

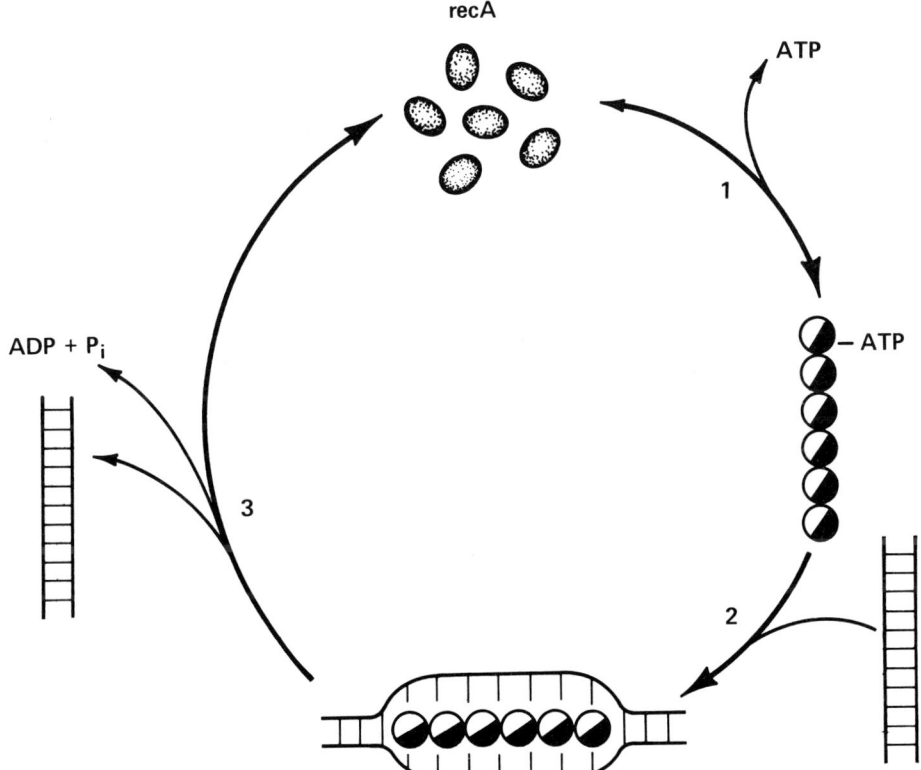

FIG. 4. Model for duplex DNA-dependent ATP hydrolysis catalyzed by recA protein at pH 6.2. In step 1 recA protein binds ATP and oligomerizes without ATP hydrolysis. These recA protein multimers contain bound ATP and are in a conformation that efficiently binds duplex DNA (step 2), and promotes local unwinding of the helix. The single-stranded regions that are produced stimulate recA protein to hydrolyze ATP and to dissociate from the DNA (step 3).

cofactor in the unwinding reaction and neither analog is hydrolyzed by recA protein. Unlike DNA unwinding by DNA helicases, where a single-stranded region contiguous with a duplex region is required for unwinding, recA protein unwinds fully duplex DNA molecules. *In vivo*, the combined action of recA protein and a nicking-closing enzyme could function like a DNA gyrase in supertwisting DNA molecules.

Our observations indicate that pH has a profound effect upon the interaction of recA protein with duplex DNA. By way of contrast, the interaction of recA protein with ssDNA (as measured by DNA binding and ssDNA-dependent ATP hydrolysis) is relatively insensitive to pH between 6.2 and 9.0. Although we do not fully understand the effects of pH on recA protein–DNA interaction, we believe that pH-induced structural changes in recA protein are important for this activity (unpublished observations). In Fig. 4, we present a model that summarizes our results regarding the interaction of recA protein and duplex DNA in the presence of ATP.

III. Nucleoside Triphosphate Binding by recA Protein

RecA protein has a high affinity for ATP as measured in the ssDNA dependent ATP hydrolysis reaction ($K_m \sim 20$ μM). Furthermore, the ATP analog, ATPγS, binds exceedingly tightly to recA protein in the presence of DNA as indicated by its ability to inhibit recA protein-dependent ATPase at very low concentrations ($K_I = 0.6$ μM). A direct determination of ATP binding to recA protein is provided by photoaffinity labeling with the ATP derivative 8-azido-[γ-^{32}P]ATP. Figure 5 is an autoradiograph of a polyacrylamide gel containing recA protein labeled with this analog under a variety of conditions. The recA protein is efficiently labeled by this ATP derivative in the absence of any DNA. If either ATPγS or UTPγS is present in the labeling reaction (c and d), the amount of azido-[^{32}P]ATP associated with recA protein is drastically reduced. At the same concentration (f) ATP reduces labeling with azido-ATP although less effectively than the ATPγS or UTPγS analogs. dTTP is more effective than ATP in reducing azido-ATP labeling of recA protein (e). We conclude that, in the absence of DNA, these nucleoside triphosphates, as well as the ATPγS and UTPγS analogs, bind tightly to the same site on the recA protein.

RecA protein is efficiently labeled with 8-azido-[γ-^{32}P]ATP in crude extracts (P. Higgins, personal communication). Partially purified fractions of the *tif-1* mutant form of recA protein and the cold labile *recA629* mutant protein are also efficiently labeled with little or

DNA AND NUCLEOSIDE TRIPHOSPHATE BINDING 277

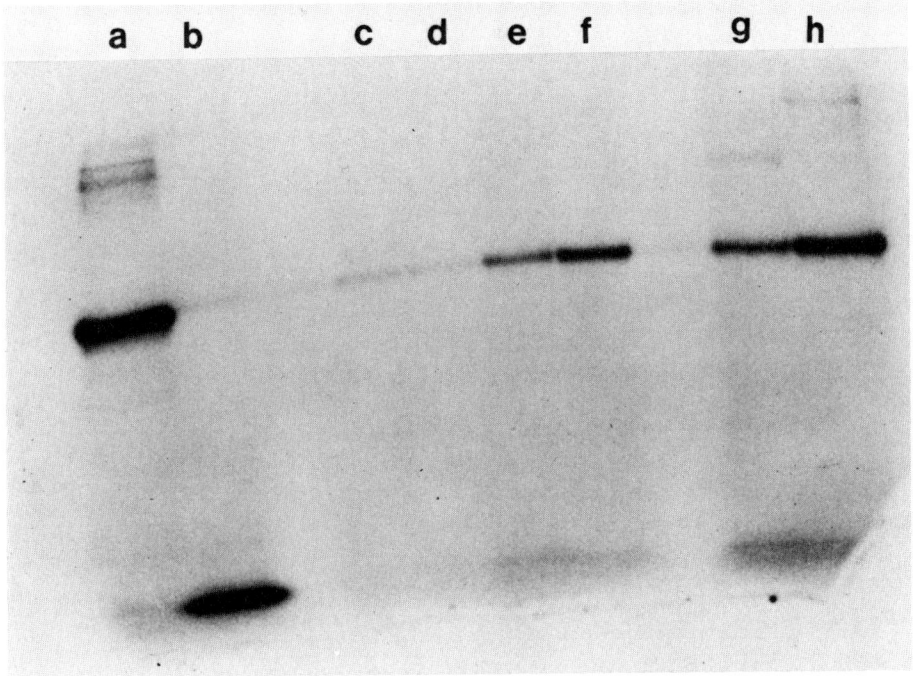

FIG. 5. Photoaffinity labeling of recA protein. Effects of other nucleoside triphosphates. RecA protein (~3 μg) was incubated with 15 μM 8-azido-[γ-^{32}P]ATP in buffer containing 5 mM Tris, pH 7.5, 5 mM MgCl$_2$, 10 mM KCl, and the indicated nucleoside triphosphate or analog. Reactions were irradiated with short-wavelength UV light for 15 minutes at 22°C and analyzed by polyacrylamide gel electrophoresis. (a) No addition; (b) after labeling, the sample was treated with trypsin; (c) 200 μM ATPγS; (d) 200 μM UTPγS; (e) 200 μM dTTP; (f) 200 μM ATP; (g) *tif-1* mutant protein (~1 μg); and (h) cold-labile *recA629* mutant protein (~1 μg).

no background labeling (g and h, respectively). Photoaffinity labeling should prove to be a particularly useful technique for studying aspects of nucleoside triphosphate binding to recA protein.

IV. Summary

The DNA-binding and nucleoside triphosphate-binding properties of recA protein are summarized below.

1. RecA protein binds to ssDNA without a nucleoside triphosphate cofactor. Addition of ATPγS results in enhanced binding of recA protein to ssDNA and removal of secondary structures in the DNA. In these recA protein · ssDNA complexes, the ATPγS is tightly bound

and is not exchangeable. Similarly the DNA is irreversibly bound. In the absence of DNA, ATPγS binds tightly but reversibly to recA protein. In the absence of ATPγS, ssDNA dissociates from recA protein.

2. RecA protein binds to polyribo- and polydeoxyribonucleotides, showing a strong affinity for poly(dT) and poly(dC) over natural DNAs and much weaker affinity for poly(dG) and poly(dA). Short oligonucleotides such as $(dT)_{12}$ do not compete for recA protein binding to natural DNA and do not stimulate ATPase activity of recA protein.

3. RecA protein removes secondary structure from φX174 DNA in the presence of ATPγS as determined by ethidium bromide fluorescence quenching. This unwinding saturates at 1 recA monomer per 4-5 nucleotides.

4. Binding of recA protein to duplex DNA requires a nucleoside triphosphate but does not require hydrolysis of this NTP. ATP and ATPγS work most efficiently to promote duplex binding by recA protein. The binding reaction is pH-sensitive and is cooperative with respect to recA protein. The binding of recA protein to duplex DNA results in unwinding of the duplex. Neither ATP hydrolysis nor a single-stranded region adjacent to the duplex is needed for recA-dependent unwinding.

5. ATPγS and ATP bind tightly to recA protein in the absence of ssDNA. Evidence from ATP hydrolysis experiments, equilibrium dialysis and photoaffinity labeling indicate that ATP, ATPγS, UTP, and UTPγS bind to the same site on the recA protein. In the presence of DNA, ATPγS binding is enhanced and stable recA protein · DNA · ATPγS complexes are formed. Neither the DNA nor the [γ-thio]triphosphate cofactor appears to be covalently linked to recA protein in these complexes.

Acknowledgments

This work was supported by grants from the National Institutes of Health and the National Science Foundation. Kevin McEntee was a recipient of an American Cancer Society Senior Fellowship. George M. Weinstock was supported by the Bank of America—Giannini Foundation.

References

1. G. M. Weinstock, K. McEntee and I. R. Lehman, *PNAS* **76**, 126 (1979).
2. T. Shibata, C. DasGupta, R. P. Cunningham and C. M. Radding, *PNAS* **76**, 1638 (1979).
3. K. McEntee, G. M. Weinstock and I. R. Lehman, *PNAS* **76**, 2615 (1979).
4. W. K. Holloman and C. M. Radding, *PNAS* **73**, 3910 (1976).
5. A. K. Ganesan, *JMB* **87**, 103 (1974).
6. T. Ogawa, H. Wabiko, T. Tsurimoto, T. Horii, H. Masutaka and H. Ogawa, *CSHSQB* **43**, 909 (1978).

7. T. Shibata, R. P. Cunningham, C. DasGupta and C. M. Radding, *PNAS* **76**, 5100 (1979).
8. J. W. Roberts, C. W. Roberts, N. L. Craig and E. M. Phizicky, *CSHSQB* **43**, 917 (1978).
9. A. Kornberg, J. F. Scott and L. L. Bertsch, *JBC* **253**, 3298 (1978).
10. N. L. Craig and J. W. Roberts, *Nature* **283**, 26 (1980).
11. R. P. Cunningham, T. Shibata, C. DasGupta and C. M. Radding, *Nature* **281**, 191 (1979).
12. W. A. Baase and J. C. Wang, *Bchem* **13**, 4299 (1974).
13. J. D. Griffith, *in* "Methods in Cell Biology," (D. M. Prescott, ed.), Vol. 7, p. 129. Academic Press, New York, 1973.

Molecular Mechanism for the Induction of "SOS" Functions

MICHIO OISHI
ROBERT M. IRBE AND
LEE M. E. MORIN

*The Public Health Research Institute of New York
New York*

Inhibition of bacterial DNA replication triggers a sequence of events that leads to the induction of various cellular functions, ("SOS" functions),[1] including prophage induction, inducible mutagenesis and DNA repair, and filamentous growth (1). It is now well established that recA protein is involved in these processes and the inhibition of DNA replication induces the synthesis of recA protein. Roberts et al. demonstrated ATP-dependent cleavage of λ repressor by purified recA protein (2) and, more recently, that the proteolysis requires not only ATP but also polynucleotides (3). Since recA protein also promotes the formation of "D-loops"[2] from superhelical DNA and homologous single-stranded DNA (4, 5), the direct involvement of recA protein in the proteolytic inactivation of repressor molecules suggests a pleiotropic and complex role for recA protein in the induction of SOS functions and in recombination.

Using prophage induction as a probe, we studied the early molecular events leading to the induction of SOS functions. We devised a biochemical assay procedure based on the phage-controlled "read-through"[3] transcription and translation of a bacterial operon (*trp* operon) integrated into an early transcribed region of the bacteriophage $\phi 80$ or λ genome (6). One of the advantages of this assay procedure is the ease with which the kinetics of phage repressor inactivation can be followed by simple assaying one enzyme activity (anthranilate synthase, EC 4.1.3.27) after subjecting the cells to various inducing conditions. Using this procedure, we found that, at optimum doses, various inducing agents exhibit distinctly different pat-

---

[1] Defined in footnote 2 in Hanawalt *et al*, this volume [Ed.].

[2] "D-loops" consist of single strands of DNA hybridized by base-pairing to short, unwound sections of duplex DNAs (5) [Ed.].

[3] "Read-through" indicates ability for transcription to proceed even in the presence of a repressor (6) [Ed.].

terns in the kinetics of repressor inactivation, suggesting that the number of steps leading to repressor inactivation varies according to the primary DNA structure produced by the agent (7). There appear to be at least three classes of inducing agents with respect to the length of time required to complete the inactivation of the repressor molecules. Agents that cause strand scissions in DNA molecules exhibit early inactivation of the repressor molecules, whereas agents that modify DNA bases, such as UV irradiation (pyrimidine dimer formation), or mitomycin C (alkylation), show an intermediate type of inactivation. Inactivation due to inhibition of DNA replication by precursor deprivation (thymine deprivation) or by inactivation of the replication machinery (DNA *ts* mutants) is a distinctly slower process.

We also found that agents that cause strand scissions, such as bleomycin and streptonigrin, cause immediate degradation of DNA, whereas DNA degradation induced by UV irradiation or mitomycin C treatment is a relatively slower process with a short lag period. Degradation of the DNA following thymine deprivation, or inactivation of the DNA replication apparatus (DNA *ts* mutants), starts to appear at a very late stage after initiation of such treatments. There was a distinct correlation between the timing of DNA degradation following the inducing treatment and the timing of repressor inactivation. Furthermore, both the synthesis of anthranilate synthase and the degradation of DNA triggered by these inducing treatments are delayed considerably when a *recB* mutation is introduced in the recipient strain, implicating *recBC*-DNase in prophage induction. The dependence of repressor inactivation upon functional $recB^+$ allele was further confirmed by prophage induction in permeabilized cells (8). dNTPs (and ATP) destabilize active DNA replication forks in the permeabilized cells and degradation of nascent DNA by *recBC*-DNase triggers the induction process. Prophage induction by thymine starvation of a thymine-requiring strain was also completly dependent on a functional gene *recB* product, suggesting a similar mechanism for induction (9). A summary of the kinetic analysis of the induction processes and induction pathways is shown in Table I.

Based on these results, we proposed the following sequence of events as the early molecular mechanisms that lead to the induction of prophage and of SOS functions in *E. coli* (9).

1. A variety of initially altered DNA structures are converted to final, possibly common, predegradative DNA structures by enzyme systems specific to each altered DNA structure.
2. The predegradative DNA is then degraded by specific DNases, mainly *recBC*-DNase. When the predegradative DNA structure

TABLE I
SUMMARY OF KINETIC ANALYSES OF INDUCTION PROCESSES

Inducing agents (conditions)	Primary mode of action on DNA	Inactivation of repressor		
		Pathways		Stage (min)
		recB	Hypothetical	
In vivo				
Bleomycin	Scissions	+	+	Early (10–15)
Streptonigrin	Scissions	+	+	Early (10–15)
Colicin E2	Scissions	NT[a]	NT	Early (10–15)
UV	Pyrimidine dimer formation	+	+	Intermediate (30)
Mitomycin C	Alkylation (cross-linking)	+	+	Intermediate (30)
Thymine starvation	Inhibition of replication	+	–	Late (60)
Nalidixic acid	Inhibition of replication	+		Intermediate-late (30–60)
DNA ts (chain elongation)[b]	Inhibition of replication	NT	NT	Late (45–70)
DNA ts (precursor)[c]	Inhibition of replication	NT	NT	Late (60)
DNA ts (initiation)[d]	Inhibition of replication	(–)	(–)	No induction
Novobiocin	Inhibition of replication	(–)	(–)	No induction
Permeabilized cells *in vitro*				
dNTPs + ATP	Degradation at replication fork	+	–	
Micrococcal DNase	?	+	–	

[a] NT, not tested.
[b] DNA ts mutants examined: E107(dnaB107ts), PC2(dnaC2ts), E486(dnaE486ts), and BT308(dnaG308ts).
[c] E101(dnaF101ts).
[d] E508(dnaA508ts).

originates from a DNA replication fork, *recBC*-DNase is solely responsible for the degradation.

3. The degradation products, either specific oligodeoxynucleotides or double-stranded DNA with single-stranded DNA tails, then act as an induction signal on an unidentified target protein to transmit the SOS (see footnote 1) message.

We have obtained preliminary evidence that specific oligodeoxynucleotides, such as d(A-G-G), inactivate phage (φ80 and λ) repressors in the permeabilized cells (9). This strongly suggests that specific oligodeoxynucleotides or single-stranded DNAs with specific base sequences are involved in the induction of SOS functions. We present here further characterization of the early molecular events that lead to the induction of prophages and SOS functions.

I. Materials and Methods

A. Bacterial Strains

The following *E. coli* K12 strains were used throughout this study: MO1512 (a derivative of AB1157) which carries *trp-E9758* and *end*, in addition to the genetic markers derived from AB1157; MO1518, the same as MO1512 but carrying *recB21*; MO1513, a φ80 lysogen of MO1512; MO1519, a φ80 lysogen of MO1518; MO3013, the same as MO1513 but carrying plasmid PTM-2 (colE1::Tn3 *recA*$^+$) (10); MO1574, a φ80*ptrp(19Oh)* lysogen of MO1512; MO1763, a φ80*ind* lysogen of MO1512; MO1542, a φ80*ind-55s* lysogen of MO1512; MO590, a φ80 lysogen of MO639; MO639, a *trp*$^-$(*trpB9579*) derivative of JC7623 (*recB21 recC22 sbcB15*); MO586, a φ80 lysogen of JC5519 (*recB21 recC22*); MO574, a φ80 lysogen of AB1157; MO591, a φ80 lysogen of MO649 (*sbcB15*); MO649, derived from MO639 by Hfr(K16) mating. MO1574 is a φ80*ptrp(19Oh)* lysogen of MO1512. MO933 is a plasmid (R13*cm*$^+$)-carrying strain of MRPl (11). φ80*ind* was kindly supplied by Dr. A. Matsushiro at Osaka University. φ80*ind-55s* was isolated in this laboratory.

B. Chemicals and Chemical Synthesis of Oligonucleotides

d(G-G), d(I-G), d(G-I), d(A-G), d(pG-G), d(pI-G), d(pG-I), d(pA-G), d(A-G-G), d(pA-G-G), d(pT-G-G), and d(pT-G-T) were synthesized by the method of Khorana *et al.* (12). Precursors for the oligodeoxynucleotides, 5'-dGMP, 5'-dAMP, 5'-dCMP; 5'-dIMP and 5'-dTMP, were purchased from Sigma except for 5'-dIMP (P. L. Biochemicals). They were desalted through a Dowex 50 W (8X) (Bio-Rad) column (40 mm × 300 mm, eluted with 20% aqueous pyridine) before modification. Acetic anhydride (Fisher), hydracrylonitrile (Eastman), isobutyryl chloride (Aldrich), and benzoyl chloride (Aldrich) were freshly distilled before use. Dicyclohexylcarbodiimide (Eastman) and ethyldiisopropylamine (Eastman) were used without further purification. Pyridine and anhydrous pyridine were prepared as described by Kössel et al. (13). Triisopropylbenzenesulfonyl chloride (Aldrich) was crystallized once from pentane (14). Blocked mononucleotide synthesis and condensation reactions were performed by the methods of Weber and Khorana (15) and Büchi and Khorana (16), except for the deblocking steps (17). The oligonucleotides were purified from the reaction mixtures first by Sephadex G-25 column chromatography (25 mm × 900 mm), eluted with 0.1 M triethylammonium bicarbonate, pH 7.5, at 4° (18). The product peak was further purified by a DE-23 (Whatman, 20 mm × 450 mm) column with an Et_3NHCO_3 gradient (0.1 M to 0.3 M in 10% ethyl alcohol). The purity of the synthesized oligonucleotides was examined by cellulose thin-layer chromatography (Brinkmann) by the methods by Narang et al. (19, 20). If necessary, the samples were further purified by preparative paper chromatography (Whatman No. 3) with one of the following solvent systems (15): (A) isopropanol/conc. ammonia/water, 7:1:2; (B) ethanol/ammonium acetate (1 M, pH 7.5), 7:3; (C) n-propanol/conc. ammonia/water, 55:10:35. r(A-G-G) was synthesized from r(A-G) and GDP with polynucleotide phosphorylase. Thymidylyl-thymidine photodimer was prepared as described by Sztump and Shugar (21). d(G-G-G) was a generous gift from Dr. E. Ohtsuka at Osaka University. d(G-G), d(A-G-A), d(C-C-G-G), d(G-G-C-C), d(A-G-A-G), $d(pG)_2$, $d(pG)_4$, d(A-A), d(A-G), d(A-C), d(T-A), r(G-G), $d(pG)_6$, $d(pG)_8$, $d(pG)_{10}$, $d(pG)_{12-18}$, $d(pA)_4$, $d(pC)_4$, and $d(pT)_4$ were purchased from Collaborative Research. d(A-T), d(G-A), d(G-C), d(G-T), d(C-A), d(C-G), d(C-C), d(C-T), d(T-G), d(T-C), d(T-T), 2'-GMP, 3'-GMP, 5'-GMP, deoxyguanosine 3',5'-bisphosphate, and poly(dG) were supplied from P. L. Biochemicals. 3'-dGMP, 5'-dGMP, cGMP, cdGMP, GDP, dGDP, GTP, and dGTP were obtained from Sigma.

Pancreatic DNase, micrococcal nuclease, pancreatic RNase, bacterial alkaline phosphatase, and venom diesterase were purchased from Worthington Biochemical Co. Polynucleotide phosphorylase was obtained from P. L. Biochemicals. Actinomycin D was supplied by Merck, Sharp and Dohme. Sources of the components in the reaction mixture have been described (8).

C. Preparation of Plasmolyzed Cells and Induction of Prophage

Cells grown to a density of 4×10^8 cells per milliliter in Luria broth (22) (25 ml) at 35°C were chilled and harvested by centrifugation at 3000 g for 5 minutes. After washing twice with buffer 3A (10 mM TrisCl, pH 7.7; 10 mM $MgCl_2$; 50 mM KCl; 1 mM EGTA; and 0.5 mM L-tryptophan), the cells were resuspended in 2.5 ml of plasmolysis buffer (10 mM TrisCl, pH 7.7; 13 mM $MgCl_2$; 100 mM KCl; 5 mM spermidine; and 1.6 M sucrose). After incubation for 10 minutes at 0°C, 5 µl of the cell suspension were directly transferred to 100 µl of reaction mixture AM and incubated at 35°C for 60 minutes with constant shaking. The standard components and their concentrations in reaction mixture AM were as follows: 20 mM TrisCl, pH 7.7; 100 mM KCl; 13 mM $MgCl_2$; 1 mM $CaCl_2$; 4 mM spermidine; 0.25 mM (each) of all L-amino acids (except L-tryptophan); 2.5 mM L-tryptophan; 0.1 mM β-NAD; 0.03 mM β-NADP; 25 µg/ml folinic acid; 10 µg/ml p-aminobenzoic acid; 2.5 mM ATP; 0.5 mM GTP; 0.5 mM CTP; 0.5 mM UTP; 25 µg/ml pyridoxine; 25 µg/ml flavin adenine dinucleotide; 10 mM phospho*enol*pyruvate; 25 µg/ml pyruvate kinase; 100 µg/ml *E. coli* tRNA mixture; 5 mM glutathione (reduced form); 2 mM dithiothreitol; and 0.34 mM phenethylalcohol. The ATP, GTP, CTP, UTP, and phospho*enol*pyruvate were all sodium salts. For UV irradiation, the cell suspension in buffer 3A (10 ml in a plastic petri dish 100 mm in diameter) was irradiated under a germicidal lamp (GE 15t8, 15 W) with constant gentle shaking.

For free phage assay, the cells (100 µl) in reaction mixture AM were diluted with 900 µl of warm Luria broth (22) with 10 mM $MgSO_4$ and 50 mM KCl, and the incubation was continued for another 120 minutes (unless specified) at 35°C with constant shaking. The reaction was stopped by mixing 100 µl of the samples with 900 µl of modified buffer T1 (6 mM TrisCl, pH 7.7; 10 µg/ml gelatin; 1 mM $MgSO_4$; 0.5 mM $CaCl_2$) containing chloramphenicol at 100 µg/ml. After appropriate dilution with the same buffer, samples were mixed with an indicator strain (MO933) and soft agar (2.5 ml of Luria broth with 0.6% agar and chloramphenicol at 100 µg/ml) and overlayed on bottom agar (Luria broth with 1% agar). Plaques were counted after overnight incubation at 35°C.

For infectious centers assay, immediately after incubation in reaction mixture AM, the samples were diluted with the modified buffer T1 and treated with antiserum against ϕ80. The infectious centers were then assayed by the same procedure for free phage assay as described above, except that no chloramphenicol was present in the dilution buffer or in the soft agar.

II. Results

A. Prophage Induction in Permeabilized Cells

In order to investigate further the nature of the predegradative DNA derived from the damaged DNA or the stalled replication fork, and also to identify the hypothetical signal substance, we sought conditions in which lysogenized cells are permeable to proteins or oligonucleotides, yet are capable of producing intact phage particles after induction. Such conditions would make it possible to analyze the effect of various macromolecules on the induction of SOS functions (see footnote 1) by biological means, which are generally more sensitive than biochemical assays. Although the earlier system (8), with permeabilized cells, demonstrated phage repressor inactivation, it failed to produce intact phage particles even after prolonged incubation. However, permeabilized and biologically active *E. coli* cells were obtained by a slight modification of the method for plasmolysis (8) and by a prior incubation of the plasmolyzed cells in a reaction mixture that allowed protein synthesis. When UV-irradiated and plasmolyzed cells, prepared from a lysogenic (ϕ80) strain (MO1513), were incubated in reaction mixture AM and further incubated in a growth medium [Luria broth, (22)], infectious phage particles were produced. The reaction mixture (AM) used for the first incubation contains most of the components necessary for *in vitro* protein synthesis. Induction of the prophage was dependent on the length of the first incubation and reached a maximum at 45–60 minutes (Table II). Direct transfer of the plasmolyzed cells into the growth medium without incubation in reaction mixture AM resulted either in no induction or in very little induction (data not shown).

Approximately 30 infectious phage particles per cell originally present were induced by the UV treatment. Since a single plasmolyzed cell produced an average of 150–300 infectious phage particles upon induction, as assayed by a single cell-burst experiment (data not shown), 30 particles would be equivalent to the induction of 10–20% of the originally present plasmolyzed cells. In separate experiments, in which we assayed infectious centers rather than

TABLE II
UV-INDUCED PROPHAGE INDUCTION IN
PLASMOLYZED CELLS[a]

Incubation in growth medium (min)	Plaques ($\times 10^{-6}$/ml)	
	Control	UV
0	0.5	10
15	0.9	100
30	2.4	150
45	2.5	460
60	2.8	500

[a] Strain MO1513 (ϕ80 lysogen) cells were irradiated with UV light (40 J/m^2) and made permeable by plasmolysis. Reaction mixtures (AM, 100 μl) were incubated at 35°C for 60 minutes. After incubation in growth medium [Luria broth, (22)] free phages were assayed as described in Section I.

free phage, up to 50% of the plasmolyzed cells were induced upon UV irradiation. The apparent discrepancy is probably due to the production of phage after 120 minutes of incubation in the growth medium. Although in some critical experiments we assayed both infectious centers and free infectious phage, in most of the experiments we assayed only free phage because of the simplicity and sensitivity of the free phage assay compared to the infectious center assay.

Figure 1 shows that the UV-triggered induction of prophage was completely eliminated in the presence of RNase (10 μg/ml) in the reaction mixture. On the other hand, prophage induction was triggered when nonirradiated permeabilized cells were exposed to d(A-G-Gp). Essentially the same results were also obtained when infectious centers rather than free phage were assayed (data not shown). The sensitivity of the UV-induced induction to RNase and the induction by d(A-G-Gp) indicates that the cells become permeable to proteins of low molecular weight and to oligodeoxynucleotides, yet maintain the capacity to induce prophage to almost the same degree as the intact nonpermeable cells.

The requirement for incubating the plasmolyzed cells before transferring them to the growth medium is apparently related to the need for protein synthesis in the plasmolyzed cells. Omission of a single amino acid (leucine) or all amino acids from the AM reaction mixture reduced the efficiency of induction to approximately 25% of the control value (see Table III). The presence of ATP was also es-

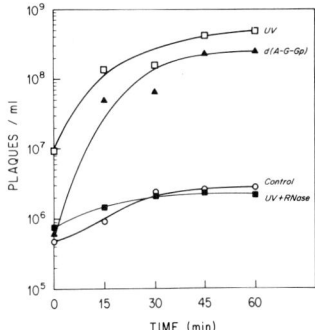

FIG. 1. Prophage induction by plasmolyzed cells. UV-irradiated (40 J/m^2) (□) or non-irradiated cells (○) (MO1513) were plasmolyzed and incubated in reaction mixture AM (100 μl, 2 × 10^7 cells) at 35°C with d(A-G-Gp) (0.4 mM) (▲) or RNase (10 μg/ml) (■). At 60 minutes 900 μl of Luria broth (22) were added, and the number of free phages was assayed at various time intervals as shown in the figure. For details, see Section I.

sential for the successful induction of prophage in the permeabilized cells. Table III also shows the effect of actinomycin D and of pancreatic RNase and DNase on prophage induction. Pancreatic RNase gave the strongest inhibition, whereas inhibition by pancreatic DNase required a higher concentration of the enzyme. This is

TABLE III
CHARACTERISTICS OF INDUCTION IN PLASMOLYZED CELLS[a]

Reaction mixture	Infectious centers (× 10^{-5})	%
Complete	18	100
Minus leucine	4.4	25
Minus all amino acids	3.5	19
Minus ATP and PePyruvate	1.2	7
Plus actinomycin D (50 μg/ml)	<0.1	0
Plus RNase, 0.1 μg/ml	2.3	14
1 μg/ml	<0.1	0
10 μg/ml	<0.1	0
Plus DNase, 1 μg/ml	7.0	39
10 μg/ml	1.7	9
100 μg/ml	1.4	7

[a] Plasmolyzed cells (2 × 10^7 cells) from UV-irradiated (40 J/m^2) MO1513 (φ80 lysogen) were incubated in reaction mixture AM (100 μl) at 35°C for 60 minutes under various conditions as described in the table. Immediately after incubation, infectious centers were assayed as described in Section I. The number of infectious centers produced by a control sample (no UV irradiation) was subtracted from each value shown in the table.

due either to (a) the difference in molecular weights of the enzymes (RNase 13,700; DNase 31,000), which might affect the uptake of these enzymes through membranes, or (b) the induction of prophage by pancreatic DNase, which might offset the inhibition by the enzyme.

B. Prophage induction by DNases

Permeabilized cells prepared by plasmolysis are sensitive to certain enzymes (8, 23). This fact prompted us to examine permeabilized lysogenic cells for prophage induction by DNase treatment instead of by conventional agents, such as UV irradiation or mitomycin C. When a $\phi 80$ lysogen (MO1513) was permeabilized by plasmolysis, incubated in the reaction mixture (AM) with small nucleases, such as pancreatic DNase (MW 31,000) or micrococcal nuclease (MW 16,800), and then incubated again in the growth medium [Luria broth (22)], phage particles were produced. The number of induced phage particles in the medium was a function of the enzyme concentration in the reaction mixture (Fig. 2). Figure 2 shows that micrococcal nuclease is a much better inducing agent than pancreatic DNase, and that the induction level obtained by micrococcal nuclease is approximately 15% of that obtained by UV treatment. We could not detect any induction with pancreatic RNase over a wide range of enzyme concentrations. Although micrococcal nuclease is known to attack RNA as well as DNA *in vitro*, these nucleases probably attack DNA in the permeabilized cells, producing nicks in the chromosome. We believe that these nicks are equivalent to the damage caused by UV irradiation or mitomycin C treatment, and that they trigger the early reactions in the SOS pathway, leading to prophage induction.

One striking characteristic of DNase-triggered induction is its absolute dependence upon the presence of a functional $recB^+$ allele. As seen in Table IV, the DNases failed to induce prophage in $recB^-$ (MO1519, *recB21*) cells as well as $recA^-$ (MO1522, *recA13*) cells. The absolute dependence of induction by DNases on *recBC* gene product was confirmed by a "read-through" (see footnote 3) biochemical assay (data not shown) (6). On the other hand, bleomycin treatment of the $recB^-$ cells induced prophage, though to a much lesser extent than $recB^+$ (MO1513) cells. This residual induction, which occurred in the absence of the *recB* gene product, was more clearly demonstrated by biochemical assays of cells induced by UV or mitomycin C. This fact indicates the existence of an unidentified alternative induction pathway (or pathways) that can substitute for the *recBC*-DNase.

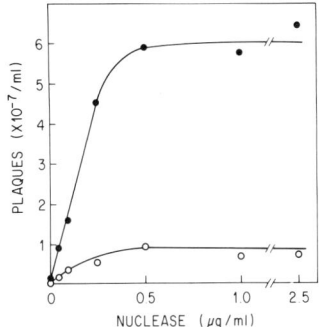

FIG. 2. Prophage induction by micrococcal nuclease and pancreatic DNase. The plasmolyzed cells (2×10^7) from strain MO1513 were incubated in reaction mixture AM (100 µl) with various concentrations of micrococcal nuclease (●) and pancreatic DNase (○) for 60 minutes at 35°C. The reaction mixtures were then diluted with 900 µl of Luria broth (22) and free phage particles were assayed after 120 minutes of incubation at 35°C. For details, see Section I.

Taking advantage of the complete dependence of DNase induction upon a functional *recBC* gene product without the complications of the alternative pathway, we further studied the mechanisms that initiate prophage induction processes. First, prophage induction is dependent on a *recBC* gene product, but is apparently independent of the recombination proficiency of the lysogens. Table V compares the prophage inducibility of four isogenic strains with all combinations of the *recBC* mutation and the *sbcB* mutation (the latter is an indirect supressor mutation of the *recBC* mutation that

TABLE IV
EFFECT OF *recA* AND *recB* MUTATIONS ON ENZYMIC INDUCTION[a]

		Plaques ($\times 10^{-5}$/ml)				
Strains	Genotype	Control	Micrococcal nuclease	Pancreatic DNase	Bleomycin	UV
MO1513	$recA^+ recB^+$	9.9	970	104	3100	2600
MO1519	$recA^+ recB21$	0.02	0.07	0.02	3.4	2.6
MO1522	$recA13 recB^+$	0.0002	0.002	0.002	0.001	0.001

[a] The plasmolyzed cells (2×10^7 cells) from strains MO1513, MO1519, and MO1522 were incubated in reaction mixture AM (100 µl) with micrococcal nuclease (0.5 µg/ml), pancreatic DNase (0.5 µg/ml), or bleomycin (0.1 µg/ml) for 60 minutes at 35°C. For UV treatment, cells were irradiated (40 J/m²) under a UV lamp before plasmolysis. The reaction mixture was then diluted with 900 µl of growth medium LB, (22) and free phage particles were assayed after a 120-minute incubation at 35°C. For details, see Section I.

TABLE V
EFFECT OF *rec* GENOTYPES ON PROPHAGE INDUCTION[a]

Strains	Genotype	Phenotype	Plaques ($\times 10^{-4}$/ml)				
			Control	Pancreatic DNase	Micrococcal nuclease	dNTPs	d(A-G-Gp)
MO574	$recB^+ recC^+ sbcB^+$	Rec$^+$	72.7	700	1540	3300	10,700
MO588	$recB21 recC22 sbcB^+$	Rec$^-$	0.7	0.9	1.0	0.6	3,200
MO591	$recB^+ recC^+ sbcB15$	Rec$^+$	73.1	1000	4300	6000	12,500
MO590	$recB21 recC22 sbcB15$	Rec$^+$	61.4	65	49	64	3,400

[a] The plasmolyzed cells (2×10^7 cells) from strains MO574, MO588, MO591, and MO590 were incubated in reaction mixture AM (100 μl) with pancreatic DNase (0.5 μg/ml), micrococcal nuclease (0.5 μg/ml), a mixture of four deoxynucleoside triphosphates (dNTPs, 0.05 mM each), or d(A-G-Gp) (0.4 mM). After 60 minutes of incubation, the samples were diluted with growth medium LB (22) (900 μl) and free phages were assayed as described in Section I.

restores the recombination proficiency of *recBC* mutants without restoring *recBC*-DNase activity) (24). Two strains that lack *recBC*-DNase (MO588 *recB⁻C⁻ sbcB⁺*, Rec⁻ and MO590 *recB⁻C⁻ sbcB⁻*, Rec⁺) fail to induce prophage after micrococcal nuclease, pancreatic DNase, or even dNTPs treatment. On the other hand, two strains that possess *recBC*-DNase (MO574 *recB⁺C⁺*, Rec⁺ and MO591 *recB⁺C⁺ sbcB⁻*, Rec⁺) responded to the nucleases and to dNTPs. Thus it is quite clear that the inducibility of the lysogens by the DNases and by dNTPs depends upon the presence of the *recBC* gene product regardless of the recombination proficiency of the strains. It should be noted that prophage induction by the nuclease treatment and by the dNTPs contrasts with the induction triggered by a specific oligodeoxynucleotide, d(A-G-Gp), in these four strains. With d(A-G-Gp), all four strains gave rise to a substantial number of free phages. This point is discussed below (see Table V).

We view *recBC*-DNase as the enzyme that processes the predegradative DNA and provides the signal for induction of SOS functions (7). Figure 3 shows the effect of micrococcal nuclease on the degradation of DNA. It is quite clear that micrococcal nuclease triggers DNA degradation when *recBC* gene product is present in the cells (MO1512). This is consistent with the fact that prophage induction by micrococcal nuclease is completely dependent upon

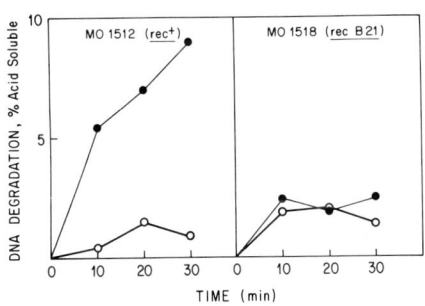

FIG. 3. Degradation of DNA by micrococcal nuclease treatment. Strains MO1512 and MO1518 were grown in "single-strength medium A" (6) at 35°C. At a cell density of 3×10^8 cells/ml, [5-*methyl*-³H]thymidine (3 µCi/ml with 0.1 µg of carrier thymidine per milliliter) was added to the medium. After 5 minutes of incubation the cells were chilled by adding ice, harvested, and plasmolyzed as described in Section I. The plasmolyzed cells (1.3×10^8 cells) were incubated in reaction mixture AM (1000 µl) with (●) and without (○) micrococcal nuclease (1 µg/ml), at 35°C. Samples (200 µl) were withdrawn intermittently and mixed with an equal volume of cold trichloroacetic acid (20% w/v) with sodium pyrophosphate (50 mM), then centrifuged (3000 rpm, 10 minutes, 4°C); the supernatant (200 µl) was withdrawn and adsorbed onto Whatman glass filters (GF/D). Radioactivities were measured by a Beckman scintillation counter.

recBC gene product. On the other hand, the degradation of DNA or the inactivation of repressor by bleomycin, UV, or mitomycin C is not completely dependent upon the presence of functional recBC gene product. Therefore, chromosomal damage caused by micrococcal nuclease is likely to be similar to the damage produced by thymine starvation (of a thy^- strain) or dNTP treatment, in which prophage induction is also completely dependent upon the presence of recBC gene product. It should be noted that pancreatic DNase treatment caused extensive DNA degradation in the plasmolyzed cells regardless of the presence of recBC gene product (data not shown). Therefore, DNA degradation per se is not sufficient to provide a signal, but rather DNA degradation by a specific DNase such as recBC-DNase is essential for induction. Further characterization of nuclease-stimulated prophage induction will be presented elsewhere.

C. Prophage Induction by Specific Oligodeoxynucleotides

Previous studies yielded evidence suggesting that phage $\phi 80$ and λ repressor molecules are inactivated when permeabilized cells are incubated with specific oligodeoxynucleotides, such as d(A-G-G). Since the biochemical assay procedure used for that study was indirect and relatively insensitive, further studies had to wait until a new, more sensitive biological assay procedure was developed, as described in Section II,A.

Using the biological assay, we undertook a series of experiments to study the oligonucleotide-dependent induction. Some oligonucleotides besides d(A-G-G), such as d(A-G) and d(G-G), trigger the induction of prophage. The effect of d(G-G) on prophage induction had been detected by read-through biochemical assay (6), but the effect of d(A-G) was apparently overlooked. As seen in Table VI, d(G-G) and d(A-G) were the only two nucleotides among the 16 possible deoxydinucleotide to trigger prophage induction. Prophage induction by d(G-G) was confirmed by infectious-centers assay (data not shown). The level of prophage induction by d(G-G) (0.4 mM) was essentially the same as that produced by other conventional inducing treatments, such as UV irradiation. We examined the effect of d(G-G) on prophage induction as a function of incubation time in the reaction mixture. A minimum of 15–30 minutes of incubation of the permeabilized cells with d(G-G) was essential for the induction of prophage (data not shown).

Figure 4 shows prophage induction as a function of d(G-G) concentration with two $\phi 80$ lysogens (MO1513 and MO3013). In both

TABLE VI
PROPHAGE INDUCTION BY
DEOXYDINUCLEOTIDES[a]

Oligonucleotides	Plaques/ml ($\times 10^{-6}$)
—	1.1
d(A-A)	1.8
d(A-G)	95.5
d(A-C)	1.4
d(A-T)	1.1
d(G-A)	0.7
d(G-G)	173
d(G-C)	0.9
d(G-T)	1.2
d(C-A)	0.5
d(C-G)	2.3
d(C-C)	0.8
d(C-T)	0.7
d(T-A)	0.9
d(T-G)	1.0
d(T-C)	0.7
d(T-T)	1.0

[a] The plasmolyzed cells (2×10^7) prepared from MO3013 were incubated in reaction mixture AM (100 µl) with deoxydinucleotides (0.1 mM) for 60 minutes at 35°C. The reaction mixtures were then diluted with 900 µl of growth medium LB (22) and free phage particles were assayed after 120 minutes of incubation at 35°C. For details, see Section I.

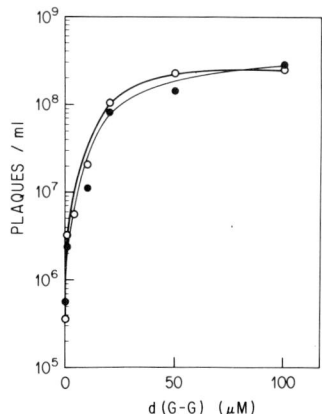

FIG. 4. Prophage induction as a function of d(G-G) concentration. Plasmolyzed cells (2×10^7 cells) from MO1513 (○) and MO3013 (●) were incubated (60 minutes, 35°C) in reaction mixture AM (100 µl) containing various concentrations of d(G-G). The samples were then diluted with 900 µl of Luria broth (22) and free phages were assayed after 120 minutes of incubation at 35°C. For details, see Section I.

cases, the induction occurred at d(G-G) concentrations as low as 10^{-7} M. As seen in Fig. 4, the induction by MO3013, which possesses a high level of recA protein due to the presence of a *recA* plasmid (pTM-2) (10), was essentially at the same level as the control strain (MO1513) throughout a wide range of d(G-G) concentrations. This suggests either that recA protein level in MO1513 is not a limiting factor for d(G-G)-dependent induction, or that recA protein is not a target for d(G-G).

In order to determine the essential structure for triggering induction, we examined the effect on induction of other dinucleotides structurally similar to d(G-G) or d(A-G) (Table VII). Although we found no significant activity with d(G-I), there was substantial induction by d(I-G). This indicates that an active dinucleotide must have guanine at the 3′ end, but the base at the 5′ end can be either guanine, adenine, or inosine. The sugar moieties of the active dinucleotide must be deoxyribose, since r(A-G) and r(G-G) have no inducing activity. The thymidylyl-thymidine photodimer, dT[-]dT, also showed no activity. Thus the essential dinucleotide structure for triggering induction is d(R-G) (R = "a purine nucleoside").

We examined a variety of guanine derivatives (mostly mononucleotides) that might be derived from d(G-G) *in vivo*. Table VIII compares the efficiency of prophage induction by such compounds. As seen in the table, none of the related compounds exhibited in-

TABLE VII
PROPHAGE INDUCTION BY DINUCLEOTIDES[a]

Oligonucleotides	No. of plaques ($\times 10^{-6}$/ml)
—	0.8
d(T[-]T)	0.3
d(I-G)	33
d(G-I)	2.6
r(A-G)	0.1
r(G-G)	1.1
d(A-G)	59
d(G-G)	150

[a] The plasmolyzed cells (2×10^7 cells) prepared from MO3013 were incubated in reaction mixture AM (100 µl) with dinucleotides (0.1 mM) for 60 minutes at 35°C. The reaction mixture was then diluted with 900 µl of growth medium LB (22), and free phage particles were assayed after 120 minutes of incubation at 35°C. For details, see Section I.

ducing activity, indicating that d(G-G) does not convert to these compounds in the cell to exhibit inducing activity. In fact, we find that radioactive d(G-G) remains unchanged in the cell even after prolonged incubation. It should be noted that d(G-G) with a phosphoric residue at the 5' position d(pG-G) did show some activity, but much less than d(G-G). We have not tested the activity with deoxyguanylyl (3'-5')deoxyguanosine 3'-phosphate, d(G-Gp).

We investigated further the role of the phosphate at the 5' position of deoxyguanosine oligonucleotides of longer chain lengths (Fig. 5). Substantial inducing activities were observed with oligodeoxyguanylyl 5'-phosphates with chain lengths from 12 to 18. The activity increased as the chain length increased up to 8 and then gradually decreased. On the other hand, when the 5'-phosphate was removed from these deoxyoligonucleotides by alkaline phosphatase, the inducing activity of the oligonucleotides was greatly reduced except for d(pG-G), which was converted to d(G-G) by the enzyme

TABLE VIII
PROPHAGE INDUCTION BY A VARIETY OF
GUANINE DERIVATIVES[a]

Compounds	Plaques/ml ($\times 10^{-6}$)
—	1.2
Guanine	0.3
Guanosine	0.9
Deoxyguanosine	0.9
2'-GMP	0.8
3'-GMP	0.5
5'-GMP	1.2
3'-dGMP	0.5
5'-dGMP	0.6
cGMP	0.7
cdGMP	0.8
GDP	1.7
dGDP	1.5
Deoxyguanosine 3',5'-bisphosphate	0.6
GTP	1.2
dGTP	0.8
d(G-G)	173

[a] The plasmolyzed cells (2×10^7 cells) prepared from MO3013 were incubated in the reaction mixture AM (100 μl) with guanine derivatives (0.1 mM) for 60 minutes at 35°C. The reaction mixtures were then diluted with 900 μl of growth medium LB, (22) and free phage particles were assayed after 120 minutes of incubation at 35°C. For details, see Section I.

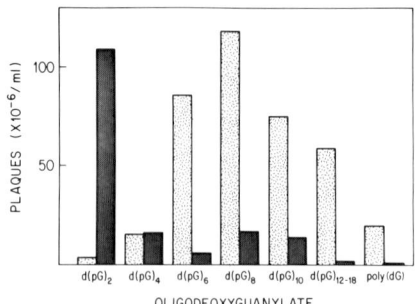

FIG. 5. Prophage induction by oligodeoxyguanylyl phosphates. The plasmolyzed cells (2×10^7 cells) prepared from MO3013 were incubated in reaction mixture AM (100 μl) with oligodeoxyguanylates (0.1 260 nm units per reaction mixture) for 60 minutes at 35°C. The reaction mixtures were then diluted with 900 μl of Luria broth (22), and free phage particles were assayed after 120 minutes of incubation at 35°C. To remove phosphates from the 5' positions, the oligodeoxyguanylates were incubated with alkaline phosphatase (0.4 unit) at 50°C for 180 minutes in buffer (10 mM TrisCl, pH 7.7, and 5 mM $MgCl_2$). Lightly shaded area is before alkaline phosphatase treatment and darkly shaded is after treatment.

treatment. The other deoxyoligomers, such as $d(pA)_4$, $d(pC)_4$, and $d(pT)_4$ exhibited no inducing activity (data not shown). The apparent requirement for a phosphoric group at the 5' end of these oligodeoxyguanylates was in sharp contrast to induction by d(pG-G) in which a phosphate at the 5' end of d(G-G) reduces its inducing activity. Thus, at least two classes of oligodeoxyguanylates, d(R-G) dinucleotides and oligodeoxyguanylyl 5'-phosphates with chain lengths between 6 and 18, seem to have prophage inducing activity.

The experiments described above clearly demonstrate that specific oligodeoxynucleotides, such as d(G-G), are able to trigger induction of prophage. The important question that emerges is, what is the biological significance of this phenomenon in light of the mechanism of induction of prophage and SOS functions? Although the ultimate answer must await further investigation, several points shed light on this question.

First, induction by d(G-G) seems to be correlated with well established biological reactions involved in prophage induction; prophage was not triggered by d(G-G) in cells prepared from a strain with a *recA* mutation (MO1522), and prophage was also not induced by d(G-G) from cells lysogenic for induction deficient (ind^-) ϕ80 (MO1763 and MO1542). Furthermore, d(G-G) induces prophage from a lysogen (MO1574) in which the ϕ80 (ϕ80*ptrp*) genome is integrated at the *trp* operon (rather than at the ϕ80 attactment site)

(25). Integration and excision of $\phi 80ptrp$ in this strain are known to be independent of phage *int* and *xis* functions, and are solely dependent on the generalized host recombination system based on *recA* gene product (25). This type of prophage is also UV inducible. These experimental results are shown in Table IX.

Second, induction triggered by d(G-G) is independent of the product of gene *recBC* (*recBC*-DNase). As shown in Table IX d(G-G) can trigger prophage induction from lysogens with *recB*(C) mutations. Furthermore, we could not detect any degradation of chromosomal DNA following d(G-G) treatment (data not shown). These results are in contrast to all other inducing treatments in which prophage induction is partially or completely dependent upon the presence of the functional *recBC* gene product, and is related to the degradation of DNA by the enzyme.

Thus it is attractive to speculate that d(G-G) or oligodeoxyguanylates are functionally equivalent to the *in vivo* signal substance, which is either a degradation product of *recBC*-DNase action on the predegradative DNA or an intermediate in a *recBC*-independent SOS pathway (7). On the other hand, it is also possible that the action of d(G-G) or oligodeoxyguanylates in prophage induction is due to some specific effect not directly related to the mechanism of *in vivo* induction of SOS functions. For example, these oligodeoxynucleotides may change the conformation of $\phi 80$ repressor molecules so that they become susceptible to proteolytic cleavage by recA protein. This may

TABLE IX
EFFECT OF HOST AND PHAGE GENOTYPE ON INDUCTION[a]

			Plaques ($\times 10^{-5}$/ml)			
Strains	Host genotype	Prophage	Control	d(G-G)	dNTPs	UV
MO1513	rec^+	$\phi 80$	4.8	1700	1110	5000
MO1519	$recB21$	$\phi 80$	0.2	1140	NT	129
MO1522	$recA13$	$\phi 80$	0.001	0.002	0.001	0.001
MO1542	rec^+	$\phi 80ind$-$55s$	0.006	0.007	0.003	NT
MO1765	rec^+	$\phi 80ind$	0.005	0.006	0.008	0.010
MO1574	rec^+	$\phi 80ptrp190h$	5.4	1900	1290	NT

[a] Permeabilized cells (2×10^7 cells) from strains MO1513, MO1519, MO1765, MO1522, MO1542, and MO1574 were incubated in reaction mixture AM (100 μl) for 60 minutes at 35°C. The concentrations of d(G-G) and dNTPs (a mixture of deoxynucleoside triphosphates) in AM were 0.1 mM and 0.05 mM (each), respectively. For UV treatment, the cells were irradiated under a UV lamp (40 J/m^2) as described in Section I. The samples (100 μl) were diluted with growth medium LB (22) (900 μl), and free phages were assayed after 120 minutes of incubation at 35°C. For details, see Section I. NT, not tested.

account for the puzzling fact that d(G-G) triggers φ80 induction much more efficiently than λ induction (data not shown). More biochemical characterization of the d(G-G)-triggered prophage induction, including analysis of the *in vitro* inactivation of repressor molecules, must be performed to more fully understand this important process.

III. Summary

Permeabilized and biologically active *E. coli* cells were obtained by plasmolysis and incubation of the cells in a reaction mixture that allows protein synthesis. These cells become permeable to low-molecular-weight proteins and oligodeoxynucleotides. Using this system, we studied two types of prophage (φ80) induction. One is the induction by low-molecular-weight DNases, such as micrococcal and pancreatic DNase, and the other by specific guanine-containing oligodeoxynucleotides. Prophage induction by the DNases was dependent on a functional *recBC*-DNase and degradation of the DNA by the enzymes. On the other hand, the oligodeoxynucleotide induction was independent of *recBC*-DNase and did not require degradation of chromosomal DNA. There are two classes of guanine-containing deoxyoligonucleotides that are effective in prophage induction. One is specific deoxydinucleotides having the sequence A(or G or I)-G; the other is oligodeoxyguanylates with a 5' phosphate and a chain length between 6 to 18.

Acknowledgments

We thank Dr. P. Margolin for helpful discussions throughout the study and Ms. D. Smith for preparation of the manuscript. We also thank Drs. A. J. Clark, T. Ogawa, A. Matsushiro, and E. Ohtsuka for sending phage and bacterial strains and d(G-G-G). The help during preparation of oligonucleotides by Drs. O. Bhanot, M. Seligman, and G. Kurr-Adler is also highly acknowledged. This work was supported by Grant GM-21073 from the National Institutes of Health to M. O.; L. M. is a Medical Science Trainee supported by Medical Science Training Grant 5T32GM07308.

References

1. E. M. Witkin, *Bacteriol. Rev.* **40**, 869 (1976).
2. J. Roberts, C. W. Roberts, N. L. Craig and E. Phizicky, *CSHSQB* **43**, 917 (1979).
3. N. L. Craig and J. W. Roberts, *Nature* **283**, 26 (1980).
4. K. McEntee, G. M. Weinstock and I. R. Lehman, *PNAS* **76**, 2615 (1979).
5. T. Shibata, R. P. Cunningham, C. Das Gupta and C. M. Radding, *PNAS* **76**, 5100 (1979).
6. C. L. Smith and M. Oishi, *Mol. Gen. Genet.* **148**, 131 (1976).
7. C. L. Smith and M. Oishi, *PNAS* **75**, 1657 (1978).
8. M. Oishi and C. L. Smith, *PNAS* **75**, 3569 (1978).
9. M. Oishi, C. L. Smith and B. Friefeld, *CSHSQB* **43**, 897 (1979).

10. T. Ogawa, H. Wabiko, T. Tsurimoto, T. Horii, H. Masukata and H. Ogawa, *CSHSQB* **43**, 909 (1978).
11. M. Yamamoto, M. Ishizawa and H. Endo, *JMB* **58**, 103 (1971).
12. H. G. Khorana, K. L. Agarwal, H. Büchi, M. H. Caruthers, N. K. Gupta, K. Kleppe, A. Kumar, E. Ohtsuka, U. L. RajBhandary, J. H. van de Sande, V. Sgaramella, T. Terao, H. Weber and T. Yamada, *JMB* **72**, 209 (1972).
13. H. Kössel, H. Büchi and H. G. Khorana, *JACS* **89**, 2185 (1967).
14. R. Lohrmann and H. G. Khorana, *JACS* **88**, 829 (1966).
15. H. Weber and H. G. Khorana, *JMB* **72**, 219 (1972).
16. H. Büchi and H. G. Khorana, *JMB* **72**, 251 (1972).
17. A. Kumar and H. G. Khorana, *JMB* **72**, 329 (1972).
18. S. A. Narang, J. J. Michniewicz and S. K. Dheer, *JACS* **91**, 936 (1969).
19. S. A. Narang and J. J. Michniewicz, *Anal. Biochem.* **49**, 379 (1972).
20. S. A. Narang, O. S. Bhanot, S. K. Dheer, J. Goodchild and J. J. Michniewicz, *BBRC* **41**, 1248 (1970).
21. E. Sztump and D. Shugar, *BBA* **61**, 555 (1962).
22. J. H. Miller, "Experiments in Molecular Genetics," p. 433. Cold Spring Harbor Laboratory, Cold Spring Harbor, New York, 1972.
23. K. Shimizu and M. Sekiguchi, *Mol. Gen. Genet.* **168**, 37 (1979).
24. S. R. Kushmer, H. Nagaishi, A. Templin and A. J. Clark *PNAS* **68**, 824 (1971).
25. A. Oka, H. Ozeki and J. Inselburg, *Virology* **46**, 556 (1971).

Induction and Enhanced Reactivation of Mammalian Viruses by Light

| Larry E. Bockstahler
| Bureau of Radiological Health
| Food and Drug Administration
| Rockville, Maryland

When certain bacteria are treated with ultraviolet radiation or other DNA-damaging agents, a group of functions, including prophage induction, "Weigle reactivation"[1] of ultraviolet-irradiated bacteriophage, mutagenesis, filamentous growth and error-prone DNA repair, are expressed. These inducible functions are known as "SOS" functions,[2] and there is evidence that suggests they belong to the same genetic regulatory unit ("SOS repair hypothesis") (1, 2). The term "SOS" (the international distress signal) implies that damage to DNA, or inhibition of DNA synthesis, initiates a regulatory signal that results in derepression of these functions (2).

Relatively little is known about mammalian cell functions that can be induced by treatment with physical or chemical agents. Several phenomena presently under investigation can be induced or activated in mammalian cells by ultraviolet or visible radiation of certain wavelengths, either alone or in combination with photosensitizing chemicals. Examples include differentiation of erythroleukemic cells (3, 4), the induction of specific enzymes (4a), tumor virus induction (5), and enhanced reactivation of mammalian viruses (6, 7).

We have been investigating the enhanced reactivation and induction of mammalian viruses because (a) they represent potential mammalian "SOS" functions, (b) they may be associated with mammalian cell oncogenic transformation; and (c) we wish to use viruses as tools to study phenomena related to DNA repair. In this report, I review certain studies that concern the induction and enhanced reactivation of DNA-containing mammalian viruses by light.

I. Photodynamic Induction of Simian Virus 40

Simian virus 40 (SV40) has proved to be useful in studies on DNA-containing mammalian virus induction [see review by Bockstahler and Hellman (5)]. This well characterized tumor virus contains a ge-

---

[1] Defined in footnote 3 in Hanawalt *et al.* in this volume [Ed.].
[2] Defined in footnote 2 in Hanawalt *et al.* in this volume [Ed.].

nome consisting of double-stranded DNA, is cytolytic in permissive (monkey kidney) cells, and transforms certain nonpermissive cells [see reviews by Fareed and Davoli (8) and Kelly and Nathans (9)]. In inducible lines of SV40-transformed cells, the complete viral genome is integrated covalently into the cellular DNA. Excision of viral SV40 DNA from SV40-transformed cell DNA is an early event in induction (10); this occurs during the induction of bacteriophage lambda from lysogenic *Escherichia coli*. Thus, although its mechanism is unknown, some aspects of SV40 induction resemble bacteriophage induction.

Our interest in using photodynamic action as a damaging treatment to study the induction of SV40 originated in part from reviewing the potential risks of clinical antiviral photodynamic therapy (11). Bacteriophage induction can occur by photodynamic treatment of lysogenic bacteria (12). We hypothesized that one risk of clinical photodynamic treatment might be the induction of latent tumor virus(es), if present in infected cells or in the surrounding uninfected cells of the patients. Some support for the hypothesis came from the demonstration that *in vitro* photodynamic induction of virus could occur in a mammalian-virus host-cell system (13). SV40 was induced by treatment of SV40-transformed hamster kidney cells with visible fluorescent light and proflavine. Our initial report (13) indicated that the relationship between light exposure and dye concentration in producing maximum photodynamic induction was complex. The purpose of the investigation described below was to further examine this relationship by observing the induction responses that occur for different values of these two parameters.

Photodynamic induction of SV40 from SV40-transformed hamster kidney cells was carried out with the clone E line, established by Kaplan *et al.* (14). Cultures of E-line cells were treated with either 1.1, 6.2, or 12.3 μM proflavine sulfate, irradiated with different exposures of visible light, and incubated for virus expression. The periods for incubation of cells with dye (1 hour) and for cells after photodynamic treatment (3 days) were selected on the basis of previous investigation (L. E. Bockstahler, J. M. Cantwell, S. A. Adams, and O. L. Ellingson, unpublished). The SV40 infectivity of cell extracts was determined by plaque assay in permissive monkey kidney cells. Response curves for induced virus infectivity as a function of light exposure are shown in Fig. 1. For a constant proflavine concentration of 1.1 μM, the induction of infectious virus increased with increasing light exposure to a maximum and remained essentially constant for higher exposures of light (Fig. 1C), in agreement with previous observations (13). The maximum level of SV40 induced represents a thousand-fold increase

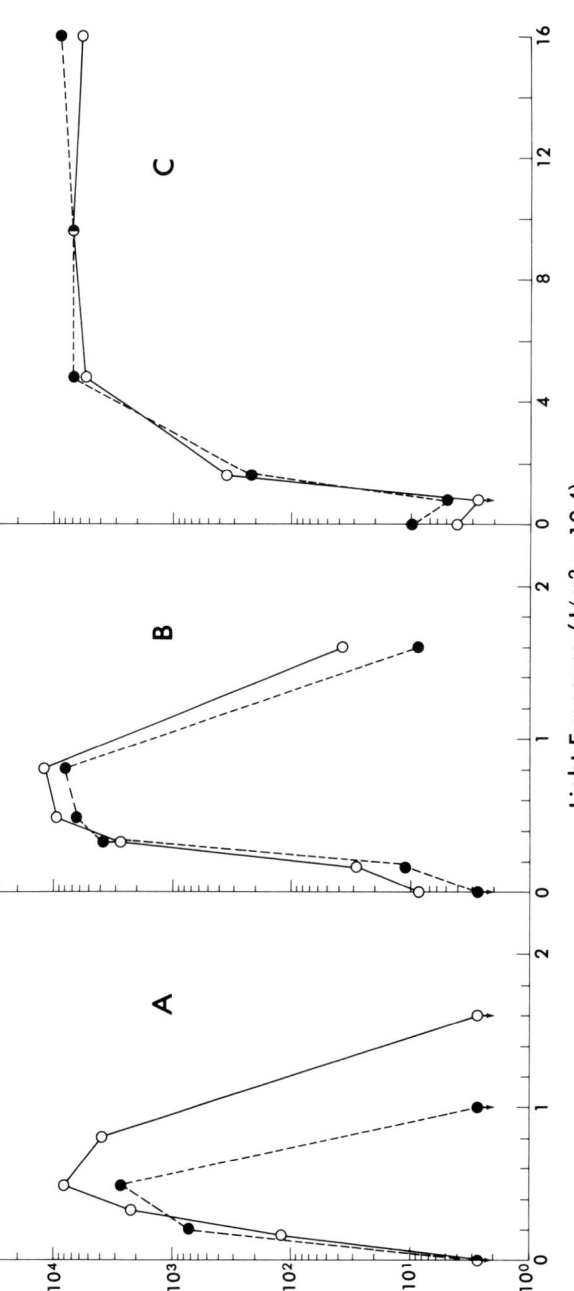

FIG. 1. Photodynamic induction of infectious simian virus 40 from SV40-transformed hamster kidney cells. The light source consisted of 12 parallel Sylvania F40T12-D-LT "Daylight" fluorescent lamps filtered with 5-mm plate glass. Irradiance under exposure conditions was 26.7 W/m² (350–750 nm). Uncertainty of the irradiance measurement due to systematic and random calibration errors was ±25% (13). For photodynamic treatment, nearly confluent cultures of clone E-line cells (14) in plastic flasks were rinsed with Dulbecco's phosphate-buffered saline (P_f/NaCl), incubated for 1 hour at 37°C in the dark with filter-sterilized proflavine sulfate (Mead Johnson) in P_f/NaCl (A, 12.3 µM; B, 6.2 µM; C, 1.1 µM), and rinsed with P_f/NaCl to remove unbound dye. Fresh medium (13) was added. The cultures were irradiated with different exposures of visible light and incubated for 72 hours at 37°C to permit virus expression. Cell-free extracts were assayed for induced SV40 infectivity by plaque determination on permissive CV-1 (TC7) monkey kidney cells. Procedures for light source calibration, irradiation, cell growth, and virus infectivity assay have been described (13). The induced virus yield is expressed as the average plaque-forming units (PFU) per 10⁶ transformed cells. Each point represents the average of three determinations (cell extracts from three replicate samples, each assayed once). Values lying below 3 PFU are lower than the detection limit of the assay and are indicated by points with attached arrows. The solid line and dashed curve) from Fig. 1B of Bockstahler and Cantwell (13) are included in C

II. Ultraviolet-Enhanced Reactivation of Herpes Simplex Virus

Enhanced reactivation consists of an increase in the survival of UV-irradiated mammalian virus when assayed for infectivity in damaged host cells, as compared with assay in untreated cells [see reviews by Lytle (6) and Sarasin (7)]. Enhanced reactivation appears to be a general phenomenon. It occurs with several nuclear-replicating viruses [including herpes simplex virus, simian virus 40, and simian adenovirus 7 (19)], in cells from a number of different normal and transformed cell lines (6), and with various DNA-damaging treatments (including UV, X-ray, carcinogen,[4] and photodynamic treatment (6, 20, 21).

The ultraviolet-enhanced reactivation of herpes simplex virus in CV-1 monkey kidney cells is demonstrated in Fig. 3. Unirradiated and UV-irradiated virus were assayed by plaque formation on monolayers of UV-irradiated cells. The capacity of the cells to support plaque formation by unirradiated virus (upper curve of Fig. 3) decreased as a function of cell UV exposure. However, the relative plaque formation of irradiated virus was increased by relatively low UV exposures. The reactivation can be quantitated in terms of a "UV reactivation factor," the normalized ratio of the two curves; the maximum amount of reactivation expressed in this experiment represents about a doubling in plaque-forming ability. The amount of reactivation increases as a function of UV dose to the virus (22). Higher values can be obtained by increasing the interval between irradiation of the cells and initiation of the virus assay (22).

Ultraviolet-enhanced reactivation resembles Weigle reactivation (see footnote 1) of bacteriophage (23, 23a); however, we prefer to use a different terminology for the phenomena until their mechanisms have been elucidated. The evidence that enhanced reactivation of herpesvirus by UV is inducible includes the facts that (a) it increases with time after cell irradiation (22); and (b) it can be prevented by inhibition of protein synthesis (6, 24).

Little is known about the mechanism of enhanced reactivation of mammalian viruses. In general, it is produced by treatments that cause breaks or gaps in cellular DNA or inhibit DNA synthesis. The wavelength dependence of this herpesvirus reactivation suggests a combined nucleic acid–protein target (25). The role of DNA repair in the reactivation is unclear. Its expression in human cells is apparently not dependent on excision repair (26). It might arise from a shift in the balance between DNA degradative and repair enzymes in irradiated

---

[4] See article by Grunberger and Weinstein, This Series 23 [Ed.]

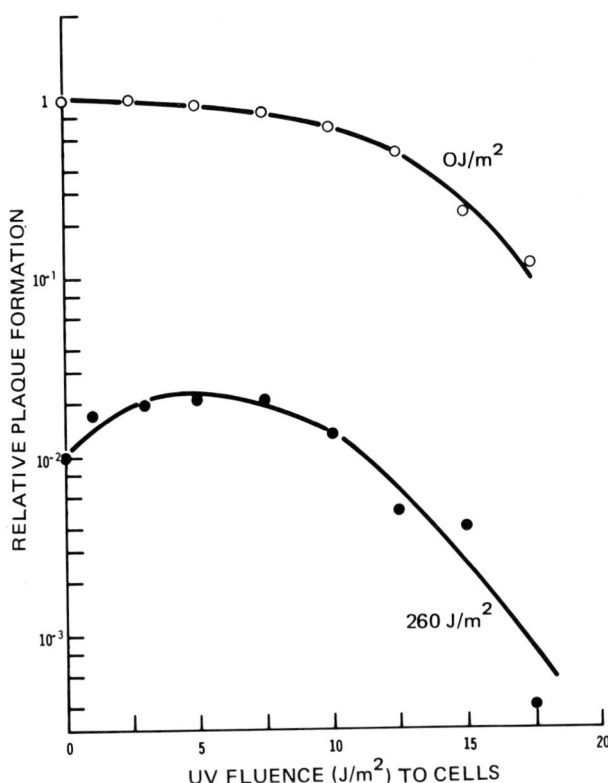

FIG. 3. Ultraviolet-enhanced reactivation of herpes simplex virus (type 1) in CV-1 monkey kidney cell monolayers. Virus and cells were irradiated separately with germicidal (principally 254 nm) ultraviolet light. Unirradiated virus (○) and irradiated virus (●; 260 J/m^2) were assayed for plaque formation immediately after UV-irradiation of cells. Procedures for cell and virus growth, virus infectivity assay, UV irradiation, and dosimetry have been described (22). Reprinted from Bockstahler *et al.* (22) with permission of *Mutation Research*.

cells; this has been proposed as a possible mechanism for Weigle reactivation (27). A possibility that the mechanism involves inducible, error-prone repair is discussed below.

DNA-damaging agents are efficient inducers of enhanced reactivation and DNA-containing latent tumor viruses, but there is little evidence suggesting that these processes are otherwise closely related. It is technically difficult to compare them using the same virus–host cell system. However, UV-enhanced reactivation of herpesvirus and UV induction of SV40 can be observed in the same line of SV40-transformed hamster cells (28). The UV dose response curves for both phenomena were remarkably similar.

Is Ultraviolet-Enhanced Reactivation Mutagenic?

The hypothesis that enhanced reactivation and induction of mammalian virus share a common "SOS" mechanism associated with inducible, error-prone DNA repair is attractive; however, it is difficult to design virus induction experiments to test it. Since UV-enhanced reactivation is similar in some aspects to mutagenic Weigle reactivation in bacteria (23), it is relevant to investigate whether or not the former is also mutagenic. We (K. B. Hellman, L. E. Bockstahler, J. M. Cantwell, and A. Strickland, unpublished observations) have attempted to answer this question by (a) growing UV-irradiated, temperature-sensitive mutants of herpes simplex virus (D9, F18, and B2 received from P. A. Schaffer, Sidney Farber Cancer Institute, Boston) at a permissive temperature in UV-irradiated and unirradiated CV-1 (TC7) monkey kidney cells; (b) preparing extracts of infected cells; and (c) measuring the reversion frequency of progeny mutants to wild-type growth by assay in host cells at nonpermissive temperature. We observed reversion of irradiated mutants that had been grown in irradiated as well as unirradiated cells; however, no conclusions could be drawn regarding the relative amount of reversion, because the data were statistically unreliable and irreproducible. Thus, we obtained no evidence for the existence of inducible, error-prone DNA repair in monkey kidney cells. However, our findings do not prove that such does not exist in mammalian cells. The results from several other relevant investigations using various virus host–cell mutagenesis-assay systems are briefly discussed below.

Sarasin and Benoit (29; see also (7)) examined the rate of back mutation (reversion) to wild-type growth of two UV-irradiated SV40 temperature-sensitive (ts) mutants (tsA58 and tsB20) grown in UV-irradiated monkey kidney cells. The progeny virus from irradiated cells had a significantly higher reversion frequency as compared with the progeny from unirradiated cells. The mutagenesis increased with increasing UV dose to the virus. No obvious correlation could be observed between the reversion frequency and UV dose to the cells, probably owing to statistical fluctuations in the data (29). These experiments were performed using a low multiplicity of infection (m.o.i.) in order to avoid multiplicity reactivation of irradiated virus. The authors concluded that UV-enhanced reactivation is mutagenic and associated with inducible error-prone repair. Reversion of a UV-irradiated SV40 ts mutant (BC245) following propagation in UV-irradiated and unirradiated monkey kidney cells has been observed (29a); however, the reversion frequencies between the two sets of cells were not significantly different.

DasGupta and Summers (24) found that the frequency of forward mutagenesis (to dBrUrd-resistant TK^- mutants) of UV-irradiated herpes simplex virus in UV-irradiated monkey kidney cells was significantly greater than that in unirradiated cells. These experiments were carried out using a high m.o.i. Control assays for the existence of possible multiplicity reactivation were not reported. The dependence of mutagenesis on UV dose to cells or virus was not examined.

Day and Ziolkowski (30) examined the reversion of a UV-irradiated adenovirus 5 temperature-sensitive mutant to wild-type growth in human fibroblasts. The frequency of reversion increased as a function of UV dose to the mutant virus; however, no greater increase in reversion frequency was observed in UV-irradiated cells compared with unirradiated cells.

Lytle et al. (30a), working in this laboratory, measured the forward mutagenesis (to iododeoxycytidine-resistant TK^- mutants) of UV-irradiated herpes simplex virus in UV-irradiated monkey kidney cells. The mutation frequency of irradiated virus (a) was not greater in irradiated cells than that in unirradiated cells, under conditions of low m.o.i. (0.2 PFU per cell) and with no multiplicity reactivation present; and (b) was greater in irradiated cells than that in unirradiated cells for high m.o.i. (2 PFU per cell) and demonstrated the presence of multiplicity reactivation. In other words, the reactivation was not associated with increased mutagenesis, except under conditions where multiplicity reactivation was also present. These results suggest that the increased herpes virus mutagenesis observed (24) may have been associated with multiplicity reactivation rather than enhanced reactivation.

Other relevant mammalian cell mutagenesis studies have been reported. Cleaver (31) designed mutagenesis experiments to test for X-ray-induced, error-prone repair of UV-induced DNA damage in Chinese hamster ovary cells, and found no evidence for such repair. Chang et al. (32) reported that enhanced postreplication repair in Chinese hamster cells is error-free.

In conclusion, there is at present relatively little supportive experimental evidence for the existence of an inducible error-prone DNA repair system in mammalian cells, and for virus induction and enhanced reactivation being mammalian "SOS" functions (see footnote 2). Further research is needed to elucidate the mechanisms of inducible functions in mammalian cells. Perhaps it is time that we design original models as a basis for this research, instead of relying on models concerned with mechanisms of inducible functions in prokaryotes.

III. Summary

Several processes can be induced in mammalian cells by light or other DNA-damaging agents; however, relatively little is known about their induction mechanisms. Two of these processes, enhanced reactivation and induction of mammalian viruses, resemble the bacterial "SOS" functions, Weigle reactivation, and induction of bacteriophage. In this report, several studies on the characterization of these potential mammalian "SOS" functions are reviewed. Induction of simian virus 40 after treatment of SV40-transformed hamster cells with proflavine and visible fluorescent light is described. Response curves were obtained by determining induced SV40 infectivity for different values of light exposure and dye concentration. The induction was clearly a function of both these parameters, which gave partial justification for calling the effect a photodynamic induction. The maximum levels of induction found for each response curve were similar in amount and represented about a thousandfold increase above spontaneous background levels observed with untreated cells. The results suggest that photodynamic treatment also reduces the capacity of the cells to support the growth of induced virus.

The ultraviolet-enhanced reactivation of herpes simplex virus in host monkey kidney cells is described. Information from this laboratory and others concerning whether or not UV-enhanced reactivation of mammalian viruses is associated with increased viral mutagenesis is reviewed. There is relatively little evidence to indicate the existence of inducible, error-prone DNA repair in mammalian cells.

ACKNOWLEDGMENTS

The author acknowledges J. M. Cantwell for excellent technical assistance, O. Ellingson and C. W. Hench for spectral irradiance measurements. He is grateful to J. C. Kaplan, P. H. Black, and P. A. Schaffer for gifts of virus and cell stock; C. D. Lytle, K. B. Hellman, and J. Z. Beer for valuable suggestions; A. Sarasin and G. B. Zamansky for sharing their unpublished results; W. M. Leach and M. L. Shore for encouragement; and colleagues of the Bureau of Radiological Health for critical review of the manuscript.

REFERENCES

1. M. Radman, in "Molecular Mechanisms for Repair of DNA" (P. C. Hanawalt, and R. B. Setlow, eds.), p. 355. Plenum, New York, 1975.
2. E. M. Witkin, *Bacteriol. Rev.* **40**, 869 (1976).

3. M. Terada, U. Nudel, E. Fibach, R. A. Rifkind and P. A. Marks, *Cancer Res.* **38**, 835 (1978).
4. W. Scher and C. Friend, *Cancer Res.* **38**, 841 (1978).
4a. U. Lichti, E. Patterson, T. Ben, S. H. Yuspa and G. T. Bowden, *Photochem. Photobiol.* **32**, 177 (1980).
5. L. E. Bockstahler and K. B. Hellman, *Photochem. Photobiol.* **30**, 743 (1979).
6. C. D. Lytle, *Natl. Cancer Inst. Monogr.* **50**, 145 (1978).
7. A. Sarasin, *Proc. 6th Int. Congr. Radiat. Res.* 1979, p. 462 (1979).
8. G. C. Fareed and D. Davoli, *ARB* **46**, 471 (1977).
9. T. J. Kelly, Jr. and D. Nathans, *Adv. Virus Res.* **21**, 85 (1977).
10. T. Rakusanova, J. C. Kaplan, W. P. Smales, and P. H. Black, *J. Virol.* **19**, 279 (1976).
11. L. E. Bockstahler, T. P. Coohill, K. B. Hellman, C. D. Lytle, and J. E. Roberts, *Pharmacol. Ther.* **4**, 473 (1978).
12. D. Freifelder, *Virology* **30**, 567 (1966).
13. L. E. Bockstahler and J. M. Cantwell, *Biophys. J.* **25**, 209 (1979).
14. J. C. Kaplan, S. M. Wilbert, J. J. Collins, T. Rakusanova, G. B. Zamansky and P. H. Black, *Virology* **68**, 200 (1975).
15. C. D. Lytle and L. D. Hester, *Photochem. Photobiol.* **24**, 443 (1976).
16. G. B. Zamansky, L. F. Kleinman, J. B. Little, P. H. Black and J. C. Kaplan, *Virology* **73**, 468 (1976).
17. L. E. Bockstahler and S. A. Adams, in "Symposium on Biological Effects and Measurement of Light Sources" (D. G. Hazzard, ed.), p. 203. Health, Education, and Welfare Publication (FDA) No. 77-8002, Bureau of Radiological Health, Rockville, Maryland, 1976.
18. G. B. Zamansky, J. B. Little, P. H. Black and J. C. Kaplan, *Mutat. Res.* **51**, 109 (1978).
19. L. E. Bockstahler and C. D. Lytle, *Photochem. Photobiol.* **25**, 477 (1977).
20. C. D. Lytle, J. Coppey and W. D. Taylor, *Nature* **272**, 60 (1978).
21. A. Sarasin and P. C. Hanawalt, *PNAS* **75**, 346 (1978).
22. L. E. Bockstahler, C. D. Lytle, J. E. Stafford and K. F. Haynes, *Mutat. Res.* **35**, 189 (1976).
23. J. J. Weigle, *PNAS* **39**, 628 (1953).
23a. R. Devoret, in this volume.
24. U. B. DasGupta and W. C. Summers, *PNAS* **75**, 2378 (1978).
25. T. P. Coohill, L. C. James, and S. P. Moore, *Photochem. Photobiol.* **27**, 725 (1978).
26. C. D. Lytle, R. S. Day, III, K. B. Hellman and L. E. Bockstahler, *Mutat. Res.* **36**, 257 (1976).
27. C. S. Rupert and W. Harm, *Adv. Radiat. Biol.* **2**, 1 (1966).
28. G. B. Zamansky, L. F. Kleinman, P. H. Black and J. C. Kaplan, *Mutat. Res.* **70**, 1 (1980).
29. A. Sarasin and A. Benoit, *Mutat. Res.* **70**, 71 (1980).
29a. J. J. Cornelis, J. H. Lupker and A. J. van der Eb, *Mutat. Res.* **71**, 139 (1980).
30. R. S. Day, III and C. Ziolkowski, in "DNA Repair Mechanisms" (P. C. Hanawalt, E. C. Friedberg and C. F. Fox, eds.), p. 535. Academic Press, New York, 1978.
30a. C. D. Lytle, J. G. Goddard and C. Lin, *Mutat. Res.* **70**, 139 (1980).
31. J. E. Cleaver, *Mutat. Res.* **52**, 247 (1978).
32. C. C. Chang, S. M. D'Ambrosio, R. Schultz, J. E. Trosko and R. B. Setlow, *Mutat. Res.* **52**, 231 (1978).

Comparative Induction Studies

ERNEST C. POLLARD
D. J. FLUKE AND
DENO KAZANIS

Zoology Department
Duke University
Durham, North Carolina

Seeking to determine whether all of the induced repair (S.O.S.)[1] functions induced by a single agent in a controlled and measurable way actually are manifested together (i.e., are expressed coordinately), we studied the responses to UV light (monochromatic, 265 nm) as a function of the UV dose (1, 2). The yields of the product of gene recA (p-recA), which we consider primary, were determined by electrophoresis on gradient polyacrylamide gels, in the presence of dodecyl sulfate, to separate the p-recA band. Figure 1 shows the induction process as a function of UV dose. It follows approximately the single-hit, all-or-nothing behavior suggested by the equation

$$\text{Total yield} - \text{constitutive yield} = 1 - \exp(-D/D_0)$$

where D is UV dose and D_0 is a reciprocal sensitivity measure. While we have a good many yield curves for several strains of *Escherichia coli*, in this article the emphasis is on a comparison of yield curves for various phenomena induced in *E. coli* AB1157, a strain we employ to examine wild-type behavior in repair-related genes. The results are listed in Table I.

Figure 2 shows information from this table plotted as ratios to the relative amount of p-recA and shown as a function of UV dose. If the relative amounts for the various manifestations are really all coordinate expressions of one phenomenon, this ratio should be unity. It can be seen that three of the phenomena (Weigle reactivation,[2] septum inhibition, and the inhibition of postradiation DNA degradation) conform reasonably to this proposition. The variations at the low doses are understandable because of difficulties in measuring low doses and low ratios. The ratio for induced radioresistance, while basically varying around unity, is not the same. The most significant observation, however, is that the induction of λ shows no agreement; in

---

[1] Defined in footnote 2 of Hanawalt *et al.* in this volume [Ed.].
[2] Defined in footnote 3 of Hanawalt *et al.*, in this volume [Ed.].

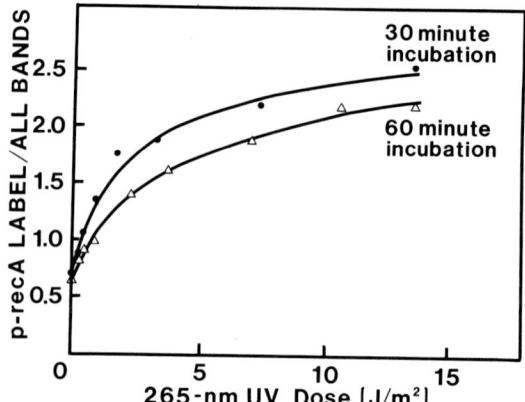

FIG. 1. Cells of strain AB1157 were given various inducing doses of UV, and the yield of the product of gene *recA* (p-recA) was assayed by gel electrophoresis. After UV, the cells were given $^{35}SO_4^{2-}$ and incubated for 30 or 60 minutes. After separating the cell proteins by electrophoresis, the p-recA band was cut out and dissolved in hydrogen peroxide (30%), and its ^{35}S activity was compared with that in all other bands. The relative yield for different inducing doses is plotted against the UV dose.

fact, it is clear that a dose effect must be present in addition to the induction of p-recA.

This apparent anomaly for λ induction may be explained as follows. If the induction needs the activated *recA* gene product, the activation probably occurs in single-stranded sections of DNA, and so the concentration governing the induction is not simply that of p-recA as seen on gels. The induction of λ represents both the concentration

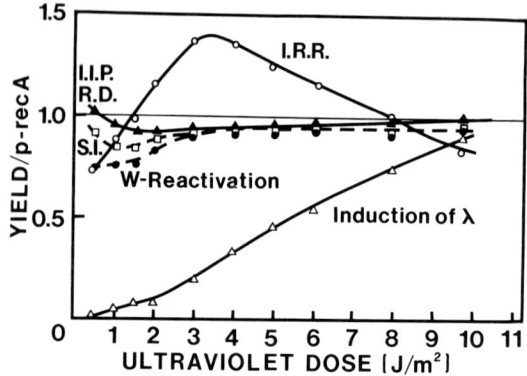

FIG. 2. The yield of various phenomena associated with induced repair, relative to the yield of the *recA* gene product in strain AB1157 plotted from information from Table I.

TABLE I
AVERAGE RELATIVE YIELDS FOR p-recA (PROTEIN X) AND FIVE OTHER MANIFESTATIONS OF INDUCED REPAIR IN *Escherichia coli* STRAIN AB1157

265-nm UV DOSE (J m^{-2})	Relative p-recA[a]	Relative W-reactivation[b]	Ratio	IRR[c]	Ratio	SI[d]	Ratio	IIPRD[e]	Ratio	λ-ind[f]	Ratio
0.5	0.175	0.128	0.74	0.13	0.74	0.16	0.91	0.18	1.03	0.0018	0.01
1.0	0.310	0.236	0.76	0.28	0.90	0.26	0.84	0.29	0.95	0.011	0.04
1.5	0.415	0.314	0.76	0.41	0.99	0.34	0.82	0.38	0.92	0.027	0.07
2.0	0.485	0.413	0.85	0.56	1.15	0.42	0.87	0.45	0.93	0.050	0.08
3.0	0.605	0.547	0.90	0.83	1.37	0.55	0.91	0.57	0.94	0.116	0.19
4.0	0.700	0.650	0.93	0.95	1.36	0.66	0.94	0.66	0.94	0.23	0.33
5.0	0.785	0.75	0.94	0.98	1.25	0.74	0.94	0.74	0.94	0.35	0.19
6.0	0.855	0.82	0.96	1.00	1.17	0.80	0.94	0.82	0.96	0.46	0.54
8.0	0.950	0.88	0.93	0.96	1.01	0.89	0.94	0.93	0.98	0.70	0.74
10.0	0.990	0.93	0.94	0.87	0.88	0.95	0.96	1.00	1.01	0.92	0.93

[a] The fraction of the *recA* gene product, p-recA, observed as described in Fig. 1, at the inducing doses shown.

[b] The fraction of maximum Weigle reactivation observed as described in Fig. 5. Weigle reactivation (see footnote 2), or W-reactivation, is an increased yield of UV irradiated phage when absorbed by an irradiated host.

[c] The fraction of maximum radioresistance (IRR) induced by a dose of UV followed by an incubation period of 40 minutes, adding rifampin to stop further induction and testing sensitivity with X-rays (1).

[d] The fraction of the maximum number of cells of length outside the normal range (septum inhibition; SI) after the dose of UV shown in column 1.

[e] The fraction of maximum inhibition of DNA degradation (IIPRD) produced by X-rays for each inducing dose in column 1 (1).

[f] The fraction of the maximum development of infectious centers and free phage in lysogens of AB1157 as observed for the doses in column 1 (2).

of p-recA *and* of the agent (amount of single-stranded region) that activates it. This two-component process would require more action by UV than is needed to induce p-recA alone.

Returning to the induced radioresistance, Fig. 3 shows the complicating factor. In this experiment, the cells are given an inducing dose of UV light, are incubated for the times shown in the figure, and are then given rifampin to prevent any further induction upon being tested for X-ray sensitivity by graded doses of X-rays. The cells were also examined for septum inhibition. The percentage of cells immediately exposed to rifampin and X-rays and showing septum inhibition was 3%. After growth for 60 minutes, a marked shoulder develops with 69% of the cells septum inhibited and an indication of a sensitive component as low doses. After incubation for 120 minutes, this sensitive component has grown considerably. The septum-inhibited component has also grown, but is now a smaller percentage of the total. An extra 30 minutes of growth leaves the shouldered, resistant component still readily identifiable, and the number of septum-inhibited cells is now at 22% of the total. Thus the culture includes a population of cells that is resistant, a population that is still sensitive and growing,

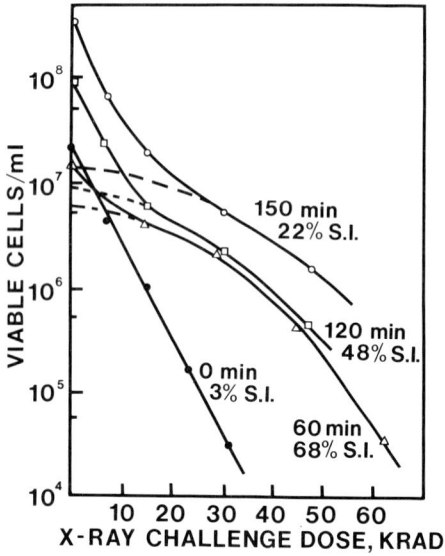

FIG. 3. The consequence of incubation of a culture partly induced for radioresistance. The presence of two classes of cell can be seen: resistant and not growing rapidly; sensitive and growing. A third class can be inferred by the initial yield in the 60-minute incubation case: cells damaged to the point of failing to give rise to colonies upon plating. S.I. = septum inhibition.

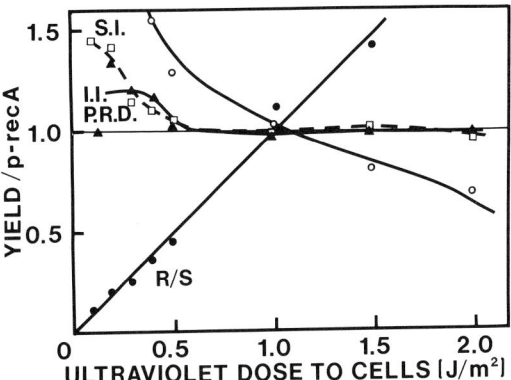

FIG. 4. This is similar to Fig. 2 for the uvr^- strain WU3516-89. Here the ratio of leucine revertants per survivor is included. It shows a proportional increase, suggesting that, in addition to the need for the *recA* gene product, a premutational lesion is required. The number of these is proportional to the dose. S.I. = septum inhibition; I.I.P.R.D. = inducible inhibition of postradiation DNA degradation; R/S = revertants per survivor.

and a population that has been damaged by the inducing dose so that it will not form colonies in the plating (shown by the lower initial number in the 60-minute incubation case). Separation and quantitative measurement of these three components is necessary, and has been done only in rough approximation in the past.

This experimental finding is significant in extrapolating from high-dose observations to low-dose estimates of effect. If human cells are capable of induced repair, it would be wise to be cautious about the linear extrapolation. Ionizing radiation can induce, so that the survival at high doses will be largely of the fraction of cells that are induced and resistant. Extrapolation to noninducing dose levels will not reveal the other two classes, both of which might be important in the long-term effects of radiation on people.

Figure 4 shows a continuation of such comparative work with a uvr^- strain, WU3610-89, for which the observation of the production of leucine revertants by UV has been included. It can be seen that induced radioresistance does not fit with coordinate expression, but that septum inhibition and inducible inhibition of post-radiation DNA degradation again do fit reasonably well. The ratio is mostly unity. Note that the leucine revertants are particularly understandable: they should not fit and they do not, but as plotted, the ratio of the revertants per survivor to the yield of p-recA at that dose produces a very nice proportionality to the dose. This confirms that the premutagenic lesion occurs with a yield proportional to the dose. So we see that the premutagenic lesion, which by itself produces no mutation, is converted into a mutagenic lesion by the *recA* gene product—a nice and quite understandable result.

Figures 5–7 are concerned with quantitation of Weigle (W-) reactivation (see footnote 2) (3). The data of Fig. 5 show the survival of UV-irradiated λ phage on AB1157 cells that have been given various doses of UV. For doses up to 25 J/m², there is a clear increase in the survival of irradiated phage at all levels of phage survival. A dose-modifying factor (DMF) reasonably describes the measurements. The maximum DMF for AB1157 is 1.8.

The results for strain AB1886 (*uvrA*⁻) are presented in Fig. 6. The effect of UV on the phage produces a characteristic survival curve with changing slope and a point of inflection at about 20 J/m². There is Weigle reactivation observable, to a very much lesser extent than in AB1157 and not conforming to any type of dose-modifying factor. A 7% figure might be justified in this strain, at the dose level involving maximal reactivation.

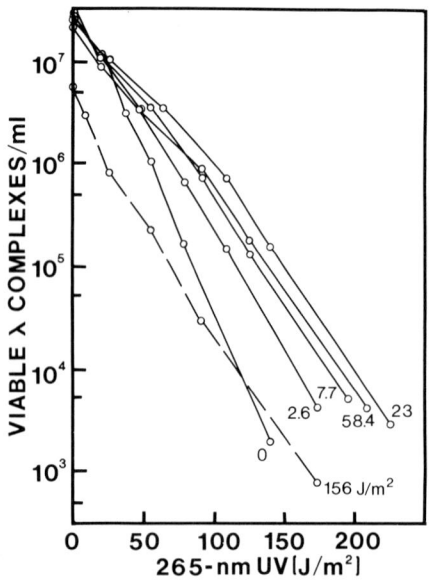

FIG. 5. Plaque survival of λ phage subjected to various 265-nm UV doses and infecting *Escherichia coli* AB1157 host cells that had already received various doses of the same UV light. Survival of the phage is progressively higher with cell UV doses up to 25 J/m². At larger doses, the phage survival curve is depressed without substantial further change in slope, an impairment of cell capacity to support phage growth. The cell UV doses are indicated next to the lowest point on each curve.

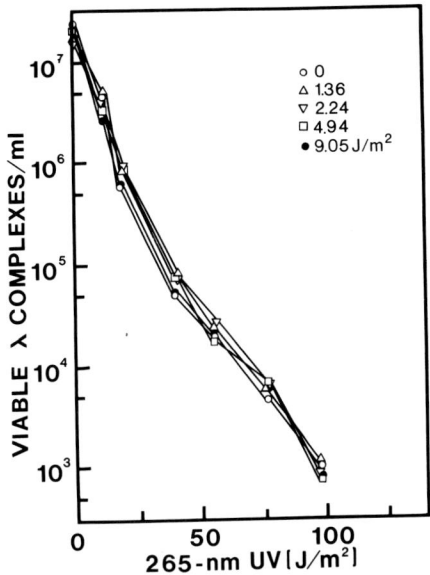

FIG. 6. Similar observations to Fig. 5 in the $uvrA^-$ host strain AB1886. A very much smaller increased survival of the phage in the irradiated cells is seen.

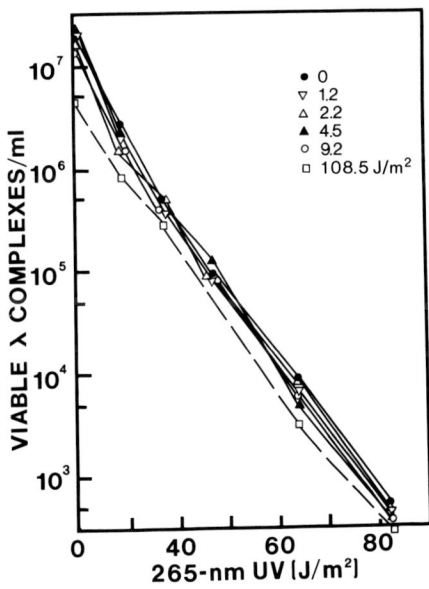

FIG. 7. Phage plaque survival in *Escherichia coli* AB2463, *recA*, that had received the various doses shown. The reactivation phenomenon associated with Weigle reactivation (see footnote 2), in which irradiated phage survive better on irradiated cells, is not seen.

Figure 7 is for the same kind of experiment performed with strain AB 2470, which is $recA^-$. No Weigle reactivation can be seen, justifying the conclusion that Weigle reactivation requires the $recA^+$ gene.

In summary, we can state that the degree of UV exposure is coordinate with the induction of the $recA$ gene product for Weigle reactivation, septum inhibition, and inhibition of postirradiation DNA degradation. It is not related by a simple function to λ induction, and not simply so for induced radioresistance. The production of leucine revertants by UV is coordinate, with the added requirement of a premutagenic lesion proportional to the dose. Weigle reactivation in strain AB1157 can be described as a dose-modifying factor of value up to 1.8. In AB1886 ($uvrA^-$), much less Weigle reactivation is seen, not describable as a DMF, and reaching only about 7% at the most favorable phage inactivation level. No Weigle reactivation at all is seen in strain AB 2463 $recA^-$.

Acknowledgment

We acknowledge support from D.O.E. Contract No. DE-AS05-76EVO3631. We were ably assisted in the Weigle reactivation work by Mr. Eric Bronner.

References

1. E. C. Pollard, S. Person, M. Rader and D. J. Fluke, *Radiat. Res.* **72** 519 (1977).
2. D. J. Fluke and E. C. Pollard, *in* "DNA Repair Mechanisms" (P. C. Hanawalt, E. C. Friedberg and C. F. Fox, eds.), p. 387. Academic Press, New York, 1978.
3. J. J. Weigle, *PNAS* **39** 628 (1953).

VII. Conclusion

Concluding Remarks   325
ERNEST C. POLLARD

Concluding Remarks

ERNEST C. POLLARD

Zoology Department
Duke University
Durham, North Carolina

The first reaction of anyone who listened to the papers in the early part of this symposium in which DNA sequences and their relation to the functioning control processes were described would, I think, be that this part of science is in the midst of an explosive expansion. What is most interesting is that the sheets of new sequences can be held in the hand and the discussion of their significance can be followed and understood. There is an underlying reason for this, namely, that the regulatory processes in the cell have a simplicity and elegance, not to say beauty, that is becoming apparent. This creates an underlying excitement, readily detectable at this symposium; and taking part in the revelation of the nature of the regulatory processes is immensely rewarding.

In participating in this excitement we have to remember that we also have a responsibility to future science. We do face a major educational problem ahead, and we should give some thought to the language we are developing. It may not be easy to do so in this period of rapid advance, but future generations of workers will thank us if we show some care and restraint.

In making these concluding remarks I would like to comment on the importance of the work that has been described. As far as the first part of the symposium is concerned, what is involved is a basic understanding, leading to the ability to control (a strong word, but not too strong) genetics and differentiation. That means the basic understanding, and possibly control, of considerable parts of living things. I think we are going to see expansion into a set of industries: something of great potential. As far as the advances in understanding of DNA repair are concerned, there is a different potential, again far from trivial, because there is latent the potential for the understanding of cancer and, with it, of mutagenesis. From there, it is only a short march to a deeper understanding of evolution. So we do indeed face an educational problem: new departments are shaping out of the material here. If I were a dean, I would be worrying about the developments demonstrated here.

Now I have the problem of unifying the two sections of the symposium, the section on transcription and replication, and the one on repair. I have found it to be a challenge, but I now see how one could think of the symposium in a unifying way—and it is very basic. In my thinking about science, I have been able to see three great unifications—and, incidentally, these unifications are peculiar to science. They are not amalgamations, nor accretions, and they are not to be seen in art or anywhere else. They are true unifications. The first, a giant, is the unification between light and electricity. Maxwell achieved it: he put the two together and they became one subject. In my lifetime, there has been the unification between atomic physics and chemistry, a unification that has changed both subjects, particularly chemistry. Now the third great unification is the unification of genetics and biochemistry, and that is coming in the lifetime of everyone here. The unity between the two parts of this symposium is part of that great unification, and in all of this conference we have been concerned, not with the overlap, but the oneness of genetics and biochemistry, so that it has made good sense to put the two parts together. The result has been a magnificent symposium, and it is fascinating to think where all this will go.

Index

A

Acinar cell, genes of, 174–175

B

Bacterial strains, induction of SOS functions and, 284
Bacteriophage lambda
 decision by, 113–115
 lysogenic response, 109–111
 prophage induction and, 115–116
 prophage maintenance, 111–113
 regulatory circuits of
  early development, 103–107
  lytic development, 107–109
Bacteriophage T7
 deoxyribonucleic acid
  repair *in vitro*, 229–231
  replication and phage production *in vitro*, 231–233
  in vitro packaging of. 228–229
  response to DNA damage, 233–235
  in vitro recombination, 233
Base selection, mechanisms of, 51–54

C

Chromatin, replicating, conformation of, 143–146
Chromatin replication
 discussion of, 146–148
 materials and methods, 136–137
Complementary strand
 distribution of RNA primers on, 28–30
 replication, origin of, 20–23
Complexes of (ϕXDNA) · (dnaB protein) · primase, 16–17

D

Deoxyribonucleases, prophage induction by, 290–294
Deoxyribonucleic acid
 alkylated, inducible repair of
  adaptive response of mammalian cells, 242–243
  chemistry of adaptive response, 240–242
  genetics of adaptive response, 238–239
  substrate for adaptive response, 239–240
 bacteriophage T7
  repair *in vitro*, 229–231
  replication and phage production *in vitro*, 231–233
  response to damage, 233–235
  in vitro packaging of, 228–229
 damaged, repair responses to, 181–184
 damage, regulation of cellular functions induced by
  generalized mechanism for, 260
  major regulatory role of *lex* A repressor, 260
  SOS repair hypothesis, 259
 duplex, binding and unwinding by *rec* A protein, 272–276
 excision repair of pyrimidine dimers
  current model, 207–212
  original model, 198–200
 inducible error-prone repair
  in bacteria, 258–259
  targeted versus nontargeted mutagenesis, 259
  third repair process, 256–257
  for UV damaged phage, 257–258
  UV reactivation, 256
 linear duplex, replication of, 54–61
 long patch repair in *E. coli*
  inducibility and genetic control, 189–192
  possible mechanisms, 193–195
  time course, 187–188
 new, evidence against preferential nucleosome assembly on, 125–126
 newly replicated, evidence for nucleosome assembly on, 123–125
 origin-containing fragment of *E. coli*, nucleotide sequence of, 34–37
 ϕX protein n' and, 11–12
 repair patch size distribution, 185–187
 single stranded, characterization of

327

complexes with rec A protein, 266–272
uncoated φX, protein dnaB and, 13–14
Deoxyribonucleic acid glycosylases, base excision repair and, 200–203
Deoxyribonucleic acid polymerase, replicative, properties of, 49–51
Deoxyribonucleic acid polymerase α, primer-template recognition by, 71–79
Deoxyribonucleic acid polymerase(s) α and β, primer-template utilization by, 64–67
Deoxyribonucleic acid polymerase β, primer-template recognition by, 68–71
Deoxyribonucleic acid polymerase δ
 inhibitor studies, 94–95
 proofreading function of 3'-to-5' exonuclease activity of, 92–94
 structural properties of, 87–92
 3'-to-5' exonuclease activity of, 84–87
Deoxyribonucleic acid replication origins of replicons in E. coli, diversity and homology with S. typhimurium, 41–42
Deoxyribonucleic acid-ribonucleic acid junctions, sites within the 245 base-pair fragment and directionality of DNA replication, 41

E

Editing, mechanisms of, 51–54
Endonuclease(s)
 apurinic/apyrimidinic, 203
  degradative pathway of repair of sites of base loss, 204–207
  insertion pathway of repair of sites of base loss, 207
 repair, reconstituted in vitro from partially purified uvr+ gene products, 218–221

G

GATC Sequence, origin-containing region and, 41

H

Herpes simplex virus, ultraviolet-enhanced reactivation of, 308–310
Histone(s)
 acetylation, gene activity and, 163–165
 newly synthesized, deposition of, 137–143
 site of assembly
  discussion, 130–133
  results, 126–130

I

Induction of SOS functions
 materials and methods, 281–284
  bacterial strains, 284
  chemicals and chemical synthesis of oligonucleotides, 284–286
  preparation of plasmolyzed cells and induction of prophage, 286–287
 results
  prophage induction
   by DNases, 290–294
   in permeabilized cells, 287–290
   by specific oligodeoxynucleotides, 294–300
Induction studies, comparative, 315–322
Insulin, genes for, 171–173

M

Mutagenicity, ultraviolet enhanced reactivation and, 310–312

N

Nucleoside triphosphates, binding by rec A protein, 276–277
Nucleosome(s)
 absence in a fraction of SV40 chromatin, 155–159
 evidence against preferential assembly on new DNA, 125–126
 evidence for assembly on newly replicated DNA, 123–125
Nucleotide sequence, of origin-containing DNA fragment of E. coli, 34–37
 functional analysis of, 43–45

INDEX

O

Oligodeoxynucleotides, prophage induction by, 294–300
Oligonucleotides, chemical synthesis of, 284–286
Origin fragment, cloned, as replication origin of *E. coli*, 40
Origin region
  of *E. coli* DNA
    minimal size and location, 37–40
    potentials for secondary structure, 42–43

P

Pancreas, endocrine and exocrine genes, structure and expression of, 171–175
Polyoma mutants grown in embryonal carcinoma cells, sequence rearrangements in, 159–163
Primase, priming system and, 14–16
Priming system, employing dna B protein and primase, 14–16
Prophage induction
  in permeabilized cells, 287–290
  preparation of plasmolyzed cells and, 286–287
Prophage lambda, induction of
  dormancy and, 252–253
  gene products involved, 253–254
  mechanism of, 255–256
  primary target of inducing agents, 253
  rec A protein cleaves λ repressor, 254–255
  rec A protein cleaves *lex* A protein, 255
Protein(s)
  prepriming, required for SS → RF on SSB-coated φX DNA, 10–11
  purified, reconstitution of RF-RF system from, 27–28
Protein dnaB
  priming system and, 14–16
  processivity in replication intermediate, 19–20
  as replication promoter, 17–19
  uncoated φX DNA and, 13–14

Protein n'
  recognition of a specific sequence on φX DNA, 11–12
  role in mobile replication promoter, 25

R

Reactivation, ultraviolet-enhanced, mutagenicity and, 310–312
rec A Protein, nucleoside triphosphate binding by, 276–277
Replication
  of complementary strand, origin of, 20–23
  promotion, dna B protein and, 17–19
Replication intermediate, processivity of dnaB protein in, 19–20
Replication promoter
  mobile
    polarity of, 23–24
    role of protein n' in, 25
rep Protein, holoenzyme and association with (gene-A protein) · (RF-II) complex in a looped, rolling-circle intermediate, 26–27
Ribonucleic acid polymerases II and III, factors involved in transcription of eukaryotic genes by, 176–177
Ribonucleic acid primers, distribution on lagging complementary strand, 28–30

S

Somatostatin(s), genes for, 173–174
SV40, photodynamic induction of, 303–307
SV40 chromatin, absence of nucleosomes in a fraction of, 155–159

T

Tetrahymena pyriformis, chromatin replication in, 135–148

U

uvrA⁺ Gene product, molecular properties of, 221–223
uvrABC Endonuclease, mechanism of action of, 224–225
uvrB⁺ Gene product, properties of, 223–224
uvrC⁺ Gene product, properties of, 223–224

X

Xenopus, 5 S RNA, control of transcription of, 175–176

Y

Yeast, tRNA gene, mutations blocking expression of, 178

Contents of Previous Volumes

Volume 1
"Primer" in DNA Polymerase Reactions—*F. J. Bollum*
The Biosynthesis of Ribonucleic Acid in Animal Systems—*R. M. S. Smellie*
The Role of DNA in RNA Synthesis—*Jerard Hurwitz and J. T. August*
Polynucleotide Phosphorylase—*M. Grunberg-Manago*
Messenger Ribonucleic Acid—*Fritz Lipmann*
The Recent Excitement in the Coding Problem—*F. H. C. Crick*
Some Thoughts on the Double-Stranded Model of Deoxyribonucleic Acid—*Aaron Bendich and Herbert S. Rosenkranz*
Denaturation and Renaturation of Deoxyribonucleic Acid—*J. Marmur, R. Rownd, and C. L. Schildkraut*
Some Problems Concerning the Macromolecular Structure of Ribonucleic Acids—*A. S. Spirin*
The Structure of DNA as Determined by X-Ray Scattering Techniques—*Vittorio Luzzati*
Molecular Mechanisms of Radiation Effects—*A. Wacker*

Volume 2
Nucleic Acids and Information Transfer—*Liebe F. Cavalieri and Barbara H. Rosenberg*
Nuclear Ribonucleic Acid—*Henry Harris*
Plant Virus Nucleic Acids—*Roy Markham*
The Nucleases of *Escherichia coli*—*I. R. Lehman*
Specificity of Chemical Mutagenesis—*David R. Krieg*
Column Chromatography of Oligonucleotides and Polynucleotides—*Matthys Staehelin*
Mechanism of Action and Application of Azapyrimidines—*J. Skoda*
The Function of the Pyrimidine Base in the Ribonuclease Reaction—*Herbert Witzel*
Preparation, Fractionation, and Properties of sRNA—*G. L. Brown*

Volume 3
Isolation and Fractionation of Nucleic Acids—*K. S. Kirby*
Cellular Sites of RNA Synthesis—*David M. Prescott*
Ribonucleases in Taka-Diastase: Properties, Chemical Nature, and Applications—*Fujio Egami, Kenji Takahashi, and Tsuneko Uchida*
Chemical Effects of Ionizing Radiations on Nucleic Acids and Related Compounds—*Joseph J. Weiss*
The Regulation of RNA Synthesis in Bacteria—*Frederick C. Neidhardt*
Actinomycin and Nucleic Acid Function—*E. Reich and I. H. Goldberg*
De Novo Protein Synthesis *in Vitro*—*B. Nisman and J. Pelmont*
Free Nucleotides in Animal Tissues—*P. Mandel*

Volume 4
Fluorinated Pyrimidines—*Charles Heidelberger*
Genetic Recombination in Bacteriophage—*E. Volkin*
DNA Polymerases from Mammalian Cells—*H. M. Keir*
The Evolution of Base Sequences in Polynucleotides—*B. J. McCarthy*
Biosynthesis of Ribosomes in Bacterial Cells—*Syozo Osawa*
5-Hydroxymethylpyrimidines and Their Derivatives—*T. L. V. Ulbricht*
Amino Acid Esters of RNA, Nucleotides, and Related Compounds—*H. G. Zachau and H. Feldmann*
Uptake of DNA by Living Cells—*L. Ledoux*

Volume 5

Introduction to the Biochemistry of D-Arabinosyl Nucleosides—*Seymour S. Cohen*
Effects of Some Chemical Mutagens and Carcinogens on Nucleic Acids—*P. D. Lawley*
Nucleic Acids in Chloroplasts and Metabolic DNA—*Tatsuichi Iwamura*
Enzymatic Alteration of Macromolecular Structure—*P. R. Srinivasan and Ernest Borek*
Hormones and the Synthesis and Utilization of Ribonucleic Acids—*J. R. Tata*
Nucleoside Antibiotics—*Jack J. Fox, Kyoichi A. Watanabe, and Alexander Bloch*
Recombination of DNA Molecules—*Charles A. Thomas, Jr.*
  Appendix I. Recombination of a Pool of DNA Fragments with Complementary Single-Chain Ends—*G. S. Watson, W. K. Smith, and Charles A. Thomas, Jr.*
  Appendix II. Proof that Sequences of A, C, G, and T Can Be Assembled to Produce Chains of Ultimate Length, Avoiding Repetitions Everywhere—*A. S. Fraenkel and J. Gillis*
The Chemistry of Pseudouridine—*Robert Warner Chambers*
The Biochemistry of Pseudouridine—*Eugene Goldwasser and Robert L. Heinrikson*

Volume 6

Nucleic Acids and Mutability—*Stephen Zamenhof*
Specificity in the Structure of Transfer RNA—*Kin-ichiro Miura*
Synthetic Polynucleotides—*A. M. Michelson, J. Massoulié, and W. Guschlbauer*
The DNA of Chloroplasts, Mitochondria, and Centrioles—*S. Granick and Aharon Gibor*
Behavior, Neural Function, and RNA—*H. Hydén*
The Nucleolus and the Synthesis of Ribosomes—*Robert P. Perry*
The Nature and Biosynthesis of Nuclear Ribonucleic Acids—*G. P. Georgiev*
Replication of Phage RNA—*Charles Weissmann and Severo Ochoa*

Volume 7

Autoradiographic Studies on DNA Replication in Normal and Leukemic Human Chromosomes—*Felice Gavosto*
Proteins of the Cell Nucleus—*Lubomir S. Hnilica*
The Present Status of the Genetic Code—*Carl R. Woese*
The Search for the Messenger RNA of Hemoglobin—*H. Chantrenne, A. Burny, and G. Marbaix*
Ribonucleic Acids and Information Transfer in Animal Cells—*A. A. Hadjiolov*
Transfer of Genetic Information during Embryogenesis—*Martin Nemer*
Enzymatic Reduction of Ribonucleotides—*Agne Larsson and Peter Reichard*
The Mutagenic Action of Hydroxylamine—*J. H. Phillips and D. M. Brown*
Mammalian Nucleolytic Enzymes and Their Localization—*David Shugar and Halina Sierakowska*

Volume 8

Nucleic Acids—The First Hundred Years—*J. N. Davidson*
Nucleic Acids and Protamine in Salmon Testes—*Gordon H. Dixon and Michael Smith*
Experimental Approaches to the Determination of the Nucleotide Sequences of Large Oligonucleotides and Small Nucleic Acids—*Robert W. Holley*
Alterations of DNA Base Composition in Bacteria—*G. F. Gause*
Chemistry of Guanine and Its Biologically Significant Derivatives—*Robert Shapiro*
Bacteriophage φX174 and Related Viruses—*Robert L. Sinsheimer*
The Preparation and Characterization of Large Oligonucleotides—*George W. Rushizky and Herbert A. Sober*
Purine N-Oxides and Cancer—*George Bosworth Brown*
The Photochemistry, Photobiology, and Repair of Polynucleotides—*R. B. Setlow*
What Really is DNA? Remarks on the Changing Aspects of a Scientific Concept—*Erwin Chargaff*
Recent Nucleic Acid Research in China—*Tien-Hsi Cheng and Roy H. Doi*

Volume 9

The Role of Conformation in Chemical Mutagenesis—*B. Singer and H. Fraenkel-Conrat*
Polarographic Techniques in Nucleic Acid Research—*E. Paleček*
RNA Polymerase and the Control of RNA Synthesis—*John P. Richardson*
Radiation-Induced Alterations in the Structure of Deoxyribonucleic Acid and Their Biological Consequences—*D. T. Kanazir*
Optical Rotatory Dispersion and Circular Dichroism of Nucleic Acids—*Jen Tsi Yang and Tatsuya Samejima*
The Specificity of Molecular Hybridization in Relation to Studies on Higher Organisms—*P. M. B. Walker*
Quantum-Mechanical Investigations of the Electronic Structure of Nucleic Acids and Their Constituents—*Bernard Pullman and Alberte Pullman*
The Chemical Modification of Nucleic Acids—*N. K. Kochetkov and E. I. Budowsky*

Volume 10

Induced Activation of Amino Acid Activating Enzymes by Amino Acids and tRNA—*Alan H. Mehler*
Transfer RNA and Cell Differentiation—*Noboru Sueoka and Tamiko Kano-Sueoka*
N^6-(Δ^2-Isopentenyl)adenosine: Chemical Reactions, Biosynthesis, Metabolism, and Significance to the Structure and Function of tRNA—*Ross H. Hall*
Nucleotide Biosynthesis from Preformed Purines in Mammalian Cells: Regulatory Mechanisms and Biological Significance—*A. W. Murray, Daphne C. Elliott, and M. R. Atkinson*
Ribosome Specificity of Protein Synthesis *in Vitro*—*Orio Ciferri and Bruno Parisi*
Synthetic Nucleotide-peptides—*Zoe A. Shabarova*
The Crystal Structures of Purines, Pyrimidines and Their Intermolecular Complexes—*Donald Voet and Alexander Rich*

Volume 11

The Induction of Interferon by Natural and Synthetic Polynucleotides—*Clarence Colby, Jr.*
Ribonucleic Acid Maturation in Animal Cells—*R. H. Burdon*
Liporibonucleoprotein as an Integral Part of Animal Cell Membranes—*V. S. Shapot and S. Ya. Davidova*
Uptake of Nonviral Nucleic Acids by Mammalian Cells—*Pushpa M. Bhargava and G. Shanmugam*
The Relaxed Control Phenomenon—*Ann M. Ryan and Ernest Borek*
Molecular Aspects of Genetic Recombination—*Cedric I. Davern*
Principles and Practices of Nucleic Acid Hybridization—*David E. Kennell*
Recent Studies Concerning the Coding Mechanism—*Thomas H. Jukes and Lila Gatlin*
The Ribosomal RNA Cistrons—*M. L. Birnstiel, M. Chipchase, and J. Speirs*
Three-Dimensional Structure of tRNA—*Friedrich Cramer*
Current Thoughts on the Replication of DNA—*Andrew Becker and Jerard Hurwitz*
Reaction of Aminoacyl-tRNA Synthetases with Heterologous tRNA's—*K. Bruce Jacobson*
On the Recognition of tRNA by Its Aminoacyl-tRNA Ligase—*Robert W. Chambers*

Volume 12

Ultraviolet Photochemistry as a Probe of Polyribonucleotide Conformation—*A. J. Lomant and Jacques R. Fresco*
Some Recent Developments in DNA Enzymology—*Mehran Goulian*
Minor Components in Transfer RNA: Their Characterization, Location, and Function—*Susumu Nishimura*
The Mechanism of Aminoacylation of Transfer RNA—*Robert B. Loftfield*
Regulation of RNA Synthesis—*Ekkehard K. F. Bautz*
The Poly(dA-dT) of Crab—*M. Laskowski, Sr.*

The Chemical Synthesis and the Biochemical Properties of Peptidyl-tRNA—*Yehuda Lapidot and Nathan de Groot*

Volume 13

Reactions of Nucleic Acids and Nucleoproteins with Formaldehyde—*M. Ya. Feldman*
Synthesis and Functions of the -C-C-A Terminus of Transfer RNA—*Murray P. Deutscher*
Mammalian RNA Polymerases—*Samson T. Jacob*
Poly(adenosine diphosphate ribose)—*Takashi Sugimura*
The Stereochemistry of Actinomycin Binding to DNA and Its Implications in Molecular Biology—*Henry M. Sobell*
Resistance Factors and Their Ecological Importance to Bacteria and to Man—*M. H. Richmond*
Lysogenic Induction—*Ernest Borek and Ann Ryan*
Recognition in Nucleic Acids and the Anticodon Families—*Jacques Ninio*
Translation and Transcription of the Tryptophan Operon—*Fumio Imamoto*
Lymphoid Cell RNA's and Immunity—*A. Arthur Gottlieb*

Volume 14

DNA Modification and Restriction—*Werner Arber*
Mechanism of Bacterial Transformation and Transfection—*Nihal K. Notani and Jane K. Setlow*
DNA Polymerases II and III of *Escherichia coli*—*Malcolm L. Gefter*
The Primary Structure of DNA—*Kenneth Murray and Robert W. Old*
RNA-Directed DNA Polymerase—Properties and Functions in Oncogenic RNA Viruses and Cells—*Maurice Green and Gary F. Gerard*

Volume 15

Information Transfer in Cells Infected by RNA Tumor Viruses and Extension to Human Neoplasia—*D. Gillespie, W. C. Saxinger, and R. C. Gallo*
Mammalian DNA Polymerases—*F. J. Bollum*
Eukaryotic RNA Polymerases and the Factors That Control Them—*B. B. Biswas, A. Ganguly, and D. Das*
Structural and Energetic Consequences of Noncomplementary Base Oppositions in Nucleic Acid Helices—*A. J. Lomant and Jacques R. Fresco*
The Chemical Effects of Nucleic Acid Alkylation and Their Relation to Mutagenesis and Carcinogenesis—*B. Singer*
Effects of the Antibiotics Netropsin and Distamycin A on the Structure and Function of Nucleic Acids—*Christoph Zimmer*

Volume 16

Initiation of Enzymic Synthesis of Deoxyribonucleic Acid by Ribonucleic Acid Primers—*Erwin Chargaff*
Transcription and Processing of Transfer RNA Precursors—*John D. Smith*
Bisulfite Modification of Nucleic Acids and Their Constituents—*Hikoya Hayatsu*
The Mechanism of the Mutagenic Action of Hydroxylamines—*E. I. Budowsky*
Diethyl Pyrocarbonate in Nucleic Acid Research—*L. Ehrenberg, I. Fedorcsák, and F. Solymosy*

Volume 17

The Enzymic Mechanism of Guanosine 5', 3'-Polyphosphate Synthesis—*Fritz Lipmann and Jose Sy*
Effects of Polyamines on the Structure and Reactivity of tRNA—*Ted T. Sakai and Seymour S. Cohen*

Information Transfer and Sperm Uptake by Mammalian Somatic Cells—*Aaron Bendich, Ellen Borenfreund, Steven S. Witkins, Delia Beju, and Paul J. Higgins*
Studies on the Ribosome and Its Components—*Pnina Spitnik-Elson and David Elson*
Classical and Postclassical Modes of Regulation of the Synthesis of Degradative Bacterial Enzymes—*Boris Magasanik*
Characteristics and Significance of the Polyadenylate Sequence in Mammalian Messenger RNA—*George Brawerman*
Polyadenylate Polymerases—*Mary Edmonds and Mary Ann Winters*
Three-Dimensional Structure of Transfer RNA—*Sung-Hou Kim*
Insights into Protein Biosynthesis and Ribosome Function through Inhibitors—*Sidney Pestka*
Interaction with Nucleic Acids of Carcinogenic and Mutagenic N-Nitroso Compounds—*W. Lijinsky*
Biochemistry and Physiology of Bacterial Ribonuclease—*Alok K. Datta and Salil K. Niyogi*

Volume 18

The Ribosome of *Escherichia coli*—*R. Brimacombe, K. H. Nierhaus, R. A. Garrett and H. G. Wittmann*
Structure and Function of 5 S and 5.8 S RNA—*Volker A. Erdmann*
High-Resolution Nuclear Magnetic Resonance Investigations of the Structure of tRNA in Solution—*David R. Kearns*
Premelting Changes in DNA Conformation—*E. Paleček*
Quantum-Mechanical Studies on the Conformation of Nucleic Acids and Their Constituents—*Bernard Pullman and Anil Saran*

Volume 19 (Symposium on mRNA: The Relation of Structure to Function)

I. The 5′-Terminal Sequence ("Cap") of mRNAs
Caps in Eukaryotic mRNAs: Mechanism of Formation of Reovirus mRNA 5′-Terminal m^{7}GpppGm-C—*Y. Furuichi, S. Muthukrishnan, J. Tomasz and A. J. Shatkin*
Nucleotide Methylation Patterns in Eukaryotic mRNA—*Fritz M. Rottman, Ronald C. Desrosiers and Karen Friderici*
Structural and Functional Studies on the "5′-Cap": A Survey Method of mRNA—*Harris Busch, Friedrich Hirsch, Kaushal Kumar Gupta, Manchanahalli Rao, William Spohn and Benjamin C. Wu*
Modification of the 5′-Terminals of mRNAs by Viral and Cellular Enzymes—*Bernard Moss, Scott A. Martin, Marcia J. Ensinger, Robert F. Boone and Cha-Mer Wei*
Blocked and Unblocked 5′ Termini in Vesicular Stomatitis Virus Product RNA *in Vitro:* Their Possible Role in mRNA Biosynthesis—*Richard J. Colonno, Gordon Abraham and Amiya K. Banerjee*
The Genome of Poliovirus Is an Exceptional Eukaryotic mRNA—*Yuan Fon Lee, Akio Nomoto and Eckard Wimmer*
II. Sequences and Conformations of mRNAs
Transcribed Oligonucleotide Sequences in Hela Cell hnRNA and mRNA—*Mary Edmonds, Hiroshi Nakazato, E. L. Korwek and S. Venkatesan*
Polyadenylylation of Stored mRNA in Cotton Seed Germination—*Barry Harris and Leon Dure III*
mRNAs Containing and Lacking Poly(A) Function as Separate and Distinct Classes during Embryonic Development—*Martin Nemer and Saul Surrey*
Sequence Analysis of Eukaryotic mRNA—*N. J. Proudfoot, C. C. Cheng and G. G. Brownlee*
The Structure and Function of Protamine mRNA from Developing Trout Testis—*P. L. Davies, G. H. Dixon, L. N. Ferrier, L. Gedamu and K. Iatrou*
The Primary Structure of Regions of SV40 DNA Encoding the Ends of mRNA—*Kiranur N.*

Subramanian, Prabhat K. Ghoshi, Ravi Dhar, Bayar Thimmappaya, Sayeeda B. Zain, Julian Pan and Sherman M. Weissman

Nucleotide Sequence Analysis of Coding and Noncoding Regions of Human β-Globin mRNA—*Charles A. Marotta, Bernard G. Forget, Michael Cohen/Solal and Sherman M. Weissman*

Determination of Globin mRNA Sequences and Their Insertion into Bacterial Plasmids—*Winston Salser, Jeff Browne, Pat Clarke, Howard Heindell, Russell Higuchi, Gary Paddock, John Roberts, Gary Studnicka and Paul Zakar*

The Chromosomal Arrangement of Coding Sequences in a Family of Repeated Genes—*G. M. Rubin, D. J. Finnegan and D. S. Hogness*

Mutation Rates in Globin Genes: The Genetic Load and Haldane's Dilemma—*Winston Salser and Judith Strommer Isaacson*

Heterogeneity of the 3′ Portion of Sequences Related to Immunoglobulin κ-Chain mRNA—*Ursula Storb*

Structural Studies on Intact and Deadenylylated Rabbit Globin mRNA—*John N. Vournakis, Marcia S. Flashner, MaryAnn Katopes, Gary A. Kitos, Nikos C. Vamvakopoulos, Matthew S. Sell and Regina M. Wurst*

Molecular Weight Distribution of RNA Fractionated on Aqueous and 70% Formamide Sucrose Gradients—*Helga Boedtker and Hans Lehrach*

III. Processing of mRNAs

Bacteriophages T7 and T3 as Model Systems for RNA Synthesis and Processing—*J. J. Dunn, C. W. Anderson, J. F. Atkins, D. C. Bartelt and W. C. Crockett*

The Relationship between hnRNA and mRNA—*Robert P. Perry, Enzo Bard, B. David Hames, Dawn E. Kelley and Ueli Schibler*

A Comparison of Nuclear and Cytoplasmic Viral RNAs Synthesized Early in Productive Infection with Adenovirus 2—*Heschel J. Raskas and Elizabeth A. Craig*

Biogenesis of Silk Fibroin mRNA: An Example of Very Rapid Processing?—*Paul M. Lizardi*

Visualization of the Silk Fibroin Transcription Unit and Nascent Silk Fibroin Molecules on Polyribosomes of *Bombyx mori*—*Steven L. McKnight, Nelda L. Sullivan and Oscar L. Miller, Jr.*

Production and Fate of Balbiani Ring Products—*B. Daneholt, S. T. Case, J. Hyde, L. Nelson and L. Wieslander*

Distribution of hnRNA and mRNA Sequences in Nuclear Ribonucleoprotein Complexes—*Alan J. Kinniburgh, Peter B. Billings, Thomas J. Quinlan and Terence E. Martin*

IV. Chromatin Structure and Template Activity

The Structure of Specific Genes in Chromatin—*Richard Axel*

The Structure of DNA in Native Chromatin as Determined by Ethidium Bromide Binding—*J. Paoletti, B. B. Magee and P. T. Magee*

Cellular Skeletons and RNA Messages—*Ronald Herman, Gary Zieve, Jeffrey Williams, Robert Lenk and Sheldon Penman*

The Mechanism of Steroid-Hormone Regulation of Transcription of Specific Eukaryotic Genes—*Bert W. O'Malley and Anthony R. Means*

Nonhistone Chromosomal Proteins and Histone Gene Transcription—*Gary Stein, Janet Stein, Lewis Kleinsmith, William Park, Robert Jansing and Judith Thomson*

Selective Transcription of DNA Mediated by Nonhistone Proteins—*Tung Y. Wang, Nina C. Kostraba and Ruth S. Newman*

V. Control of Translation

Structure and Function of the RNAs of Brome Mosaic Virus—*Paul Kaesberg*

Effect of 5′-Terminal Structures on the Binding of Ribopolymers to Eukaryotic Ribosomes—*S. Muthukrishnan, Y. Furuichi, G. W. Both and A. J. Shatkin*

Translational Control in Embryonic Muscle—*Stuart M. Heywood and Doris S. Kennedy*
Protein and mRNA Synthesis in Cultured Muscle Cells—*R. G. Whalen, M. E. Buckingham and F. Gros*
VI. Summary: mRNA Structure and Function—*James E. Darnell*

Volume 20
Correlation of Biological Activities with Structural Features of Transfer RNA—*B. F. C. Clark*
Bleomycin, an Antibiotic That Removes Thymine from Double-Stranded DNA—*Werner E. G. Müller and Rudolf K. Zahn*
Mammalian Nucleolytic Enzymes—*Halina Sierakowska and David Shugar*
Transfer RNA in RNA Tumor Viruses—*Larry C. Waters and Beth C. Mullin*
Integration versus Degradation of Exogenous DNA in Plants: An Open Question—*Paul F. Lurquin*
Initiation Mechanisms of Protein Synthesis—*Marianne Grunberg-Manago and François Gros*

Volume 21
Informosomes and Their Protein Components: The Present State of Knowledge—*A. A. Preobrazhensky and A. S. Spirin*
Energetics of the Ribosome—*A. S. Spirin*
Mechanisms in Polypeptide Chain Elongation on Ribosomes—*Engin Bermek*
Synthetic Oligodeoxynucleotides for Analysis of DNA Structure and Function—*Ray Wu, Chander P. Bahl and Saran A. Narang*
The Transfer RNAs of Eukaryotic Organelles—*W. Edgar Barnett, S. D. Schwartzbach, and L. I. Hecker*
Regulation of the Biosynthesis of Aminoacid:tRNA Ligases and of tRNA—*Susan D. Morgan and Dieter Söll*

Volume 22
The -C-C-A End of tRNA and Its Role in Protein Biosynthesis—*Mathias Sprinzl and Friedrich Cramer*
The Mechanism of Action of Antitumor Platinum Compounds—*J. J. Roberts and A. J. Thomson*
DNA Glycosylases, Endonucleases for Apurinic/Apyrimidinic Sites, and Base Excision-Repair—*Thomas Lindahl*
Naturally Occurring Nucleoside and Nucleotide Antibiotics—*Robert J. Suhadolnik*
Genetically Controlled Variation in the Shapes of Enzymes—*George Johnson*
Transcription Units for mRNA Production in Eukaryotic Cells and Their DNA Viruses—*James E. Darnell, Jr.*

Volume 23
The Peptidyltransferase Center of Ribosomes—*Alexander A. Krayevsky and Marina K. Kukhanova*
Patterns of Nucleic Acid Synthesis in *Physarum polycephalum*—*Geoffrey Turnock*
Biochemical Effects of the Modification of Nucleic Acids by Certain Polycyclic Aromatic Carcinogens—*Dezider Grunberger and I. Bernard Weinstein*
Participation of Modified Nucleosides in Translation and Transcription—*B. Singer and M. Kröger*
The Accuracy of Translation—*Michael Yarus*
Structure, Function, and Evolution of Transfer RNAs (with Appendix Giving Complete Sequences of 178 tRNAs)—*Ram P. Singhal and Pamela A. M. Fallis*

Volume 24

Structure of Transcribing Chromatin—*Diane Mathis, Pierre Oudet, and Pierre Chambon*
Ligand-Induced Conformational Changes in Ribonucleic Acids—*Hans Günter Gassen*
Replicative DNA Polymerases and Mechanisms at a Replication Fork—*Robert K. Fujimura and Shishir K. Das*
Antibodies Specific for Modified Nucleosides: An Immunochemical Approach for the Isolation and Characterization of Nucleic Acids—*Theodore W. Munns and M. Kathryn Liszewski*
DNA Structure and Gene Replication—*R. D. Wells, T. C. Goodman, W. Hillen, G. T. Horn, R. D. Klein, J. E. Larson, U. R. Müller, S. K. Neuendorf, N. Panayotatos, and S. M. Stirdivant*

Volume 25

Splicing of Viral mRNAs—*Yosef Aloni*
DNA Methylation and Its Possible Biological Roles—*Aharon Razin and Joseph Friedman*
Mechanisms of DNA Replication and Mutagenesis in Ultraviolet-Irradiated Bacteria and Mammalian Cells—*Jennifer D. Hall and David W. Mount*
The Regulation of Initiation of Mammalian Protein Synthesis—*Rosemary Jagus, W. French Anderson, and Brian Safer*
Structure, Replication, and Transcription of the SV40 Genome—*Gokul C. Das and Salil K. Niyogi*